, the

MEI structured mathematics

Mechanics 5 & 6

PAT BRYDEN
FF PAVELIN

Series Editor: Roger Porkess

MEI Structured Mathematics is supported by industry:
BNFL, Casio, GEC, Intercity, JCB, Lucas, The National Grid Company,
Sharp Texas Instruments, Thorn EMI

Hodder & Stoughton

A MEMBER OF THE HODDER HEADLINE GROUP

Acknowledgements

The authors and publishers would like to thank the following companies, institutions and individuals who have given permission to reproduce copyright material for this book. Every effort has been made to acknowledge ownership of copyright. The publishers will be happy to make suitable arrangements with any copyright holders whom it has not been possible to contact.

Illustrations were drawn by Ken Ovington of Precision Art and Jeff Edwards Illustration and Design.

Photographs: Addison-Wesley Longman Ltd/Francis Weston Sears and Mark W. Zemansky 192; Lawrence Berkeley/Science Photo Library 71 (left); Pat Bryden 146, 183; CNES Diffusion 241; Roddy Paine 71 (right), 101, 126 (all), 162, 218 (3 right), 250, 287; Tanya Piejus 218 (left); Carl Purcell/Science Photo Library 45; TRIP/RACAL 87.

The Associated Examining Board.
The University of Cambridge Local Examinations Syndicate
The University of London Examinations and Assessment Council
The University of Oxford Delegacy of Local Examinations

The above examination boards cannot accept any responsibility for the accuracy or method of working in the answers given.

British Library Cataloguing in Publication Data

Mechanics. – Book 5–6. – (MEI Structured Mathematics Series)
  I. Bryden, Pat II. Pavelin, Clive
  III. Series
  530

ISBN–0–340–57860 2

First published 1996
Impression number   10 9 8 7 6 5 4 3 2 1
Year   1999 1998 1997 1996

Typeset by J. W. Arrowsmith Ltd., Bristol.
Printed in Great Britain for Hodder & Stoughton Educational, a division of Hodder Headline Plc, 338 Euston Road, London NW1 3BH by Bath Press, Avon.

MEI Structured Mathematics

Mathematics is not only a beautiful and exciting subject in its own right but also one that underpins many other branches of learning. It is consequently fundamental to the success of a modern economy.

MEI Structured Mathematics is designed to increase substantially the number of people taking the subject post-GCSE, by making it accessible, interesting and relevant to a wide range of students.

It is a credit accumulation scheme based on 45-hour Components which may be taken individually or aggregated to give:

3 Components   AS Mathematics
6 Components   A Level Mathematics
9 Components   A Level Mathematics + AS Further Mathematics
12 Components   A Level Mathematics + A Level Further Mathematics

Components may alternatively be combined to give other A or AS certifications (in Statistics, for example) or they may be used to obtain credit towards other types of qualification.

The course is examined by the Oxford and Cambridge Schools Examination Board, with examinations held in January and June each year.

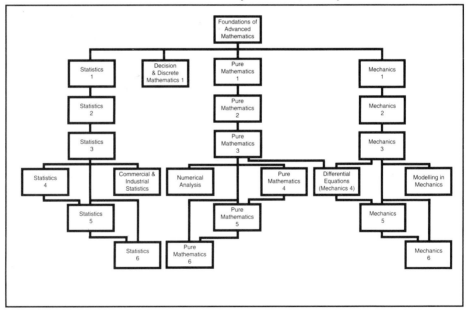

This is one of the series of books written to support the course. Its position within the whole scheme can be seen in the diagram above.

Mathematics in Education and Industry is a curriculum development body which aims to promote the links between Education and Industry in Mathematics, and to produce relevant examination and teaching syllabuses and support material. Since its foundation in the 1960s, MEI has provided syllabuses for GCSE (or previously O Level), Additional Mathematics and A Level.

For more information about MEI Structured Mathematics or other syllabuses and materials, write to MEI Office, 11 Market Street, Bradford-on-Avon BA115 1LL.

Introduction

This is the fifth in the series of books written to support the Mechanics components of MEI Structured Mathematics and covers the last two of these components. As well as supporting the MEI scheme, Mechanics 5 & 6 is suitable if you are studying Mechanics or Applied Mathematics as part of any A level Further Mathematics course or as part of a degree course.

Like the other books in the series, Mechanics 5 & 6 presents the subject as a powerful and exciting means of modelling real situations, and contains several experiments and investigations for you to carry out. In the belief that an informed and intuitive understanding of mathematical results is necessary to practising applied mathematicians, engineers and scientists, we have taken every opportunity to apply techniques to interesting problems in the real world while at the same time encouraging you to develop an understanding of the mathematics involved.

Mechanics 5 & 6 builds on the basic mechanical principles introduced in the earlier books and develops and exploits the use of calculus and vector methods. These are applied to a range of topics including the motion of bodies under variable forces, rotation of rigid bodies, relative velocity, variable mass and the application of energy methods to stability problems.

We would like to thank the many people who have given help and advice with this book as it has developed, including David Edsall, Robin Grayson, Val Hanrahan, David Holland, Mike Jones, Alan Bryden, and our general editor Roger Porkess. Acknowledgements are also due to the examination boards who have allowed us to use their questions in the exercises.

Pat Bryden and Cliff Pavelin

Contents

Vectors review

This grand book – the Universe... is written in the language of mathematics

Galileo Galilei (1623)

Vectors are important in modelling and solving two- and three-dimensional problems in mechanics. You have already used vector methods, particularly in *Mechanics 1* and *Pure Mathematics 3*. This chapter will help you to review and consolidate your knowledge of vectors with particular reference to their application in mechanics.

Vector basics

A vector is a quantity with both magnitude (also known as length, size or modulus) and direction. In mechanics, vectors are used to represent quantities such as displacement, velocity, acceleration, force and momentum.

A vector is usually written $\overrightarrow{PQ}$ or a and is typeset as $\overrightarrow{PQ}$ or as bold **PQ** or **a**.

In figure 1.1, $\overrightarrow{PQ} = \overrightarrow{OA}$ because their lengths and directions are the same.

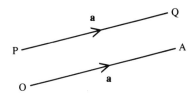

Figure 1.1

Vectors can be added: $\mathbf{a} + \mathbf{b} = \mathbf{c}$

and subtracted: $\mathbf{a} - \mathbf{b} = \mathbf{d}$

and multiplied by a scalar: $\lambda\mathbf{a}$ has magnitude $\lambda \times$ magnitude of a

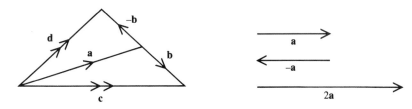

Figure 1.2

The length of the vector **a** is denoted $|\mathbf{a}|$ or a. When $a = 1$, **a** is a *unit vector*. A unit vector is denoted by having a hat **â**.

The *position vector* of a point A relative to a point O is the vector $\overrightarrow{OA}$. Normally, O is the origin of Cartesian axes and the position vector of a point is usually denoted by the corresponding lower-case letter, thus $\overrightarrow{OA} = \mathbf{a}$, $\overrightarrow{OB} = \mathbf{b}$, and so on. Then $\overrightarrow{AB} = \mathbf{b} - \mathbf{a}$, where **b** and **a** are the position vectors of A and B with respect to an assumed origin (figure 1.3) which is not necessarily drawn on the diagram.

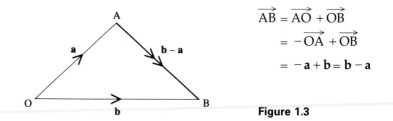

$$\overrightarrow{AB} = \overrightarrow{AO} + \overrightarrow{OB}$$
$$= -\overrightarrow{OA} + \overrightarrow{OB}$$
$$= -\mathbf{a} + \mathbf{b} = \mathbf{b} - \mathbf{a}$$

Figure 1.3

Cartesian components

Vectors can also be expressed in Cartesian component form, with respect to some appropriate origin and axes. In three dimensions:

$$\overrightarrow{OA} = \begin{pmatrix} x \\ y \\ z \end{pmatrix} = x\mathbf{i} + y\mathbf{j} + z\mathbf{k}$$

where x, y, z are the displacements represented by the vector in the x, y and z directions and **i**, **j** and **k** are the unit vectors in the directions of the co-ordinate axes.

Vector addition, subtraction and multiplication by a scalar are easy when vectors are given in Cartesian component form, as you simply apply the operation to each component.

Given $\overrightarrow{OA} = \begin{pmatrix} a_1 \\ a_2 \\ a_3 \end{pmatrix}$ $\overrightarrow{OB} = \begin{pmatrix} b_1 \\ b_2 \\ b_3 \end{pmatrix}$ then $2\overrightarrow{OA} + \overrightarrow{OB} = \begin{pmatrix} 2a_1 + b_1 \\ 2a_2 + b_2 \\ 2a_3 + b_3 \end{pmatrix}$

Vector equation of a line

A line is specified by giving its direction and a point through which it passes. The vector equation of a line is

$$\mathbf{r} = \mathbf{a} + \lambda\mathbf{u}$$

or in component form $\mathbf{r} = (a_1 + \lambda u_1)\mathbf{i} + (a_2 + \lambda u_2)\mathbf{j} + (a_3 + \lambda u_3)\mathbf{k}$

$$
\text{or} \quad \begin{pmatrix} x \\ y \\ z \end{pmatrix} = \begin{pmatrix} a_1 \\ a_2 \\ a_3 \end{pmatrix} + \lambda \begin{pmatrix} u_1 \\ u_2 \\ u_3 \end{pmatrix} = \begin{pmatrix} a_1 + \lambda u_1 \\ a_2 + \lambda u_2 \\ a_3 + \lambda u_3 \end{pmatrix}
$$

The line passes through the point A with position vector **a** and it is in the direction of the vector **u** (figure 1.4). λ is a scalar variable. The vector **r** is the position vector of the point R with co-ordinates (x, y, z). As λ varies, R always lies on the line through A in the direction of **u**. Since A can be chosen as *any* point on the line and since any multiple of **u** (e.g. 3**u** or −10**u**) can be used instead of **u**, the same line has many different forms of equation.

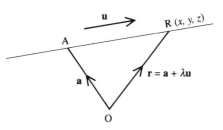

Figure 1.4

Particle moving with constant velocity

In *Mechanics 1* you saw that a particle moving with a constant velocity **v** passing through a point with position vector r_0 will have at a time t later a position

$$
\mathbf{r} = \mathbf{r}_0 + t\mathbf{v}
$$

This is the equation of a line through r_0 in the direction of **v**. In other words, it is the vector equation of the path of the particle, just as you would expect.

Line of action of a force

Although a force is often given as a vector, direction and magnitude alone are not enough fully to determine the effect of a force. For example, given that the force $\mathbf{F} = \mathbf{i} + 2\mathbf{j}$ is acting on a body, you have sufficient information to resolve **F** in any direction, but not sufficient to find its *moment* about any point. For this, you need to know a point on the *line of action* of the force. Suppose the force is acting at the point P with position vector 2**i** (figure 1.5). Then the equation of the line of action of **F** is

$$
\mathbf{r} = 2\mathbf{i} + \lambda(\mathbf{i} + 2\mathbf{j})
$$

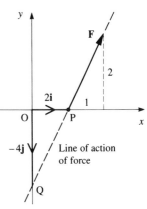

Figure 1.5

Notice that you may just as well have been given any other point on the line of action of **F**. The point Q with position vector $-4\mathbf{j}$ is also on the line of action of the force (put $\lambda = -2$ in the equation). This means that the effect of this force will be exactly the same whether it is acting at $2\mathbf{i}$ or $-4\mathbf{j}$.

The example illustrates the use of the vector equation of a line in describing a force.

EXAMPLE

In the xy plane, a force $\mathbf{F} = F_x \mathbf{i} + F_y \mathbf{j}$ of magnitude 10 acts along the line

$$\mathbf{r} = \mathbf{i} + \mathbf{j} + \lambda\,(3\mathbf{i} + 4\mathbf{j})$$

in the direction of increasing λ.
(i)   Find F_x and F_y.
(ii) **F** acts through the point A$(a, 0)$. Find a.
(iii) Deduce the moment of **F** about O.

Solution

(i)   **F** is in the same direction as $3\mathbf{i} + 4\mathbf{j}$.

$$|3\mathbf{i} + 4\mathbf{j}| = \sqrt{(3^2 + 4^2)} = 5$$

So the unit vector in the direction of **F** is $\tfrac{1}{5}(3\mathbf{i} + 4\mathbf{j})$. Also, $|\mathbf{F}| = 10$, hence

$$\mathbf{F} = \frac{10}{5}(3\mathbf{i} + 4\mathbf{j}) = 6\mathbf{i} + 8\mathbf{j}$$

$$\Rightarrow \quad F_x = 6 \quad \text{and} \quad F_y = 8$$

(ii)

$$\mathbf{r} = \mathbf{i} + \mathbf{j} + \lambda\,(3\mathbf{i} + 4\mathbf{j})$$
$$= (1 + 3\lambda)\mathbf{i} + (1 + 4\lambda)\mathbf{j}$$
$$\Rightarrow \quad x = 1 + 3\lambda \quad \text{and} \quad y = 1 + 4\lambda$$

where (x, y) is a point on the line.
When $y = 0$, $\lambda = -\tfrac{1}{4} \Rightarrow x = \tfrac{1}{4}$, hence $a = \tfrac{1}{4}$.

(iii) To find the moment about O, regard the force as acting at A and resolved into its two components F_x and F_y. F_x does not contribute, so the result is

$$a F_y = \tfrac{1}{4} \times 8 = 2 \quad \text{anticlockwise}$$

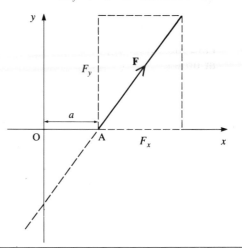

Use of vectors to describe motion

1

Two and three-dimensional motion is often described using *vector* notation.

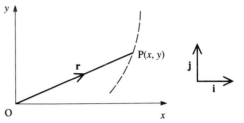

Figure 1.6

Figure 1.6 shows a particle P moving in the xy plane. Its position at any time t can be given by its x and y co-ordinates. These co-ordinates can also be regarded as the components of the position vector of P relative to the origin. Denote this vector OP by **r**. Then

$$\mathbf{r} = x\mathbf{i} + y\mathbf{j}$$

where **i** and **j** are, as usual, the unit vectors in the x and y directions.

In the x direction, the speed of the particle is $\dot{x}$ (i.e. dx/dt) and in the y direction $\dot{y}$, so the velocity of the particle, denoted by $\dot{\mathbf{r}}$, is the vector

$$\mathbf{v} = \dot{\mathbf{r}} = \dot{x}\mathbf{i} + \dot{y}\mathbf{j}$$

Similarly, the acceleration of P can be found by differentiating the components again:

$$\frac{d\mathbf{v}}{dt} = \ddot{x}\mathbf{i} + \ddot{y}\mathbf{j}$$

The following examples demonstrate the use of vectors to represent velocity and acceleration.

EXAMPLE

A particle moves in the xy plane so that its position vector with respect to the origin at a given time t is

$$\mathbf{r} = 3t^2\mathbf{i} + (4t - 6)\mathbf{j}$$

Find the vector expressions for the velocity and acceleration of the particle at time t.

Solution

Differentiate once for velocity

$$\mathbf{v} = \frac{d\mathbf{r}}{dt} = 6t\mathbf{i} + 4\mathbf{j}$$

And again for acceleration

$$\frac{d\mathbf{v}}{dt} = 6\mathbf{i}$$

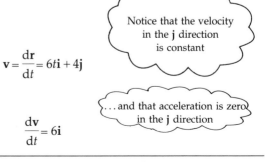

Notice that the velocity in the **j** direction is constant

...and that acceleration is zero in the **j** direction

EXAMPLE

The acceleration of a particle at time t is given by

$$\mathbf{a} = \cos t\,\mathbf{i} + \sin t\,\mathbf{j}$$

At time $t = 0$, its position relative to the origin is $-\mathbf{i}$ and its velocity is $-\mathbf{j}$. Prove that the particle always lies on the circle $x^2 + y^2 = 1$.

Solution

The velocity and then the position of the body are obtained by successive integration of

$$\mathbf{a} = \cos t\,\mathbf{i} + \sin t\,\mathbf{j}$$

The velocity is therefore

$$\mathbf{v} = (\sin t + c_1)\mathbf{i} + (-\cos t + c_2)\mathbf{j}$$

where c_1 and c_2 are the constants of integration.

When $t = 0$, $\mathbf{v} = c_1\mathbf{i} + (-1 + c_2)\mathbf{j}$. But the velocity is $-\mathbf{j}$ when $t = 0$, so c_1 and c_2 are both zero. Hence

$$\mathbf{v} = \sin t\,\mathbf{i} - \cos t\,\mathbf{j}$$

Integrating again gives

$$\mathbf{r} = (-\cos t + k_1)\mathbf{i} + (-\sin t + k_2)\mathbf{j}$$

Again, the constants of integration k_1 and k_2 are zero, since $\mathbf{r}$ is $-\mathbf{i}$ when t is zero. So

$$\mathbf{r} = -\cos t\,\mathbf{i} - \sin t\,\mathbf{j}$$

The co-ordinates of the particle at time t are given by $x = -\cos t$, $y = -\sin t$. This always lies on the circle $x^2 + y^2 = 1$ (since $\sin^2 t + \cos^2 t = 1$).

Differentiating vectors

The result of differentiating a vector has been assumed to be that vector formed by differentiating its x and y components. This is also consistent with the general definition of differentiation as a rate of change. In a small period of time δt, the particle P moves from a point A, with position vector $\mathbf{r}(t)$, to B with position vector $\mathbf{r}(t + \delta t)$.

From figure 1.7 you can see that the vector AB is $\mathbf{r}(t + \delta t) - \mathbf{r}(t)$. This is the change in position of P:

$$\delta\mathbf{r} = \mathbf{r}(t + \delta t) - \mathbf{r}(t)$$

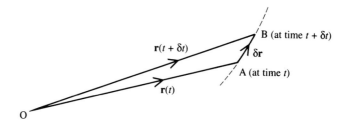

Figure 1.7

The average velocity of P over this small period is the change in position divided by time, $\dfrac{\delta \mathbf{r}}{\delta t}$.

As δt tends to zero, $\dfrac{\delta \mathbf{r}}{\delta t}$ becomes the instantaneous velocity at A. This is written $\dfrac{d\mathbf{r}}{dt}$, where

$$\frac{d\mathbf{r}}{dt} = \lim_{\delta t \to 0} \frac{\mathbf{r}(t + \delta t) - \mathbf{r}(t)}{\delta t}$$

This definition is independent of any particular co-ordinate system. But $\delta \mathbf{r}$ can be written in terms of its components (figure 1.8):

$$\delta \mathbf{r} = \delta x\, \mathbf{i} + \delta y\, \mathbf{j}$$

$$\Rightarrow \quad \frac{\delta \mathbf{r}}{\delta t} = \frac{\delta x}{\delta t}\, \mathbf{i} + \frac{\delta y}{\delta t}\, \mathbf{j}$$

As $\delta t \to 0$

$$\frac{d\mathbf{r}}{dt} = \frac{dx}{dt}\, \mathbf{i} + \frac{dy}{dt}\, \mathbf{j}$$

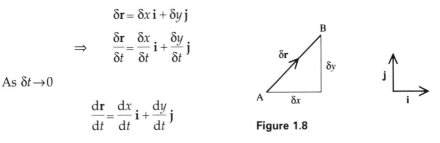

Figure 1.8

This is the result you obtain simply by differentiating the x and y components separately.

Newton's Second Law in vector form

If $\mathbf{F} = F_x \mathbf{i} + F_y \mathbf{j}$ is a force acting on a body of mass m, then the acceleration along each component axis is derived from the component of force along that axis:

$$F_x = m\ddot{x} \quad \text{and} \quad F_y = m\ddot{y}$$

where (x, y) are the co-ordinates of the body. Thus Newton's Second Law can be expressed in vector form:

$$\mathbf{F} = m\mathbf{a}$$

as you saw in *Mechanics 1*. The acceleration $\mathbf{a}$ may be written as $\dfrac{d\mathbf{v}}{dt}$ or $\dfrac{d^2\mathbf{r}}{dt^2}$.

The vectors $\mathbf{F}$ and $\mathbf{a}$ may, of course, be three dimensional, in which case there will be a z component also:

$$\mathbf{F} = F_x \mathbf{i} + F_y \mathbf{j} + F_z \mathbf{k} = m\ddot{x}\mathbf{i} + m\ddot{y}\mathbf{j} + m\ddot{z}\mathbf{k}$$

EXAMPLE

A particle P of mass 2 kg moves under the action of a force, which at time t is given by

$$\mathbf{F} = 8\cos 2t\, \mathbf{i} + 8\sin 2t\, \mathbf{j}$$

At time $t = 0$, P is at rest at the origin.
(i)   Find the position vector of P relative to O at time t.
(ii) Find at what times the kinetic energy is a maximum.

Solution

(i) By Newton's Second law

$$m \frac{d\mathbf{v}}{dt} = \mathbf{F}$$

So

$$2 \frac{d\mathbf{v}}{dt} = 8 \cos 2t\mathbf{i} + 8 \sin 2t\mathbf{j}$$

$$\frac{d\mathbf{v}}{dt} = 4 \cos 2t\mathbf{i} + 4 \sin 2t\mathbf{j}$$

Integrating gives:

$$\mathbf{v} = 2 \sin 2t\mathbf{i} - 2 \cos 2t\mathbf{j} + \mathbf{p}$$

> Each component on the right-hand side has been integrated and the vector $\mathbf{p}$ is the constant of integration. The right-hand side might equally have been written with separate constants for each component of $\mathbf{p}$:
> $$\mathbf{v} = (2 \sin 2t + p_1)\mathbf{i} - (2 \cos 2t + p_2)\mathbf{j}$$

The body is at rest when $t = 0$, so

$$0 = -2\mathbf{j} + \mathbf{p} \quad \Rightarrow \quad \mathbf{p} = 2\mathbf{j}$$

So

$$\mathbf{v} = 2 \sin 2t\mathbf{i} + (2 - 2 \cos 2t)\mathbf{j}$$

Writing $\mathbf{v} = \dfrac{d\mathbf{r}}{dt}$ and integrating again gives

$$\mathbf{r} = -\cos 2t\mathbf{i} + (2t - \sin 2t)\mathbf{j} + \mathbf{q}$$

The constant of integration $\mathbf{q}$ is again determined by looking at the initial conditions. When $t = 0$, the particle is at the origin, so

$$0 = -\mathbf{i} + \mathbf{q} \quad \Rightarrow \quad \mathbf{q} = \mathbf{i}$$

The position at time t is thus

$$\mathbf{r} = (1 - \cos 2t)\mathbf{i} + (2t - \sin 2t)\mathbf{j}$$

(ii) The kinetic energy is

$$\tfrac{1}{2}m v^2 = \tfrac{1}{2} \times 2 \times |\mathbf{v}|^2$$

> Sum of the squares of the components of v

$$= (2 \sin 2t)^2 + (2 - 2 \cos 2t)^2$$

$$= 4 \sin^2 2t + 4 - 8 \cos 2t + 4 \cos^2 2t$$

$$= 8(1 - \cos 2t) \qquad (\text{since } \sin^2 2t + \cos^2 2t = 1)$$

$$= 16 \sin^2 t$$

This is maximum when $\sin t = \pm 1$, that is $t = \pi/2,\ 3\pi/2,\ 5\pi/2, \ldots$

HISTORICAL NOTE

It was more than 60 years after Newton's discoveries before it was realised that his Second Law could be conveniently expressed in the form of components of acceleration and force along Cartesian axes. Euler was the first to present it in this way, in 1736. The vector notation was invented much later, by 19th century mathematical physicists (notably the American Willard Gibbs and the Scot James Clerk Maxwell) who used vectors to give a much simpler formulation of the equations describing electric and magnetic fields. The elegant and powerful notation then spread to mechanics and many other branches of physics.

Why use vectors?

In the foregoing examples, the x and y components of the vectors were treated independently. So you may wonder why write, for example,

$$\mathbf{a} = \cos t\mathbf{i} + \sin t\mathbf{j}$$

rather than

$$a_x = \cos t \qquad a_y = \sin t$$

The vector form is more compact, but is this the only reason to use vectors? The answer is no. There are vector operations which make vector methods a much simpler way of dealing with certain problems. The *scalar product*, a reminder of which is given below, is one such operation. The *vector product*, which you will meet later, is another example, almost essential in three-dimensional work.

The scalar product

In *Pure Mathematics 3* you met the *scalar* or *dot* product between two vectors. Its properties are summarised here together with some simple applications in mechanics.

The scalar product is written $\mathbf{p}.\mathbf{q}$, where $\mathbf{p}$ and $\mathbf{q}$ are vectors. The result is a number (or scalar), *not* a vector. It is defined as the product of the magnitudes of the vectors and the cosine of the angle between them (figure 1.9):

$$\mathbf{p}.\mathbf{q} = |\mathbf{p}||\mathbf{q}|\cos\theta$$

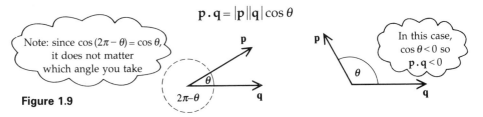

Note: since $\cos(2\pi - \theta) = \cos\theta$, it does not matter which angle you take

In this case, $\cos\theta < 0$ so $\mathbf{p}.\mathbf{q} < 0$

Figure 1.9

From the definition, it is clear that $\mathbf{p}.\mathbf{q} = \mathbf{q}.\mathbf{p}$ (commutative law). It can also be shown that

$$\mathbf{p}.(\mathbf{q} + \mathbf{r}) = \mathbf{p}.\mathbf{q} + \mathbf{p}.\mathbf{r} \quad \text{(distributive law)}$$

so that the scalar product on vectors can be used rather like ordinary multiplication on numbers.

The value of the scalar product is easy to work out if the components of the vectors in the direction of x and y (and z in the three-dimensional case) are known. Given that

$$\mathbf{p} = p_x\mathbf{i} + p_y\mathbf{j}$$

$$\mathbf{q} = q_x\mathbf{i} + q_y\mathbf{j}$$

then $\qquad\qquad\qquad\qquad \mathbf{p} \cdot \mathbf{q} = p_x q_x + p_y q_y$

In the three-dimensional case, when each vector has a z component, the result is

$$\mathbf{p} \cdot \mathbf{q} = p_x q_x + p_y q_y + p_z q_z$$

The scalar product of two non-zero vectors is zero if and only if the two vectors are perpendicular (because $\cos 90°$ is zero).

The scalar product of a vector with itself is the square of its magnitude:

$$\mathbf{p} \cdot \mathbf{p} = |\mathbf{p}|^2 = p_x^2 + p_y^2 \quad (p_x^2 + p_y^2 + p_z^2 \text{ in three dimensions})$$

The component of a force in a given direction

One use of the scalar product is the calculation of the component of a force in any given direction. If $\mathbf{F}$ is a force and $\hat{\mathbf{n}}$ is a *unit* vector, then $\mathbf{F} \cdot \hat{\mathbf{n}}$ is the component (or resolved part) of a force $\mathbf{F}$ in the direction of $\hat{\mathbf{n}}$. You can see from figure 1.10 that $\mathbf{F} \cdot \hat{\mathbf{n}}$ has the value $|\mathbf{F}| \cos \theta$ (since $|\hat{\mathbf{n}}| = 1$), that is the resolved part in the direction $\hat{\mathbf{n}}$.

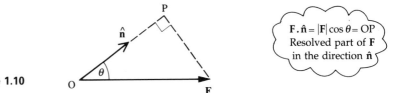

$\mathbf{F} \cdot \hat{\mathbf{n}} = |\mathbf{F}| \cos \theta = OP$
Resolved part of $\mathbf{F}$
in the direction $\hat{\mathbf{n}}$

Figure 1.10

EXAMPLE

A small ring of mass 100 g slides on a smooth straight wire in the direction $\mathbf{i} + 2\mathbf{j} + 2\mathbf{k}$. Apart from the normal reaction of the wire, forces on the ring are its weight $-\mathbf{k}$ N and a constant force due to the wind of $\frac{1}{2}\mathbf{i} + \frac{1}{2}\mathbf{j}$ N. The ring has an initial speed of 2 m s^{-1}.

(i) Find the unit vector in the direction of the wire.
(ii) Find the resolved part of the total forces on the ring in the direction of motion.
(iii) Hence find the acceleration of the ring and the distance travelled before coming to rest.

Solution

(i) $\qquad\qquad\qquad\qquad |\mathbf{i} + 2\mathbf{j} + 2\mathbf{k}| = \sqrt{(1^2 + 2^2 + 2^2)} = 3$

Hence the unit vector in the direction of motion is

$$\hat{\mathbf{n}} = \tfrac{1}{3}(\mathbf{i} + 2\mathbf{j} + 2\mathbf{k}) = \tfrac{1}{3}\mathbf{i} + \tfrac{2}{3}\mathbf{j} + \tfrac{2}{3}\mathbf{k}$$

(ii) The normal reaction R of the wire is perpendicular to the wire and has no component parallel to it. The resultant of the other forces is

$$\mathbf{F} = \tfrac{1}{2}\mathbf{i} + \tfrac{1}{2}\mathbf{j} - \mathbf{k}$$

The resolved part in the direction of motion is

$$\mathbf{F}.\hat{\mathbf{n}} = (\tfrac{1}{6} + \tfrac{1}{3} - \tfrac{2}{3}) = -\tfrac{1}{6}\,\text{N}$$

(iii) Applying $F = ma$ in the direction of motion gives

$$-\tfrac{1}{6} = \tfrac{1}{10}a \quad (\text{mass is } \tfrac{1}{10}\,\text{kg})$$

$$\Rightarrow \quad a = -\tfrac{5}{3}$$

The ring decelerates at $\tfrac{5}{3}\,\text{m s}^{-2}$.

To find when the ring comes to rest, apply $v^2 = u^2 + 2as$:

$$0 = 4 - 2 \times \tfrac{5}{3}s$$

$$\Rightarrow \quad s = \tfrac{6}{5}$$

The ring comes to rest after $\tfrac{6}{5}$ m.

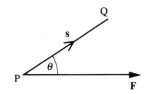

Work and power as scalar products

When a constant force $\mathbf{F}$ acts on a body which moves through a displacement $\mathbf{s}$ (figure 1.11), the work done is the component of $\mathbf{F}$ in the direction of motion, multiplied by the distance travelled:

$$\text{work done} = |\mathbf{F}|\cos\theta|\mathbf{s}| = \mathbf{F}.\mathbf{s}$$

Figure 1.11

In general, forces are variable and objects move through curves, rather than straight lines, in two and three dimensions; the calculation of the work done is then more difficult. However, it still involves the scalar product, as you will see in Chapter 3.

When a force $\mathbf{F}$ moves a body through a displacement $\delta\mathbf{s}$ in time δt, the work done is

$$\delta W = \mathbf{F}.\delta\mathbf{s}$$

The rate of working, i.e. power, is thus

$$\frac{\delta W}{\delta t} = \mathbf{F}.\left(\frac{\delta\mathbf{s}}{\delta t}\right)$$

In the limit as $\delta t \to 0$, this gives

$$\text{power} = \frac{\mathrm{d}W}{\mathrm{d}t} = \mathbf{F}.\mathbf{v}$$

The power expended by a force on a moving body is its speed multiplied by the component of the force in the direction of motion. This is the equivalent of the Fv formula you use when the force is in the same direction as the motion.

With reference to figure 1.12, the power at P is

$$|\mathbf{F}|\cos\theta|\mathbf{v}| = \mathbf{F}.\mathbf{v}$$

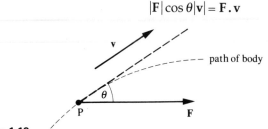

Figure 1.12

EXAMPLE

A particle of mass 2 kg, starting from the rest at the origin, moves under the action of a force, which at time t is given by

$$\mathbf{F} = 8\cos 2t\mathbf{i} + 8\sin 2t\mathbf{j}$$

In the example on page 8, it was shown that at time t, the velocity is

$$\mathbf{v} = 2\sin 2t\mathbf{i} + (2 - 2\cos 2t)\mathbf{j}$$

the acceleration is

$$\mathbf{a} = 4\cos 2t\mathbf{i} + 4\sin 2t\mathbf{j}$$

and that the kinetic energy is

$$E = 8(1 - \cos 2t)$$

(i)   Find the first time $t(t > 0)$ that the velocity is perpendicular to the acceleration.

(ii) Find the rate of working at time t.

(iii) Verify that the power is equal to the rate of change of kinetic energy.

Solution

(i)   When the acceleration is perpendicular to the velocity, $\mathbf{a}.\mathbf{v} = 0$.

$$\mathbf{a}.\mathbf{v} = (4\cos 2t)(2\sin 2t) + (4\sin 2t)(2 - 2\cos 2t)$$

$$= 8\sin 2t\cos 2t + 8\sin 2t - 8\sin 2t\cos 2t$$

$$= 8\sin 2t$$

The first time for which $\mathbf{a}.\mathbf{v} = 0$ is thus when $\sin 2t = 0$, that is $t = \pi/2$.

(ii)   Power $= \mathbf{F}.\mathbf{v} = [8\cos 2t\mathbf{i} + 8\sin 2t\mathbf{j}].[2\sin 2t\mathbf{i} + (2 - 2\cos 2t)\mathbf{j}]$

$$= 16\sin 2t\cos 2t + 8\sin 2t(2 - 2\cos 2t)$$

$$= 16\sin 2t$$

(iii) Given the kinetic energy $E = 8(1 - \cos 2t)$, the rate of change of kinetic energy is

$$\frac{dE}{dt} = 16\sin 2t$$

which is the power.

This is a general result. Work done by all forces = increase in kinetic energy Rate of working = power = rate of increase in kinetic energy

1. The position vector of a particle relative to an origin O at time t is $t^3\mathbf{i} + 9t^2\mathbf{j}$. Find the speed of the particle at time $t = 3$.

2. A particle P initially at rest on the y axis has an acceleration at time t given by $6t\mathbf{i} + 2\mathbf{j}$. Find expressions for the velocity and position vector at time t.

3. Three forces $\mathbf{F}_1 = 2\mathbf{i} + \mathbf{j}$, $\mathbf{F}_2 = -\mathbf{i} - 5\mathbf{j}$ and $\mathbf{F}_3 = 2\mathbf{i}$ act on a particle at a point with position vector $\mathbf{i} + 2\mathbf{j}$.
   (i) Find the resultant force.
   (ii) What is the vector equation of the line of action of the resultant force?
   (iii) What is the moment of the resultant force about the origin?

4. At time $t = 0$, a body of mass 1 kg is moving with a velocity of $4\mathbf{i}$ m s^{-1}. A force $\mathbf{F} = 2t\mathbf{i} + 3\mathbf{j}$ N acts on it. Find the velocity and kinetic energy when $t = 2$.

5. At time $t = 0$, particles P and Q are at points with respective position vectors $\mathbf{i} + \mathbf{j} + \mathbf{k}$ and $7\mathbf{i} + 13\mathbf{j} - 5\mathbf{k}$. P is moving with a constant velocity $\mathbf{i} + 2\mathbf{j} - \mathbf{k}$ and Q with constant velocity $-2\mathbf{i} - 4\mathbf{j} + 2\mathbf{k}$.
   (i) Write down the vector equation of the line representing the path of P.
   (ii) Show that Q moves on the same line.
   (iii) Find t when P and Q collide.

6. At time t, the position vector of a body of mass m is $t^2\mathbf{i} + \sin t\mathbf{j}$.
   (i) Calculate the force $\mathbf{F}$ acting on the body at time t.
   (ii) Use $\mathbf{F}.\mathbf{v}$ to work out the rate at which $\mathbf{F}$ is doing work at time t.
   (iii) Integrate to find the total work done between times 0 and t and verify this is the same as the increase in kinetic energy of the body.

7. At time t, a charged particle moving in an electric field has a position vector
$$\mathbf{r} = l(\omega t - \sin \omega t)\mathbf{i} + l(\omega t + \cos \omega t)\mathbf{j}$$
where ω and l are constants. Show that the acceleration has constant magnitude $l\omega^2$.

8. A charged particle of mass m is constrained to move inside a smooth straight tube whose direction is parallel to $3\mathbf{i} - 4\mathbf{j}$, where $\mathbf{i}$ is taken as horizontal and $\mathbf{j}$ is vertically upwards. As well as gravity, the particle is acted on at time t by an electrical force $-\frac{1}{3}mgt\mathbf{i}$. The particle is at rest at time $t = 0$.
   (i) Write down the resolved part of the total force in the direction of the tube.
   (ii) By applying Newton's Second Law in the direction of the tube, show that the particle is next at rest after 8 seconds.

9. A body of unit mass is initially at the origin moving with a velocity $-2\mathbf{j}$. It is acted on by a force
$$\mathbf{F} = 2\cos t\mathbf{i} + 2\sin t\mathbf{j}$$
   (i) Calculate the acceleration and velocity of the body at time t.
   (ii) Show that the velocity is always perpendicular to the force and that the speed is constant.
   (iii) Calculate the position at time t.
   (iv) Show that the body moves in a circle and finds its centre and radius.
   (v) Show that the force has a direction and magnitude consistent with this circular motion.

10. The position vector $\mathbf{r}$ metres of a particle P at time t seconds is given by
$$\mathbf{r} = \sin 3t\mathbf{i} - \cos 3t\mathbf{j}$$
   (i) Find the velocity of P at time t.
   (ii) Show that the speed of P is constant and find its value.

11. A particle P moves so that at time t its position vector $\mathbf{r}$ relative to a fixed origin O is
$$\mathbf{r} = (a - a\cot^2 \omega t)\mathbf{i} + 2a\cot \omega t\mathbf{j}, \quad 0 < \omega t \leqslant \tfrac{\pi}{2}$$
where a and ω are positive constants.
   (i) Show that $|\mathbf{r}| = a\csc^2 \omega t$.
   (ii) Show that $\mathbf{v}$, the velocity of P, is given by
$$\mathbf{v} = 2|\mathbf{r}|\omega[\cot \omega t\mathbf{i} - \mathbf{j}]$$
   and find, in terms of a and ω, the minimum speed of P.
   (iii) Find t when P is moving perpendicular to OP and show that P never moves parallel to OP.

[AEB 1994]

2

Relative motion

As we rush, as we rush, in the train,
The trees and the houses go wheeling back,
But the starry heavens above that plain
Come flying in our track.

James Thomson

For Discussion

What is the writer of the above verse saying?

When you are in a train, you know that you are moving and the trees and houses are stationary ... or are they?

Do stars move?

Do you think there is anywhere in the universe which is stationary?

For centuries mariners and other travellers have located their positions using the sun and stars. The need to understand their motion relative to the earth and to each other gave a powerful impetus to the study of astronomy and many branches of mathematics and physics.

Your perception of the motion of an object relative to you depends partly on the way you perceive distance. The apparent velocity of an object depends on such things as the time it takes to cross your field of vision and the way its position changes relative to other objects. Fortunately, however, you do not need to rely on your senses to describe relative motion; it can be described simply and precisely using the vector ideas with which you are familiar. The theory in this chapter provides a powerful tool for solving problems when two or more objects are in motion.

Definitions and notation

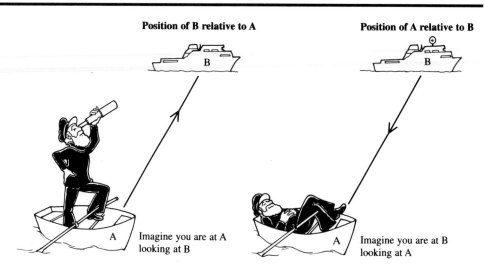

Position of B relative to A

Position of A relative to B

B

B

A — Imagine you are at A looking at B

A — Imagine you are at B looking at A

Relative position

The position of a point B relative to a point A is defined as the displacement vector $\overrightarrow{AB}$. Conversely, the position of A relative to B is the vector $\overrightarrow{BA}$.

It is useful when considering relative motion to denote the position vectors of A and B when referred to an origin O by $\mathbf{r}_A$ and $\mathbf{r}_B$ respectively. Then the position vector of B *relative to* A is given by $\overrightarrow{AB} = \mathbf{r}_B - \mathbf{r}_A$ and the position vector of A *relative to* B is $\overrightarrow{BA} = \mathbf{r}_A - \mathbf{r}_B$ as shown in figure 2.1.

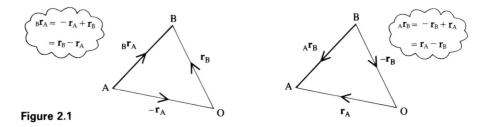

Figure 2.1

An alternative notation for the position vector of B relative to A is $_B\mathbf{r}_A$, which preserves the order of the letters. Then

$$_B\mathbf{r}_A = \mathbf{r}_B - \mathbf{r}_A$$

Similarly, the position of A relative to B is

$$_A\mathbf{r}_B = \mathbf{r}_A - \mathbf{r}_B = -_B\mathbf{r}_A$$

Notice that these are independent of the position of the origin O.

Relative velocity and acceleration

The velocity of a point is defined as the rate of change of its position vector:

$$\mathbf{v} = \frac{d\mathbf{r}}{dt} = \dot{\mathbf{r}}$$

A velocity found in this way is not an absolute velocity, however. It is the velocity of the point *relative to the origin*, which is assumed to be fixed. When the origin is not stationary, it is possible to define the velocity of any point relative to it by differentiating the equation for the relative position. Using a similar notation to that for relative position, the velocity of B relative to A is

$$_B\mathbf{v}_A = _B\dot{\mathbf{r}}_A = \dot{\mathbf{r}}_B - \dot{\mathbf{r}}_A$$

or

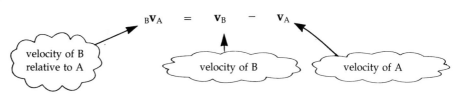

Again, this is independent of the position of the origin O.

The velocity of A relative to B is

$$_A\mathbf{v}_B = \mathbf{v}_A - \mathbf{v}_B = -_B\mathbf{v}_A$$

The relative accelerations of A and B are obtained by differentiating again with respect to time.

The acceleration of A relative to B is

$$_A\mathbf{a}_B = \mathbf{a}_A - \mathbf{a}_B$$

Relative motion in one dimension

The examples which follow illustrate how problems involving two or more moving bodies can be simplified using the foregoing definitions.

EXAMPLE

An escalator of length h m is moving upwards at a constant rate of u m s^{-1}.
(i) Sarah walks up it at a constant speed of v m s^{-1} relative to the escalator. How long will it take her to reach the top?
(ii) Tom runs down the escalator at a constant speed of w m s^{-1} relative to the escalator. How long will it take him to reach the bottom?
(iii) Sarah walks at 0.8 m s^{-1} and Tom runs at 1.4 m s^{-1} and they take the same time to negotiate the escalator. How fast is it moving?

Solution

(i) The diagram shows the velocities of Sarah and the escalator, using S for Sarah and E for the escalator.

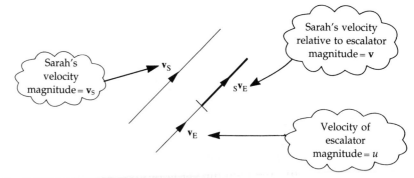

Sarah's velocity relative to the escalator is

$$_S\mathbf{v}_E = \mathbf{v}_S - \mathbf{v}_E$$

so taking upwards as the positive direction, Sarah's actual speed v_S is given by

$$v = v_S - u$$
$$\Rightarrow \quad v_S = v + u$$

The time taken by Sarah to travel a distance h at this speed is $\dfrac{h}{(u+v)}$ seconds.

(ii) Similarly for Tom

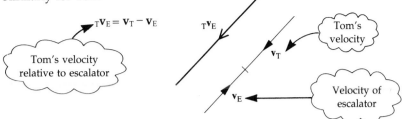

Tom's velocity relative to escalator

$_T\mathbf{v}_E = \mathbf{v}_T - \mathbf{v}_E$

$_T\mathbf{v}_E$

$\mathbf{v}_T$

$\mathbf{v}_E$

Tom's velocity

Velocity of escalator

Taking downwards as the positive direction

$$w = v_T - (-u)$$

$$\Rightarrow \quad v_T = w - u$$

The time taken by Tom is $\dfrac{h}{(w - u)}$ seconds.

(iii) When they both take the same time

$$\frac{h}{(u + v)} = \frac{h}{(w - u)}$$

$$\Rightarrow \quad u + v = w - u$$

$$\Rightarrow \quad u = \tfrac{1}{2}(w - v)$$

$$= \tfrac{1}{2}(1.4 - 0.8)$$

$$= 0.3$$

The speed of the escalator is 0.3 m s^{-1}.

EXAMPLE

A truck of length 8 m is travelling at a steady speed of 25 m s^{-1} (90 km h^{-1}) on the inside lane of a relatively quiet motorway. As it passes a junction, it overtakes a car of length 4 m coming down the slip road at 17 m s^{-1} (61.2 km h^{-1}). When the front of the car is level with the rear of the truck, the car begins to accelerate at a constant rate. After 6.5 s it reaches a speed of 30 m s^{-1} (108 km h^{-1}), which it then maintains while moving into the outer lane. After overtaking the truck, the car moves into the inner lane leaving a gap of 10 m. Ignoring the extra amount which the lane changes add to the distance moved by the car, find
(i) the time the car is furthest behind the truck and their distance apart at that time;
(ii) the time the front of the car is level with the front of the truck;
(iii) the time taken by the car in completing the manoeuvre and the distance it travels during this time.

Solution

(i) The acceleration of the car can be found using $v = u + at$:

$$30 = 17 + 6.5a$$

$$\Rightarrow \quad a = 2$$

It is now possible to solve the problem using relative motion.

The diagram shows the initial and final positions of the car *relative to the truck*. The direction of motion is taken as positive.

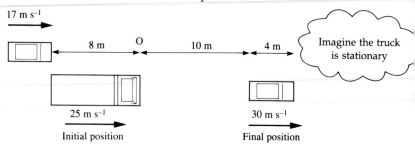

The acceleration of the car relative to the truck is given by

$$_Ca_T = a_C - a_T = 2 \quad \text{(the truck has zero acceleration)}$$

The initial relative velocity is

$$_Cu_T = u_C - u_T$$
$$= 17 - 25$$
$$= -8$$

Using $v = u + at$ for the relative motion gives

$$_Cv_T = -8 + 2t \quad (t \leqslant 6.5)$$

Integrating gives

$$_Cs_T = -8t + t^2 + k$$

The initial displacement of the front of the car relative to the front of the truck is –8, so the relative displacement after time t is

$$_Cs_T = -8t + t^2 - 8 \quad (t \leqslant 6.5)$$
$$= (t - 4)^2 - 24 \qquad \qquad \text{①}$$

The function on the right-hand side is a minimum when $t = 4$. Then

$$_Cs_T = -24$$

The car is furthest behind after 4 seconds when it is 24 m behind the front of the truck, leaving a gap of 16 m.

(ii) From ①, when $t = 6.5$, $_Cs_T = 6.25 - 24$
$$= -17.75$$

After this the speeds of the car and truck are both constant so the relative velocity is a constant $(30 - 25) = 5$ m s^{-1}. The time from this point is $(t - 6.5)$ seconds, so the relative displacement is now given by

$$_Cs_T = -17.75 + 5(t - 6.5) \quad (t \geqslant 6.5)$$

and is zero when $t = 10.05$, i.e. the car and the truck are level after a total of about 10 seconds.

(iii) The time taken by the car to move in front of the truck is given by

$$14 = -17.75 + 5(t - 6.5)$$
$$\Rightarrow \quad t = 12.85$$

The total time for the manoeuvre is 12.85 seconds and during this time the truck has travelled a distance of $25 \times 12.85 = 321.25$ m.

The car has travelled $321.25 + 8 + 10 + 4 = 343$ m (to 3 significant figures).

The diagram shows the displacement–time graphs for the motion of the truck, the car and the car relative to the truck.

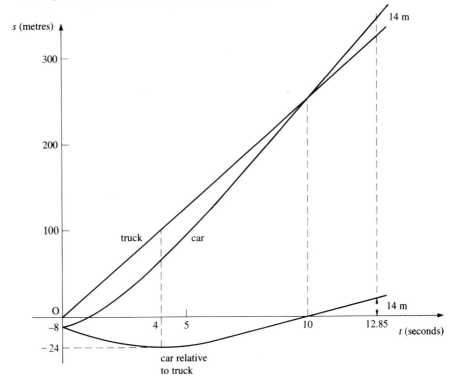

Relative motion in two dimensions

The effect of wind or current

When an object is moving in air or water, it is likely to experience resistive forces due to the presence of the medium, such as those discussed in Chapter 3, but its motion might also be affected by the motion of the air or water. The courses of aircraft and boats are altered by winds and currents; someone wishing to row a boat directly across a river, for example, must aim at a point upstream (figure 2.2). For a boat moving in a current, its velocity relative to the water is given by the equation

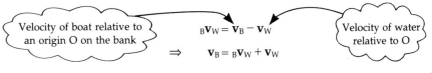

Velocity of boat relative to an origin O on the bank

$$_B\mathbf{v}_W = \mathbf{v}_B - \mathbf{v}_W$$

Velocity of water relative to O

$$\Rightarrow \quad \mathbf{v}_B = {}_B\mathbf{v}_W + \mathbf{v}_W$$

The *actual velocity* of the boat is the vector sum of the velocity of the boat relative to the water and the velocity of the current.

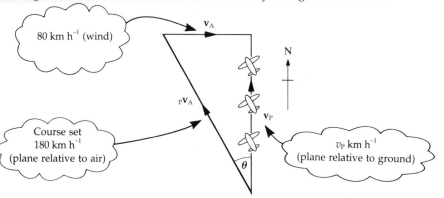

The boat is aimed upstream in the direction of $_B\mathbf{v}_W$ but actually travels straight across

$\mathbf{v}_W$

$_B\mathbf{v}_W$

$\mathbf{v}_B$

current

Figure 2.2

The velocity of the boat relative to the water, $_B\mathbf{v}_W$, is sometimes called the 'velocity in still water', the boat would have this velocity in the absence of any current.

It is interesting to note that a boat which is floating in a current without power (i.e. $_B\mathbf{v}_W = 0$) moves with the same velocity as the water and suffers no resistance to its motion due to the water.

EXAMPLE

The pilot of a plane wishes to fly due north but a wind of 80 km h^{-1} is blowing from the west. The speed of the plane in still air is 180 km h^{-1}.
(i)   In what direction should the pilot set her course?
(ii) What is the speed of the plane over the ground?

Solution

Using P and A to represent the plane and the air, the equation for relative motion is

$$_P\mathbf{v}_A = \mathbf{v}_P - \mathbf{v}_A$$

$$\Rightarrow \quad \mathbf{v}_P = {_P\mathbf{v}_A} + \mathbf{v}_A$$

The direction of $\mathbf{v}_P$ is required to be due north, $\mathbf{v}_A$ is 80 km h^{-1} from the west and the magnitude of $_P\mathbf{v}_A$ is 180 km h^{-1}, so the velocity triangle is as shown.

80 km h^{-1} (wind)

$\mathbf{v}_A$

$_P\mathbf{v}_A$

N

$\mathbf{v}_P$

Course set
180 km h^{-1}
(plane relative to air)

v_P km h^{-1}
(plane relative to ground)

θ

The direction of the course set is the direction of $_P\mathbf{v}_A$ and the plane actually moves north with speed $v_P = |\mathbf{v}_P|$ km h^{-1}.

(i) From the diagram $\sin\theta = \dfrac{80}{180}$

$$\theta = 26.39°$$

The course set should be on a bearing of 333.6°.

(ii) Using Pythagoras's theorem:

$$v_P^2 = (180)^2 - (80)^2$$

$$\Rightarrow \quad v_P = 161.25$$

The speed of the plane over the ground (its 'ground speed') is 161 km h^{-1}.

The relative velocity of the wind

In the last two examples, the velocities of the plane and the boat were changed by the wind and the current. In the next example, the wind does not blow the cyclist off course, but the cyclist has to work harder to maintain his speed when the wind is against him because the air resistance depends on the velocity of the wind relative to him. A cyclist travelling at 10 m s^{-1} in still air feels the air resistance as a wind blowing against him at 10 m s^{-1}. When there is an actual wind of 10 m s^{-1} blowing against him, he will feel a wind of 20 m s^{-1}, but when the wind is behind him, we have a situation (at last!) where there is no air resistance at all.

EXAMPLE

A cyclist is travelling due north at 10 m s^{-1} and there is a wind blowing. Using **i** and **j** as unit vectors in the east and north directions:
(i)   Find the velocity of the wind relative to the cyclist when it is blowing from the south west at a speed of 8 m s^{-1}.
(ii) Is it possible for the 8 m s^{-1} wind to appear to be coming from the west?
(iii) Show that, if the wind changes direction but maintains the speed of 8 m s^{-1}, its velocity relative to the cyclist always has a southerly (head on) component no matter in what direction it is blowing.
(iv) Under what conditions does the wind aid the cyclist in the sense that it is behind him and there is no air resistance to his forward motion?

Solution

(i)   Let C represent the cyclist and W the wind. Then

$$_W\mathbf{v}_C = \mathbf{v}_W - \mathbf{v}_C$$

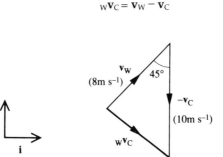

But
$$\mathbf{v_W} = 8\cos 45°\,\mathbf{i} + 8\sin 45°\,\mathbf{j}$$
$$= 4\sqrt{2}\mathbf{i} + 4\sqrt{2}\mathbf{j}$$

Also
$$\mathbf{v_C} = 10\mathbf{j}$$

$$\Rightarrow \quad {}_W\mathbf{v_C} = 4\sqrt{2}\mathbf{i} + (4\sqrt{2} - 10)\mathbf{j}$$

Correct to 3 significant figures, the velocity of the wind relative to the cyclist is $5.66\mathbf{i} - 4.34\mathbf{j}$ m s^{-1} or 7.13 m s^{-1} on a bearing of 128°.

(ii) The diagram illustrates the equation ${}_W\mathbf{v_C} = \mathbf{v_W} - \mathbf{v_C}$ when the wind appears to be coming from the west.

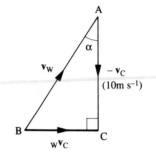

The actual speed of the wind (8 m s^{-1}) is represented by the hypotenuse BA of the right-angled triangle. As this must be greater than the length AC, which represents a speed of 10 m s^{-1}, this situation is impossible.

(iii) The northerly component of the wind is greatest when the wind is blowing from the south and it is then 8 m s^{-1}. The northerly component of the velocity of the wind relative to the cyclist is then $8\mathbf{j} - 10\mathbf{j} = -2\mathbf{j}$. The negative sign indicates that, relative to the cyclist, the wind is southerly, i.e. head on.

If the velocity of the wind is in any other direction, its northerly component is less than 8 m s^{-1}, so its velocity relative to the cyclist has a southerly component of more than 2 m s^{-1}.

(iv) The diagrams show three possible wind velocities, represented by $\overrightarrow{RP}$, and the corresponding relative velocities, represented by $\overrightarrow{RQ}$.

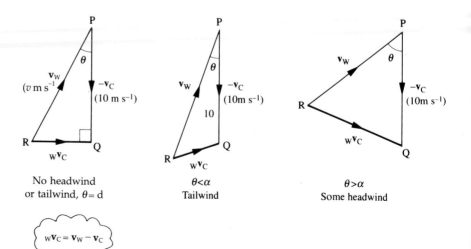

No headwind
or tailwind, $\theta = d$

$\theta < \alpha$
Tailwind

$\theta > \alpha$
Some headwind

$${}_W\mathbf{v_C} = \mathbf{v_W} - \mathbf{v_C}$$

There is no air resistance if the northerly component of the relative velocity is greater than or equal to zero. It can be seen from the first two diagrams that this is impossible unless the speed of the wind is greater than 10 m s^{-1}, i.e. greater than the speed of the cyclist. When the windspeed is $v \text{ m s}^{-1}$ and $v > 10$, it can be seen from the diagrams that the angle θ between its direction and the direction of motion of the cyclist must be less than $\alpha = \arccos(10/v)$. Then the wind aids the cyclist by having a northerly component of relative velocity.

Exercise 2A

1. A car is travelling due east at 60 km h^{-1}. A bird flies over at 45 km h^{-1}. In each of the cases (i) to (v), draw a diagram to show the velocity of the bird relative to the car and calculate this velocity. The bird flies
   (i) due east; (ii) due west; (iii) due north;
   (iv) north east; (v) on a bearing of 230°.

2. Cars A and B are driving north on a dual carriageway. A is on the inside lane travelling with speed 80 km h^{-1} and B is on the outside lane travelling at 110 km h^{-1}. B is just about to overtake A and both cars are of length 4.2 m.
   (i) Find the velocity of B relative to A.
   (ii) Write down the distance B must travel relative to A in order to overtake it.
   (iii) Find the time taken for B to overtake A.

   A third car, C, also of length 4.2 m, is driving south at 100 km h^{-1}.
   (iv) Find the time taken for B and C to pass each other.

3. A woman is walking on horizontal ground at 4 km h^{-1} and the rain is falling vertically at 7.5 km h^{-1}. At what angle to the horizontal should she hold her umbrella?

4. A girl walks across a raft which is on a river and drifting downstream at a speed of 2 m s^{-1}. She walks at right angles to the stream and a boy a short distance away on the bank sees her, apparently walking on water, cross the river at an angle of 20° to the bank. How fast is she actually walking?

5. A swift which can fly in still air at 25 m s^{-1} is caught in a wind blowing at 20 m s^{-1} from the south west. It wishes to return to its nest, which is in a direction due east of its present position. In what direction should it fly relative to the air?

6. A boy swims at a speed v relative to the water.
   (i) He wishes to swim straight across a stream which is flowing with speed u. Show that this is possible only if $v > u$ and find the direction in which he should aim.
   (ii) He decides to aim at right angles to the stream. If the width of the stream is a, how far down stream from his starting point will he land?

7. An intercity train is of length 225 m and travels at a maximum speed of 60 m s^{-1}. A local turbo train is of length 75 m and travels at a maximum speed of 40 m s^{-1}. By considering the relative motion of the trains, find the time taken for the intercity to pass the turbo in the following circumstances:
   (i) They are both travelling at maximum speed
       (a) in the same direction;
       (b) in opposite directions.
   (ii) The turbo is accelerating at a constant rate of 0.75 m s^{-2} out of a station and has a speed of 30 m s^{-1} when the intercity, travelling at maximum speed in the same direction, begins to overtake it. (Assume the turbo accelerates to maximum speed.)

   There seem to be two solutions to part (ii), but only one is correct. In what circumstances would the second solution apply? Would this be realistic?

8. At the instant that a lift starts moving downwards with constant acceleration a m s^{-2}, a child standing in the lift throws a ball upwards with a speed of u m s^{-1} from a height h m above the floor.
   (i) Show that the ball will not hit the ceiling of the lift providing it is at least H m high, where

   $$H = h + \frac{u^2}{2(g-a)}$$

   (ii) Assuming the ball does not hit the ceiling, find the time taken for it to hit the floor in the case where $a = 1.8$, $u = 4$ and $h = 1$.

9. A uniform rectangular parcel is on an open truck at a distance d m from the back. The length of the parcel measured in the direction of motion of the truck is p m and the coefficient of friction between the parcel and the truck is μ.
   (i) Show that when the truck accelerates at a constant rate a m s^{-2} from rest, the parcel slides providing $a > \mu g$.
   (ii) The truck stops accelerating after T s. Given that $a > \mu g$ show that the velocity of the truck relative to the parcel is then $(a - \mu g)T$ and find an expression for the relative displacement.
   (iii) The truck now continues with constant velocity. What happens to the parcel?
   (iv) Show that the parcel will slide completely off the truck if

   $$T^2 > \frac{\mu g(2d + p)}{a(a - \mu g)}$$

   and hence that the maximum velocity the truck can attain at constant acceleration a, if the parcel is not to slide completely off the truck, is

   $$\sqrt{\frac{\mu ga(2d + p)}{(a - \mu g)}}$$

   (v) Find the constant acceleration in the case where $\mu = 0.5$, $g = 10$, $d = 0.75$, $p = 0.5$, and the truck reaches a maximum speed of 18 m s^{-1}.

10. The stopping distance of a vehicle is the total distance travelled by the vehicle in the interval between a driver first seeing a hazard and the vehicle coming to rest. One method of calculating the stopping distance is to use the formula

    $$x = 0.2v + 0.006v^2$$

    where x is the stopping distance in metres and v the vehicle speed in km h^{-1}.
    (i) In this formula the first term represents the distance travelled at a constant speed of v km h^{-1} in the interval T seconds between a driver seeing a hazard and starting to apply the brakes. Find the value of T.
    (ii) The second term represents the distance travelled after the brakes have been applied until the vehicle comes to rest. Assuming the brakes produce a constant retardation of f m s^{-2}, find f.
    (iii) Two cars are travelling in the same direction on a single track road. The leading car is travelling at a steady speed of 50 km h^{-1} and the following car has a steady speed of u km h^{-1}. At the instant when the cars are 50 m apart, the driver of the leading car sees a hazard. The driver of the following car does not see the hazard and is only warned of it when the brake lights of the leading car go on (i.e. when the brakes of the leading car are first applied). Find the greatest possible value of u so that the cars do not collide. [It may be assumed that the above formula applies to both of the cars and both of the drivers.]

    [AEB 1988]

11. A ship is steaming north in still water at 20 km h^{-1} with the wind relative to the ship coming from the direction $330°$. The ship alters course and steams west at the same speed. The wind relative to the ship now comes from the direction $240°$.
    (i) Show that the wind is blowing from the direction $195°$.
    (ii) Find the speed of the wind.

The ship alters course again and heads in a direction due south at a speed of 15 km h^{-1} relative to the water. It encounters a current flowing from west to east at a speed of 8 km h^{-1}. The velocity of the wind is the same as before.

(iii) Find the velocity of the ship.
(iv) Find the direction of the velocity of the wind relative to the ship.

[Cambridge 1992]

12. Cars are moving along a straight road at a steady speed of v m s^{-1}. There is a 2 seconds distance between the cars and they are each of width w m.
    (i) A cat crosses the road at right angles to the direction of motion. Use the equation of relative motion to show that the minimum safe speed for the cat is $w/2$ m s^{-1}, whatever the speed of the cars.
    (ii) If the cat crosses the road at an angle to the motion, show that its minimum safe speed is $\dfrac{vw}{\sqrt{(4v^2 + w^2)}}$.

13. An aircraft flies a distance a due north at a constant airspeed V through the air, which is blowing from the north east with constant speed v ($v < V$).
    (i) (a) Show that the time, t, taken for the outward journey is a root of the equation

    $$(V^2 - v^2)t^2 - \sqrt{2}\,avt - a^2 = 0$$

    (b) Find a similar equation for the time of the return journey and hence show that the difference between the times for the outward and return journeys is $\dfrac{\sqrt{2}av}{(V^2 - v^2)}$.

An alternative approach to this problem is suggested by the next part.

(ii) (a) Draw velocity triangles, ABC and ABD, for the outward journey and the return journey, representing the velocity of the wind by $\overrightarrow{AB}$ in each case. Combine these triangles into one triangle ADC and use it to show that the outward and return speeds are given by

$$V \cos \alpha \pm \frac{v}{\sqrt{2}}$$

where $\alpha =$ angle ACB $= \arcsin(v/\sqrt{2}V)$.

(b) Show that the total time for the outward and return journeys under the same wind conditions is

$$\frac{a\sqrt{(4V^2 - 2v^2)}}{(V^2 - 2v^2)}.$$

14. A pleasure boat travels between two points, A and B, on a river which are 6 km apart. The river flows at a constant speed from B to A. The boat can travel upstream from A to B at a constant speed in 48 minutes with a power setting of 30 kW. It can do the return trip from B to A, also at a constant speed, and with the same power setting, in 36 minutes.
    (i) Determine the speed of flow of the river in kilometres per hour.

In order to do the upstream trip in 36 minutes, the power setting has to be increased to 40 kW. It is believed that the power setting P required is equal to kV^α, where V km h^{-1} is the speed of the boat relative to the water and k and α are positive constants.

(ii) By considering the power settings for the two upstream journeys show that $\alpha = 1.145$ and find k.
(iii) By considering the energy used for an upstream trip taking T minutes, determine, correct to the nearest minute, the value of T which uses least energy.

[MEI 1992]

Collisions and closest approach

Two objects *collide* when their *relative position* is zero, so consideration of their relative motion is a useful way of tackling problems involving collisions.

EXAMPLE

Two roads are modelled by straight lines with equations relative to a fixed origin:

$$\mathbf{r}_1 = 200\mathbf{j} + \lambda(3\mathbf{i} + 4\mathbf{j}) \quad \text{and} \quad \mathbf{r}_2 = 600\mathbf{i} + \mu\mathbf{j}$$

Car A is travelling along the first road towards the point of intersection of the roads with a speed of 15 m s^{-1} and car B is approaching it along the second road with speed $v \text{ m s}^{-1}$. Initially A has position vector $200\mathbf{j}$ and B's position vector is $600\mathbf{i} - 150\mathbf{j}$.
(i) Find the velocities of A and B.
(ii) Assuming the speeds are constant, find the position vector of A relative to B after t seconds.
(iii) A does not stop when it reaches the junction. Find the maximum value of v which will ensure that B reaches the junction after A without needing to slow down.

Solution

(i) The direction of the first road is $3\mathbf{i} + 4\mathbf{j}$, and a unit vector in this direction is $\frac{1}{5}(3\mathbf{i} + 4\mathbf{j})$. A's velocity is $15 \times \frac{1}{5}(3\mathbf{i} + 4\mathbf{j}) = 9\mathbf{i} + 12\mathbf{j}$.

The direction of the second road is $\mathbf{j}$, so B's velocity is $v\mathbf{j}$.

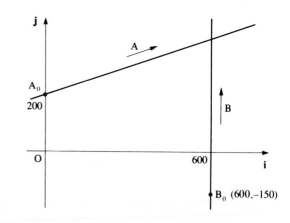

(ii) The initial position of A relative to B is

$$\mathbf{r}_A - \mathbf{r}_B = 200\mathbf{j} - (600\mathbf{i} - 150\mathbf{j})$$

$$= -600\mathbf{i} + 350\mathbf{j}$$

The velocity of A relative to B is

$$\mathbf{v}_A - \mathbf{v}_B = 9\mathbf{i} + (12 - v)\mathbf{j}$$

The position of A relative to B after time t seconds is

$$_A\mathbf{r}_B = -600\mathbf{i} + 350\mathbf{j} + [9\mathbf{i} + (12 - v)\mathbf{j}]t$$

$$= (-600 + 9t)\mathbf{i} + (350 + 12t - vt)\mathbf{j}$$

(iii) The cars will collide if $_A\mathbf{r}_B = 0$.

This can happen only if

$$(-600 + 9t) = 0 \quad \text{and} \quad (350 + 12t - vt) = 0$$

Then
$$t = \frac{600}{9} = \frac{200}{3}$$

and
$$vt = 350 + 12t$$
$$= 1150$$
$$\Rightarrow \quad v = 17.25$$

In order to be sure of avoiding A, car B must travel at no more than 17 m s^{-1}.

The next example illustrates how ideas of relative motion can be applied to find the closest distance between two objects. This is particularly relevant at sea. The closest distance can be found by using vectors or by using geometrical ideas and part (ii) of the example has been done in two ways to illustrate both methods. The vector methods used can be applied in a similar way to three-dimensional motion in the air.

Distances at sea are often measured in nautical miles. This is the length of a part of the equator which subtends an angle of 1 minute at the centre of the earth. The circumference of the earth at the equator is thus $360 \times 60 = 21\,600$ nautical miles. A knot is a speed of 1 nautical mile per hour and is approximately 1.2 mph, 1.9 km h^{-1} or 0.52 m s^{-1}, all correct to 2 significant figures.

EXAMPLE

At noon, a fishing boat is 2 nautical miles south west of a harbour entrance when the crew see a car ferry leaving the harbour at an estimated speed of 9 knots and on a bearing of 190°. The fishing boat wishes to sail due east and is required to avoid the ferry by sailing behind it. Assuming that the boat and ferry are modelled as points and that their velocities are constant:
(i) Find the maximum speed the boat can travel and still avoid the ferry.
(ii) If the boat sails at a constant speed of 6.5 knots due east, at what time is it closest to the ferry and what is the shortest distance between them?

Solution

(i) This problem can be tackled by considering the boat to be stationary and the ferry to be moving with its velocity relative to the boat.

The diagram illustrates the vector equation
$$_F\mathbf{v}_B = \mathbf{v}_F - \mathbf{v}_B.$$

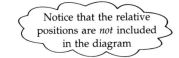

Notice that the relative positions are *not* included in the diagram

The boat and the ferry will be on a collision course if the relative velocity $_F\mathbf{v}_B$ is in a south westerly direction. In that case, the angles will be as shown. The length BC represents the speed, 9 knots, of the ferry. The speed of the boat, v_B knots can be found using the sine rule:

$$\frac{v_B}{\sin 35°} = \frac{9}{\sin 45°}$$

$$\Rightarrow \quad v_B = 7.3004$$

The fishing boat should travel at less than 7.3 knots to avoid the ferry.

(ii) The shortest distance between the fishing boat and the ferry can be found using relative displacements. When considering displacement as well as velocity two diagrams are necessary.

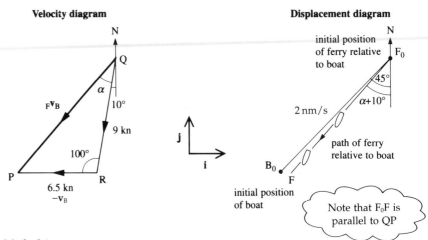

Velocity diagram

Displacement diagram

Note that F_0F is parallel to QP

Method 1

The initial position of the boat is taken as the origin and unit vectors $\mathbf{i}$ and $\mathbf{j}$ are defined to be in the directions east and north respectively.

From the velocity diagram

$$_F\mathbf{v}_B = \mathbf{v}_F - \mathbf{v}_B = (-9\sin 10°\,\mathbf{i} - 9\cos 10°\,\mathbf{j}) - 6.5\mathbf{i}$$

Integrating with respect to t gives

$$_F\mathbf{r}_B = (-9\sin 10°\,\mathbf{i} - 9\cos 10°\,\mathbf{j})t - 6.5t\mathbf{i} + \mathbf{c}$$

When $t = 0$, the position of the ferry relative to the boat is $2\cos 45°\,\mathbf{i} + 2\sin 45°)\mathbf{j}$. Hence

$$_F\mathbf{r}_B = (-9\sin 10°\,\mathbf{i} - 9\cos 10°\,\mathbf{j})t - 6.5t\mathbf{i} + 2\cos 45°\,\mathbf{i} + 2\sin 45°\,\mathbf{j}$$

$$= (\sqrt{2} - 8.0628t)\mathbf{i} + (\sqrt{2} - 8.8633t)\mathbf{j}$$

They are a distance d km apart, where

$$d^2 = (\sqrt{2} - 8.0628t)^2 + (\sqrt{2} - 8.8633t)^2$$

As d is positive (and non-zero), it is a minimum when the quadratic function of t on the right-hand side of this equation is a minimum.

Let $\qquad f(t) = (\sqrt{2} - 8.0628t)^2 + (\sqrt{2} - 8.8633t)^2$

$$f'(t) = 2(\sqrt{2} - 8.0628t)(-8.0628) + 2(\sqrt{2} - 8.8633t)(-8.8633)$$

For minimum $f(t)$, $f'(t) = 0$

$\qquad \Rightarrow \qquad 287.1337t = 47.8742$

$\qquad\qquad \Rightarrow \qquad t = 0.1667$ (giving a time of about 10 minutes)

Then $d = 0.09447$, which represents a distance of 0.09447×1.86 km or about 176 m.

Ignoring the length of the ferry, the shortest distance between the fishing boat and the ferry is about 0.094 nautical miles or 176 m at 10 minutes after noon.

Method 2
Method 1 can be applied to most problems involving variable velocities which are written as vector functions of time and can readily be extended to three dimensions.

For *constant* velocities there is an alternative method of finding the shortest distance, d, using geometry. This relies on the fact that when an object A is moving along a straight line relative to a stationary object B, A is closest to B when BA is at right angles to the path of A.

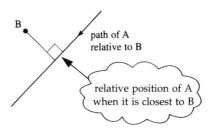

In order to compare the two methods, part (ii) of the foregoing example is repeated using the geometrical method.

First, the magnitude and direction of the relative velocity $_F\mathbf{v}_B$ are found.

Using the cosine rule in triangle PQR

$$_F v_B^2 = 6.5^2 + 9^2 - 117\cos 100°$$

$$\Rightarrow \qquad _F v_B = 11.981\,94$$

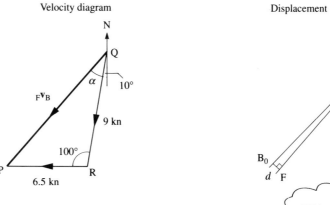

The sine rule gives

$$\sin \alpha = \frac{6.5 \sin 100°}{11.981\,94}$$

$$\Rightarrow \qquad \alpha = 32.2925°$$

The angle $B_0 F_0 F$ is $45 - (32.2925° + 10°) = 2.7075°$.

The ferry and boat are closest when the relative displacement $B_0 F$ is perpendicular to FF_0. Then

$$d = 2 \sin 2.7075$$

$$= 0.094\,475$$

The time taken is $\dfrac{2 \cos 2.7075}{11.981\,94} = 0.166\,73$ hours

As before, the shortest distance is about 0.0945 nautical miles or 176 m after 10 minutes.

HISTORICAL NOTE

The motion of the earth and planets had to be determined from observations of their motion relative to the earth. Before the 16th century, most people believed in the view propounded by Aristotle and elaborated by Ptolemy and his followers, that the earth was the centre of the universe. As more was learnt about the motion of the planets, elaborate systems involving epicycloids were invented to explain such things as the way they appeared to stop and move backwards and forwards in a loop (see Exercise 2B, Question 12). As long ago as the fourth century BC, Heraclides of Pontus (388–310 BC) and Aristarchus of Samos (310–230 BC) had suggested that the earth rotated round the sun, but Copernicus (1473–1543) was the first modern astronomer to question the Ptolemaic system. In 1514 he was invited to comment on the reform of the calendar and as a result of further study of astronomy and the theories of Greek philosophers he developed his theory that the earth rotated daily about its axis and that the planets moved round the sun. A first copy of his great work, On the Revolution of the Celestial Spheres, *written in Latin, is said to have been delivered to him on his death-bed in 1543.*

Galileo (1564–1642) invented the scientific method of combining experimentation with theory and deduced results about falling bodies, projectiles, motion on an inclined plane, and momentum, which indicated that he understood the laws of motion later stated by Newton. In 1623 he described his new scientific method in a paper which included his famous pronouncement that the 'Book of the Universe is . . . written in mathematical characters'.

Galileo developed the telescope for astronomical use and his observations convinced him that Copernicus was correct. Unfortunately, these theories had a profound effect on the prevailing view that human life was the centre of the universe and Galileo was tried for heresy and ultimately forced to spend the last eight years of his life under house arrest.

The theories expounded by Copernicus and Galileo led to two major changes in thought. Because the stars seemed to be fixed relative to the earth, it was realised

that they must be much further away than was previously thought so the universe must approach the infinite. In addition, it had been thought that bodies fell towards their natural place at the centre of the universe. Now that the earth was no longer located at this centre, a new explanation for gravity was required. This eventually led to Newton's law of universal gravitation.

Exercise 2B

1. Particles A and B are initially at points with position vectors $5\mathbf{i} + 13\mathbf{j}$ and $7\mathbf{i} + 5\mathbf{j}$ and are moving with constant velocities $3\mathbf{i} - 5\mathbf{j}$ and $2\mathbf{i} - \mathbf{j}$ respectively.
   (i)   Write down expressions for the initial position of A relative to B, and for the velocity of A relative to B. Hence show that the position of A relative to B after time t is given by

   $$-2\mathbf{i} + 8\mathbf{j} + (\mathbf{i} - 4\mathbf{j})t$$

   (ii) Show that the particles collide and find the time taken for them to do so.
   (iii) Find the position vector of A at this time and hence find the position vector of the point of collision.

2. Ships passing through the Belle Isle Straits north of Newfoundland in the summer are in danger from floating icebergs consisting of pack ice several times their own mass. A ship which is approaching the straits in a southwesterly direction notices an iceberg 5 nautical miles due west. The captain estimates that the current is carrying the iceberg at 2.5 knots on a bearing of 160°.
   (i)   Find the speed of the ship if it is on a collision course with the iceberg.
   (ii) If the ship travels at 4.5 knots, what is the shortest distance between it and the iceberg?

3. Ship A is steaming north at 30 km h^{-1} when ship B is detected at a distance of 40 km and on a bearing of 020° from A. Ship B is steaming west at 25 km h^{-1} and it is known that B's radar can detect other ships up to a maximum distance of 20 km. Without changing speed, A alters course so that it will pass behind B with as little change in direction as possible, but keeping out of the range of B's radar. B maintains the same course and speed throughout.

   (i)   Draw a space diagram to show the path of A relative to B and hence show that the direction of the velocity of A relative to B is 050° after A alters course.
   (ii) Draw a diagram to show the velocity of A relative to B and hence show that the direction of A's altered course is 018° to the nearest degree.
   (iii) Find the magnitude of the velocity of A relative to B and find, to the nearest minute, the time taken for A to get closest to B, measured from the instant that A alters course.

   [Cambridge 1993 adapted]

4. In this question the unit vectors $\mathbf{i}$ and $\mathbf{j}$ are in the directions east and north respectively. The unit of distance is the metre.

   Ships A and B each have constant velocity. Ship A is steaming north at 12 m s^{-1} and at 1000 it sights ship B at the point with position vector $4800\mathbf{i} + 12\,000\mathbf{j}$ relative to A. At 1005 ship B is at the point with position vector $1200\mathbf{i} + 4800\mathbf{j}$ relative to A.
   (i)   Show that the velocity of B relative to A is $(-12\mathbf{i} - 24\mathbf{j})$ m s^{-1} and find the velocity of B in the form $(u\mathbf{i} + v\mathbf{j})$ m s^{-1}.
   (ii) Find an expression for the displacement $\overrightarrow{AB}$, t seconds after 1000.
   (iii) Use your answer to part (ii) to find the function $f(t)$ representing the square of the distance AB.
   (iv) By finding the minimum value of $f(t)$ show that the two ships are nearest to each other at 1008 and find their distance apart at this time.

   [Cambridge 1993 adapted]

5. At noon a boat A is 9 km due west of another boat B. To an observer on B the boat A always appears to be moving on a bearing of 150° (i.e. S30°E) with constant speed 2.5 m s^{-1}.
   (i) Find the time at which the boats are closest together and the distance between them at this time.
   (ii) The velocities of B and a third boat C are given to be $(3i + 4j)$ m s^{-1} and $(5.5i + 2j)$ m s^{-1} respectively, where i and j denote unit vectors directed east and north respectively. At noon, C is directly north of A and the boats B and C are on collision course. Find
       (a) the velocity of C relative to B;
       (b) the time at which B and C would collide;
       (c) the distance between A and C at noon.

   [AEB 1987]

6. At a given instant, a cyclist A passes through a point with position vector $(-9i - 5j)$ km relative to a fixed point O and is travelling with a constant velocity of $(9i + 7j)$ km h^{-1}. At the same instant another cyclist B is passing through a point with position vector $(6i + 5j)$ km relative to O and travelling with a constant velocity of $(7i + 6j)$ km h^{-1}.
   (i) Write down the vector equations of the paths of the two cyclists and determine the position vector of the point of intersection of the two paths. Hence, or otherwise, deduce that B passes through this point one hour before A.
   (ii) Show that, initially, the position vector of B relative to A is $(15i + 10j)$ km and find the position vector of B relative to A at time t hours later. Hence, or otherwise, show that the cyclists are closest together when $t = 8$ and find this minimum separation.

   [AEB 1988]

7. An old man cycles to work each day leaving his home at a point with position vector $(-2i - 8j)$ km relative to an origin O at the centre of the village. The unit vectors i and j point east and north respectively.

One day he cycles due north at 8 km h^{-1}. At the same time, a younger man leaves his home, which has position vector $(-12i - 4j)$ km relative to O, and cycles with velocity $(6i - 6j)$ km h^{-1}. Show that after half an hour the cyclists are closest together and determine this shortest distance between them.

On that day as the men cycle to work there is a steady wind blowing. To the older man, cycling due north at 8 km h^{-1}, the wind appears to blow *from* the west. To the younger man, cycling with velocity $(6i - 6j)$ km h^{-1}, the wind appears to blow *from* the south. Find the velocity of the wind as a vector.

   [AEB 1987]

8. Unit vectors i and j are defined with i horizontal and j vertically upwards. At time $t = 0$ a particle P is projected from a fixed origin O with velocity $nu(3i + 5j)$, where n and u are positive constants. At the same instant a particle Q is projected from the point A, where $\overrightarrow{OA} = a(16i + 17j)$ with a being a positive constant, with velocity $u(-4i + 3j)$.
   (i) Find the velocity of P at time t in terms of n, u, g and t. Show also that the velocity of P relative to Q is constant and express it in the form $pi + qj$.
   (ii) Find the value of n such that P and Q collide.
   (iii) Given that P and Q do not collide and that Q is at its maximum height above A when at a point B, find, in terms of u and g, the horizontal and vertical displacements of B from A.

   [AEB 1988]

9. Two aircraft A and B are circling near an airport awaiting permission to land. The position vector of A relative to an origin O on the ground is

$$r_A(t) = a \cos \omega t i + a \sin \omega t j + 2h k$$

where a, h and ω are constants, $\omega = \dfrac{V}{a}$, where V is the speed of the aircraft and $\mathbf{k}$ is directed vertically upwards. B flies with speed V in a horizontal circle of radius a directly over the path of A at height $3h$ above O.

(i) Given that B describes its circle in the same sense as A and is $\frac{\pi}{2}$ behind A, find the position vector of B relative to O.

(ii) Show that the velocity of B relative to A has magnitude $\sqrt{2}\,V$ and makes an angle $\frac{3\pi}{4}$ with the velocity of A.

At time $t = t_0 = \dfrac{2\pi a}{V}$, A is given permission to land and leaves its circular path in the vertical plane of its tangent at $\mathbf{r}_A(t_0)$ and descends with speed V at an angle θ to the horizontal.

(iii) Find its velocity vector and position vector whilst descending at time T after the beginning of the descent.

[AEB 1987]

10. A rugby player is at the origin O (on the halfway line) and running with velocity $9\mathbf{j}$ m s^{-1}. At the same time another player is at the point $20\mathbf{i} + 25\mathbf{j}$ and runs with constant velocity $\mathbf{v}$ m s^{-1} in an effort to intercept the first player.

(i) Show that it is possible for the second player to intercept the first if

$$\mathbf{v} = -20p\mathbf{i} + (9 - 25p)\mathbf{j}$$

for some constant p.

(ii) Find an expression for the square of the speed of the second player in terms of p and differentiate it to determine the value p_m of p which will give minimum speed.

(iii) The second player runs with the minimum speed. Show that he intercepts the first after a time $1/p_m$ and find the position of the players when they meet.

11. In a clay pigeon shoot, the target is launched vertically from ground level with speed v. At a time T later the competitor fires a rifle inclined at an angle α to the horizontal. The competitor is also at ground level and a distance d from the launcher. The speed of the bullet leaving the rifle is u.

(i) Show that, if the competitor scores a hit, then

$$d\sin\alpha - (vT - \tfrac{1}{2}gT^2)\cos\alpha = (v - gT)\frac{d}{u}$$

(ii) Suppose now that $T = 0$. Show that if the competitor can hit the target before it hits the ground, then $v < u$ and

$$d < \frac{2v}{g}(u^2 - v^2)^{\frac{1}{2}}$$

[STEP 1993]

12. Mars takes approximately two Earth years to complete one revolution about the sun, is $1\frac{1}{2}$ times as far from the sun as Earth and travels in the same direction round it. Assume that both orbits are circular and lie in the same plane and that initially the sun, Earth and Mars lie in that order along a line.

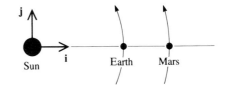

Sun   Earth   Mars

(i) Write down the position vectors of both Earth and Mars relative to the sun as functions of time t in Earth years using the directions $\mathbf{i}$ and $\mathbf{j}$ shown in the diagram and the distance from the sun to Earth as the unit of length.

(ii) Find the position vector of Mars relative to Earth as a function of time.

(iii) Show that the locus of Mars relative to Earth over the period $-1 < t \leqslant +1$ crosses an axis in the $\mathbf{i}$ direction with Earth as its origin when $t = 0$, +1 and $\pm(1/\pi)\arccos(3/4)$.

(iv) Hence sketch the locus of Mars relative to Earth over this period.

[MEI 1993]

13. Particles P_1 and P_2 of masses m_1 and m_2 are attached to the ends of a light inextensible string. They are placed on a smooth table and set in motion with the string taut and after time t have positions $x_1\mathbf{i} + y_1\mathbf{j}$ and $x_2\mathbf{i} + y_2\mathbf{j}$ relative to a fixed origin O.

(i) By considering the equations of motion of P_1 and P_2, show that their centre of mass moves with constant velocity. Will this still be the case if the string is replaced with an elastic spring?

(ii) The string is replaced with a light spring of natural length l_0 and modulus of elasticity λ which obeys Hooke's Law. The particles are placed with P_1 at O and P_2 at $l_0\mathbf{i}$. P_2 is then given an initial velocity u parallel to the x axis.

(a) Use Hooke's Law to find an expression for the acceleration of P_2 relative to P_1 in terms of their relative displacement. Hence show that, relative to P_1, P_2 moves with simple harmonic motion with central value l_0 and period $2\pi/\omega$, where

$$\omega^2 = \frac{\lambda\,(m_1 + m_2)}{m_1 m_2 l_0}$$

(b) Show that after time t the tension in the spring is $\dfrac{\lambda u}{l_0\omega}\sin\omega t$.

(c) In the case where $m_1 = m_2 = m$ and $\lambda = mg$, show that the displacements x_1 and x_2 of P_1 and P_2 after time t are given by

$$x_1 = \frac{u}{2}\left(t - \frac{1}{\omega}\sin\omega t\right)$$

and

$$x_2 = \frac{u}{2}\left(t + \frac{1}{\omega}\sin\omega t\right) + l_0$$

(d) Using the same axes, sketch graphs of x_1, x_2 and the position of the centre of mass as t varies.

14. A cog wheel of radius a and centre C, rolls without slipping inside a fixed circle of radius $2a$ and centre O. When C rotates about O with constant angular speed ω, the angular speed of the cog relative to C is ω_1.

(i) By considering the velocity of the point of contact A between the cog and the circle, show that $\omega = \omega_1$.

(ii) The diagram shows a point P such that angle POA $= \theta$. Show that the velocity of P is $2a\omega\sin\theta$ along OP.

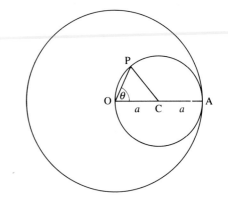

(iii) If OP $= x$, write x in terms of θ and differentiate to find $\dot{x}$ in terms of $\dot{\theta}$. Use the result in part (ii) to show that
$\dot{\theta} = -\omega$. Hence show that every point on the cog wheel moves along a diameter of the circle with simple harmonic motion of period $2\pi/\omega$ and amplitude $2a$.

Investigations

1. Imagine you are sitting in a train which is moving with constant speed in a straight line. At a certain instant you are looking in a direction at right angles to the railway track and you can see a house in the distance and a tree between you and the house. If you keep looking at the house, what will the velocity of the tree relative to the house seem to be to you? You may find it helpful to consider the distances of the tree and house from the railway track and their motion relative to you.

2. If a cutting diamond is fixed to the circumference of a cogwheel such as the one described in Exercise 2B, Question 14, the arrangement would make it possible to cut fine straight lines using a rotating drill. Investigate ways in which rotating drills are used for cutting other unlikely shapes such as square holes.

KEY POINTS

- The position of A relative to B is $_A\mathbf{r}_B = \mathbf{r}_A - \mathbf{r}_B$

- The velocity of A relative to B is $_A\mathbf{v}_B = \mathbf{v}_A - \mathbf{v}_B$

- The acceleration of A relative to B is $_A\mathbf{a}_B = \mathbf{a}_A - \mathbf{a}_B$

- Problems can be solved by assuming B is fixed and the position, velocity and acceleration of A are those relative to B.

- A and B collide when $_A\mathbf{v}_B = 0$

- To find the closest distance apart, write $|_A\mathbf{r}_B|$ or $|_A\mathbf{r}_B|^2$ as a function of t and then find the minimum value of this function

Variable forces

Is it possible to fire a projectile up to the moon?

The Earth to the Moon by Jules Verne (1865)

The answer given in Verne's book to the above question is as follows.

Yes, provided it possesses an initial velocity of 12 000 yards per second, calculations prove that to be sufficient. In proportion as we recede from the earth the action of gravitation diminishes in the inverse ratio of the square of the distance; that is to say at three times a given distance the action is nine times less. Consequently the weight of a shot will decrease and will become reduced to zero at the instant that the attraction of the moon exactly counterpoises that of the earth; at $\frac{47}{52}$ of its journey. There the projectile will have no weight whatever; and if it passes that point it will fall into the moon by the sole effect of lunar attraction.

For Discussion

If an *unpowered* projectile could be launched from the earth with a high enough speed in the right direction, it would reach the moon. What forces act on the projectile during its journey? How near to the moon will it get if its initial speed is not *quite* enough?

In Jules Verne's story, three men and two dogs were sent to the moon inside a projectile fired from an enormous gun. Although this is completely impracticable, the basic mathematical ideas in the above passage are correct. As a projectile moves further from the earth and nearer to the moon, the gravitational attraction of the earth decreases and that of the moon increases. In many of the dynamics problems you have met so far it has been assumed that forces are *constant*, whereas on Jules Verne's space missile the total force *varies* continuously as the motion proceeds.

You have already met problems involving variable force. When an object is suspended on a spring and bounces up and down, the varying tension in the spring leads to *simple harmonic motion*. If you have studied the chapter on differential equations in *Mechanics 4*, you will have seen how air resistance or fluid resistance on an oscillating object can be modelled by a force opposing the motion and proportional to the speed of the oscillator. This produces *damped* simple harmonic motion.

Gravitation, spring tension and air resistance all give rise to variable force problems, the subject of this chapter.

Newton's Second Law as a differential equation

How are variable force problems analysed? When forces are *constant*, you can find the constant acceleration using Newton's Second Law:

$$\text{force} = \text{mass} \times \text{acceleration}$$

You then calculate velocities, times and displacements using formulae such as $s = ut + \frac{1}{2}at^2$ and $v = u + at$.

When forces vary, Newton's Second Law is still valid; the force at any moment determines the *instantaneous* value of the acceleration. However, since the acceleration is varying, you have to use calculus techniques to derive details of the motion. The starting point is to write Newton's Second Law in the form of a *differential equation*:

$$F = m \frac{dv}{dt}$$

or

$$F = mv \frac{dv}{ds}$$

Notice how acceleration can be written as $v \dfrac{dv}{ds}$. This follows from the chain rule for differentiation:

$$\frac{dv}{dt} = \frac{dv}{ds} \times \frac{ds}{dt}$$

$$= v \frac{dv}{ds}$$

NOTE *Here and throughout this chapter the mass m is assumed to be constant. Jules Verne's spacecraft was a projectile fired from a gun. It was not a rocket whose mass varies, due to ejection of fuel. Rocket motion is considered in Mechanics 6.*

Deriving the constant acceleration formulae

To see the difference in use between the $\dfrac{dv}{dt}$ and $v \dfrac{dv}{ds}$ forms of acceleration, examine the case where the force, and therefore the acceleration,

$\dfrac{F}{m}$ is constant (say a). Starting from the $\dfrac{dv}{dt}$ form

$$\frac{dv}{dt} = a$$

Integrating gives

$$v = u + at$$

where u is the constant of integration ($v = u$ when $t = 0$). Since $v = \dfrac{ds}{dt}$, integrating again gives

$$s = ut + \tfrac{1}{2}at^2 + s_0$$

assuming the displacement is s_0 when $t = 0$.

These are the familiar formulae for motion under constant acceleration.

Starting from the $v\,\dfrac{dv}{ds}$ form

$$v\,\frac{dv}{ds} = a$$

Separating the variables and integrating gives

$$\int v\,dv = \int a\,ds$$

$$\Rightarrow \quad \tfrac{1}{2}v^2 = as + k \quad (k \text{ is the constant of integration})$$

Assuming $v = u$ when $s = 0$, $k = \tfrac{1}{2}u^2$, so the formula becomes

$$v^2 = u^2 + 2as$$

This is another of the standard constant acceleration formulae. Notice that *time is not involved* when you start from the $v\,\dfrac{dv}{ds}$ form of acceleration.

Solving $F = ma$ for variable force

When the force is continuously *variable*, you follow the same procedure as above. You write Newton's Second Law in the form of a differential equation and then solve it using one of the forms of acceleration $v\,\dfrac{dv}{ds}$ or $\dfrac{dv}{dt}$. The choice depends on the particular problem. Some guidelines are given below and these should be checked with the examples which follow.

Normally, the resulting differential equation can be solved by separating the variables.

When the force is a function of *time*, *F(t)*

Use $\dfrac{dv}{dt}$:

$$F(t) = m\,\frac{dv}{dt}$$

Separating the variables and integrating gives

$$m \int dv = \int F(t)\,dt$$

Assuming you can solve the integral on the right-hand side, you then have v in terms of t.

Writing v as $\dfrac{ds}{dt}$, the position as a function of time can be found by integrating again.

When *F* is given as a function of *velocity*, *F(v)*

Start from either

$$F(v) = m\,\frac{dv}{dt}$$

or

$$F(v) = mv\,\frac{dv}{ds}$$

You can separate the variables in both forms; use the first if you are interested in behaviour over time and the second when you wish to involve the displacement.

When *F* is a function of *displacement*, *F(s)*

You normally start from

$$F(s) = mv\,\frac{dv}{ds}$$

N O T E

When F is a linear function of velocity or displacement (or both), another possible starting point is to use the second derivative $\dfrac{d^2 s}{dt^2}$;

$$F = m\,\frac{d^2 s}{dt^2}$$

This gives a second order linear differential equation which can be solved by the methods described in Mechanics 4: Differential Equations. *The displacement is then given in terms of time without finding the velocity first.*

Variable force examples

The three examples that follow show the approaches when the force is given respectively as a function of time, displacement and velocity.

EXAMPLE

A crate of mass m is freely suspended at rest from a crane. When the operator begins to lift the crate further, the tension in the suspending cable increases uniformly from mg to $1.2mg$ over a period of 2 seconds.

(i)   What is the tension in the cable t s after the lifting has begun ($t \leqslant 2$)?
(ii) What is the velocity after 2 s?
(iii) How far has the crate risen after 2 s?

Assume the situation may be modelled with air resistance and cable stretching ignored. Take g as 10 m s^{-2}

Solution

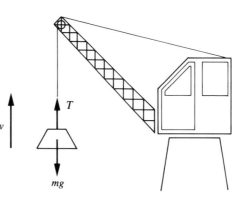

When the crate is at rest it is in equilibrium and so the tension, T, in the cable equals the weight mg of the crate. After time $t = 0$, the tension increases, so there is a net upward force and the crate rises.

(i)   The tension increases uniformly by $0.2mg$ in 2 s, i.e. increases by $0.1mg$ per second. After t s, the tension is

$$T = mg + 0.1mgt$$

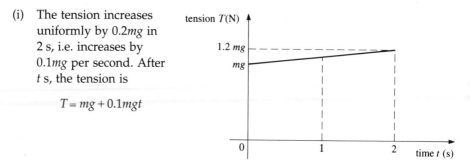

(ii) At any moment in the 2 s period, $F = ma$ gives

$$(mg + 0.1mgt) - mg = m\frac{dv}{dt}$$

$$\Rightarrow \quad \frac{dv}{dt} = 0.1gt$$

Integrating gives

$$v = 0.05\, gt^2 + k$$

where k is the constant of integration.
When $t = 0$, the crate has not quite begun to move, so $v = 0$. This gives $k = 0$ and $v = 0.05\, gt^2$

When t is 2, $v = 0.05 \times 10 \times 4 = 2$.

The velocity after 2 s is 2 m s^{-1}.

(iii) To find the displacement s, write v as $\dfrac{ds}{dt}$ and integrate again:

$$\frac{ds}{dt} = 0.05\, gt^2$$

$$\int ds = \int 0.05\, gt^2 \, dt$$

$$s = 0.05\, g \times \tfrac{1}{3} t^3 + c$$

When $t = 0$, $s = 0 \Rightarrow c = 0$.

When $t = 2$ and $g = 10$, $s = \tfrac{4}{3}$

The crate moves $\tfrac{4}{3}$ m in 2 seconds.

For Discussion

The displacement cannot be obtained by the formula $s = \tfrac{1}{2}(u + v)t$, which would give the answer 2 m. Why not?

EXAMPLE

A prototype of Jules Verne's projectile, mass m, is launched vertically upwards from the earth's surface. When it is at a height s above the surface (where s is small compared with the radius of the earth R), the magnitude of the earth's gravitational force on the projectile may be modelled as $mg\left(1 - \dfrac{2s}{R}\right)$, where g is gravitational acceleration at the earth's surface.

Assuming all other forces can be neglected

(i)   write down a differential equation of motion involving s and velocity v;
(ii) integrate this equation and hence obtain an expression for the loss of kinetic energy of the projectile between its launch and rising to a height s;
(iii) if the projectile just reaches a height of one tenth of the earth's radius before falling back, show that the launch velocity is $0.3\sqrt{(2gR)}$.

Solution

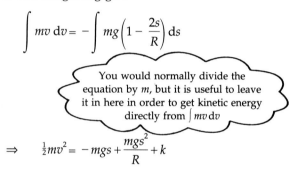

(i) Taking the upward direction as positive, the force on the projectile is $-mg\left(1-\dfrac{2s}{R}\right)$. The force is a function of s, so start from the equation of motion in the form

$$mv\frac{dv}{ds}=-mg\left(1-\frac{2s}{R}\right)$$

(ii) Separating the variables and integrating give

$$\int mv\,dv=-\int mg\left(1-\frac{2s}{R}\right)ds$$

> You would normally divide the equation by m, but it is useful to leave it in here in order to get kinetic energy directly from $\int mv\,dv$

$$\Rightarrow\quad \tfrac{1}{2}mv^2=-mgs+\frac{mgs^2}{R}+k$$

Writing v_0 for the launch velocity, $v=v_0$ when $s=0$, so $k=\tfrac{1}{2}mv_0^2$:

$$\tfrac{1}{2}mv_0^2-\tfrac{1}{2}mv^2=mgs-\frac{mgs^2}{R}$$

The left-hand side is the loss of kinetic energy, so

$$\text{loss of KE}=mgs-\frac{mgs^2}{R}$$

> You can check this is dimensionally consistent to give you confidence that your working is correct

(iii) Removing the common factor m from the above equation, and multiplying by 2 gives

$$v_0^2-v^2=2gs-\frac{2gs^2}{R}$$

If the projectile just reaches a height $s=R/10$, then the velocity v is zero at that point.

Substituting $s = R/10$ and $v = 0$ gives

$$v_0^2 = 2g\left(\frac{R}{10}\right) - \frac{2gR^2}{100R}$$

$$= \frac{18gR}{100}$$

$$\Rightarrow \quad v_0 = \frac{3}{10}\sqrt{(2gR)}$$

So the launch velocity is $0.3\sqrt{(2gR)}$.

EXAMPLE

A body of mass 2 kg, initially at rest on a smooth horizontal plane, is subjected to a force of magnitude $\dfrac{1}{(2v+1)}$ N, where v is the velocity of the body $(v > 0)$.

(i)   Find the time when the velocity is $1\ \text{m s}^{-1}$.
(ii) Find the displacement when the velocity is $1\ \text{m s}^{-1}$.

Solution

(i)   Using $F = ma = m\dfrac{dv}{dt}$

> Write acceleration in $\dfrac{dv}{dt}$ form since time is required

$$\frac{1}{(2v+1)} = 2\frac{dv}{dt}$$

Separating the variables gives

$$\int dt = \int 2(2v+1)\, dv$$

$$\Rightarrow \quad t = 2v^2 + 2v + k$$

When $t = 0$, $v = 0$ so $k = 0$ and therefore

$$t = 2v^2 + 2v$$

When $v = 1$, $t = 4$. That is, when the velocity is $1\ \text{m s}^{-1}$, the time is 4 s.

(ii)

$$F = ma = mv\frac{dv}{ds}$$

> Write acceleration in $v\dfrac{dv}{ds}$ form since displacement is required

$$\frac{1}{(2v+1)} = 2v\frac{dv}{ds}$$

Separating the variables gives

$$\int ds = \int 2v(2v+1)\, dv$$

$$\Rightarrow \quad s = \tfrac{4}{3}v^3 + v^2 + k$$

When $s = 0$, $v = 0$ so $k = 0$ and therefore

$$s = \tfrac{4}{3}v^3 + v^2$$

When $v = 1$, $s = \tfrac{7}{3}$. When the velocity is $1\ \text{m s}^{-1}$, the displacement is $\tfrac{7}{3}$ m.

Exercise 3A

1. Each of the parts (i) to (vii) of this question assumes a body mass 1 kg under the influence of a single force F N in a constant direction but with a variable magnitude given as a function of velocity v m s^{-1}, displacement s m, or time t s. In each case, express $F = ma$ as a differential equation using either $a = \dfrac{dv}{dt}$ or $a = v\dfrac{dv}{ds}$ as appropriate.

Then separate the variables and integrate giving the result in the required form and leaving an arbitrary constant in the answer.

(i) $F = 2v$ express s in terms of v

(ii) $F = 2v$ express v in terms of t

(iii) $F = 2\sin 3t$ express v in terms of t

(iv) $F = -v^2$ express v in terms of t

(v) $F = -v^2$ express s in terms of s

(vi) $F = -4s + 2$ express v in terms of s

(vii) $F = -2v - 3v^2$ express s in terms of v

2. Each of the parts (i) to (viii) of this question assumes a body of mass 1 kg under the influence of a single force F N in a constant direction but with a variable magnitude given as a function of velocity v m s^{-1}, displacement s m, or time t s. The body is initially at rest at a point O. In each case, write down the equation of motion and solve it to supply the required information.

(i) $F = 2t^2$ find v when $t = 2$

(ii) $F = -\dfrac{1}{(s+1)^2}$ find v when $s = -\frac{1}{9}$

(iii) $F = \dfrac{1}{(s+3)}$ find v when $s = 3$

(iv) $F = \dfrac{1}{(v+1)}$ find t when $v = 3$

(v) $F = 1 + v^2$ find t when $v = 1$

(vi) $F = 5 - 3v$ find t when $v = 1$

(vii) $F = 1 - v^2$ find t when $v = 0.5$

$$\left[\textbf{Hint: } \frac{1}{(1-v^2)} = \frac{1}{2}\left\{\frac{1}{(1-v)} + \frac{1}{(1+v)}\right\}\right]$$

(viii) $F = 1 - v^2$ find s when $v = 0.5$

3. A horse pulls a 500 kg cart from rest until the speed, v, is about 5 m s^{-1}. Over this range of speeds, the magnitude of the force exerted by the horse can be modelled by $500(v+2)^{-1}$ N. Neglecting resistance,

(i)   write down an expression for $v\dfrac{dv}{ds}$ in terms of v;

(ii) show by integration that when the velocity is 3 m s^{-1}, the cart has travelled 18 m;

(iii) write down an expression for $\dfrac{dv}{dt}$ and integrate to show that the velocity is 3 m s^{-1} after 10.5 s;

(iv) show that $v = -2 + \sqrt{(4 + 2t)}$;

(v)   integrate again to derive an expression for s in terms of t, and verify that after 10.5 s, the cart has travelled 18 m.

4. In simple harmonic motion, the force on a body of mass m is proportional to the distance from point O and is directed towards that point:

$$\text{force} = -ks$$

where s is the displacement from O.

(i)   Show that the equation of motion can be put in the form

$$v\frac{dv}{ds} = -\omega^2 s$$

where v is the velocity and $\omega^2 = k/m$.

(ii) Separate the variables and integrate. Given that $s = a$ when $v = 0$, write v in terms of a, s and ω.

(iii) Write v as $\dfrac{ds}{dt}$, integrate again and show that the solution can be written $s = a\sin(\omega t + \varepsilon)$, where ε is a constant.

This standard simple harmonic motion formula can also be obtained directly by writing the original equation of motion in the form of a second-order linear differential equation

$$\frac{d^2 s}{dt^2} + \omega^2 s = 0$$

and solving it using the methods described in Mechanics 4: Differential Equations.

5. A car of mass 1000 kg is travelling on level ground at $5\,\text{m s}^{-1}$ at time $t = 0$ when it begins to accelerate with its engine working at a constant 10 kW. Assume that its motion can be modelled neglecting friction and other forces.

   (i) Show that the thrust of the engine is inversely proportional to the speed of the car. Write down the equation of motion involving velocity v and time t.

  (ii) Integrate and hence express v in terms of t. Confirm that after 10 s the velocity is $15\,\text{m s}^{-1}$.

  (iii) By writing v in the form $\mathrm{d}s/\mathrm{d}t$, integrate again and show that the distance travelled after time t is given by

$$s = \frac{1}{30}\,(20t + 25)^{3/2} - \frac{25}{6}$$

  (iv) Eliminate t between the equations derived in (ii) and (iii) and hence express s in terms of v.

  (v) Show that the result in part (iv) can be derived directly from another form of the equation of motion.

Resisted motion

A strong enough wind can blow you over; the force exerted by the air obviously depends on wind speed. You feel the same effect when you cycle or ski quickly—except that it is now called *air resistance* rather than wind. Any object moving through a fluid, such as a gas or a liquid, encounters a *resistive force* opposing its motion.

Experiments have shown that the air resistance on a moving object, such as a falling pebble, is approximately proportional to the square of its speed. The resistance is caused mainly because of the force required to push the particles of air into motion, in other words to change the momentum of the air particles. It is similar when the object is at rest and the wind blows against it—the force is produced by the loss of momentum of the air as it hits the object.

However, when the object is tiny (a particle of dust, for example), is slow-moving, or when it is moving through a liquid rather than a gas, then most of the resistive force is caused by *viscosity*—friction between layers of the liquid and between the liquid and the object. The viscous force (on a falling leaf or a lump of sugar dropped into a cup of coffee, for example) is directly proportional to speed. In general, motion through a fluid will be opposed by resistive forces which depend on a mixture of both *velocity* and *velocity squared*, but it is often the case that one dominates and the other can be ignored.

Terminal velocity

Imagine that a pebble of mass m, falling through the air, is subjected to a resistance proportional to the square of its speed, say kv^2, where k is a constant. This is a reasonable modelling assumption for such an object. The net downward force is $mg - kv^2$ (figure 3.1). This is clearly a variable force. Applying Newton's Second Law at any instant gives

$$mg - kv^2 = m\frac{\mathrm{d}v}{\mathrm{d}t}$$

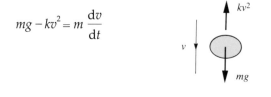

Figure 3.1

We can get some idea of the behaviour of the pebble simply by looking at the differential equation. Initially v is zero, so the acceleration is g. Thus v will begin to increase. As v increases the acceleration $\frac{\mathrm{d}v}{\mathrm{d}t}$ gets smaller. As v^2 approaches mg/k, the left-hand side of the equation approaches zero, and therefore so does the acceleration. If v ever becomes $\sqrt{(mg/k)}$, the forces balance and there is no acceleration. The pebble would then continue at the speed $\sqrt{(mg/k)}$. This is known as its *terminal velocity*: it is the limiting speed that the pebble can reach. When an object is dropped from rest, its speed gets nearer and nearer to the terminal velocity but in theory does not quite achieve it, as the later worked examples show.

For situations that can be modelled by a resistive force proportional to v—for example a tiny raindrop falling in the air—Newton's Second Law gives

$$mg - cv = m\frac{\mathrm{d}v}{\mathrm{d}t}$$

where the resistive force is cv. You can see that in this case the terminal velocity has a value mg/c.

Modelling air resistance

Air resistance is quite significant in everyday situations. For example, a 2 cm diameter pebble dropped off a mountain side (don't do it!) has a terminal velocity of about 35 m s^{-1}, and the velocity is close to this after about 7 seconds.

For Discussion

The graph shows how the speed of a granite pebble varies with time when it is dropped in air. The pebble has a diameter of 2 cm and a mass of 11.3 g. What is the gradient of the graph at the origin? How would you use the graph to estimate how far the stone has fallen before reaching 90% of its terminal velocity?

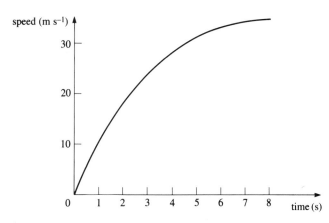

Investigation

Modelling air resistance

A sphere of radius r metres moving through the air with a velocity $v \text{ m s}^{-1}$ will encounter a resistive force of approximate magnitude

$$k_1 r v + k_2 r^2 v^2$$

It is made up of two terms, one dependent on velocity, the other on velocity squared. In SI units, approximate values for k_1 and k_2 are respectively (3.1×10^{-4}) and 0.87, and the expression gives the resistive force in newtons.

(i) Investigate the air resistance on two pieces of stone of density 2.5 g cm^{-3}, both of which can be modelled as spheres. One is a pebble of radius 2 cm, the other a grain of radius 0.02 cm. For each object, draw a graph showing how the resistive force varies for values of the velocity between zero and the terminal velocity. This is when the force is equal to the weight of the stone.

(ii) At the terminal velocity, what is the relative importance of the v term as against the v^2 term in each case? Do you think it is reasonable to ignore the v term when analysing the total motion of the pebble?

(iii) What about a 2 cm radius puff-ball of mass 0.1 g? Use the graph for the pebble to find its terminal velocity. Is it still reasonable to ignore the v term?

Solving the resistive motion equations

To solve differential equations of resistive motion, write the acceleration in the form $\dfrac{dv}{dt}$ or $v\dfrac{dv}{ds}$, depending on whether you require *time* in the answer.

You can separate the variables and then perform integration. The integrations typically result in logarithms or inverse trigonometric functions.

EXAMPLE

A raindrop of mass m starts from rest and falls vertically. When falling with velocity v, it experiences a resistive force of magnitude kv.

(i) Express the terminal velocity V_T of the raindrop in terms of m, k and g.

(ii) Show that if the raindrop starts from rest, its velocity V after time T is given by

$$V = V_T(1 - e^{-gT/V_T})$$

(iii) Assuming the raindrop has a terminal velocity of 5 m s^{-1}, how long does it fall before it has 99% of the terminal velocity (take g as 10 m s^{-2})?

(iv) Derive an expression for the distance fallen in terms of time.

Solution

(i) The forces on the raindrop are mg downwards and kv upwards. Applying Newton's Second Law at any instant:

$$mg - kv = m\frac{dv}{dt}$$

If the raindrop were moving at the terminal velocity, the acceleration would be zero, so

$$V_T = \frac{mg}{k}$$

(ii) It is helpful at this stage to divide the first equation by k and write $mg/k = V_T$ and $m/k = V_T/g$. The equation then becomes

$$V_T - v = \frac{V_T}{g}\frac{dv}{dt}$$

Now separate the variables and perform integration. At the start, $v = 0$ and $t = 0$. At time t, $v = V$. These values are used as the limits of integration in the definite integrals, so there is no need for a constant of integration.

$$\int_0^v \frac{dv}{(V_T - v)} = \int_0^t \frac{g}{V_T}\,dt$$

$$\Rightarrow \qquad [-\ln|(V_T - v)|]_0^v = \left[\frac{gt}{V_T}\right]_0^t$$

$$-\ln(V_T - v) + \ln V_T = \frac{gt}{V_T}$$

> $V_T > v$ throughout the motion so $|(V_T - v)| = (V_T - v)$

$$\ln\left[\frac{(V_T - v)}{V_T}\right] = -\frac{gt}{V_T}$$

$$-\frac{v}{V_T} = e^{-gt/v_T}$$

which gives

$$v = V_T(1 - e^{-gt/V_T})$$

$$= V_T - V_T e^{-gt/v_T}$$

This makes it clear that, as you would expect, the velocity V is equal to the terminal velocity minus a small and decreasing amount ①

(iii) When v is 99% of the terminal velocity, $v/V_T = 0.99$.

$$0.99 = 1 - e^{-gt/V_T}$$

$$= 1 - e^{-10t/5} \quad (g = 10,\ V_T = 5)$$

$$= 1 - e^{-2t}$$

$$\Rightarrow \quad e^{-2t} = 0.01$$

$$e^{2t} = 100$$

Note that the time taken to reach a given *percentage* of V_T does depend on the value of V_T. $V_T = 5$ was substituted in ②

$$\Rightarrow \quad t = \tfrac{1}{2}\ln 100 = 2.3$$ ②

The raindrop falls for 2.3 s before achieving 99% of V_T.

(iv) Having derived ①, write v as $\dfrac{ds}{dt}$ and perform integration to get

$$\int ds = \int (V_T - V_T e^{-gt/V_T})\,dt$$

$$s = V_T t + \left(\frac{V_T^2}{g}\right) e^{-gt/V_T} + c$$

where c is the constant of integration.

Since $s = 0$ when $t = 0$, $c = -\dfrac{V_T^2}{g}$, so

The first term $V_T t$ is the distance that the raindrop would fall in a time t if travelling at speed V_T the whole time. s is always less than this

$$s = V_T t - \frac{V_T^2}{g}(1 - e^{-gt/V_T})$$

Using a second-order equation

The last example involved integration twice, once to find v and again to find s. If you are not interested in v, and want s directly, you can express acceleration as $\dfrac{d^2 s}{dt^2}$ and solve the second-order differential equation using the method described in *Mechanics 4: Differential Equations*. If you are not well practised with such equations, you may still prefer the first solution.

However, it is useful to know more than one way of solving problems. The solution of the second-order equation is summarised here. You should refer to *Mechanics 4: Differential Equations* to remind yourself of the details.

Newton's Second law for motion of the raindrop gives

$$mg - kv = m \frac{d^2s}{dt^2}$$

Writing v as $\frac{ds}{dt}$ and rearranging gives

$$\frac{d^2s}{dt^2} + \frac{k}{m}\frac{ds}{dt} = g$$

This is in the standard form for a second-order linear differential equation with no term in s.

> This method would not work with a kv^2 law of resistance as the resulting second-order equation would be nonlinear and the standard methods of solution would not apply

The *auxiliary equation* is

$$\lambda^2 + \left(\frac{k}{m}\right)\lambda = 0$$

giving $\lambda = 0$ or $-k/m$.

So the *complementary function* is

$$s = A e^{0t} + B e^{-kt/m}$$
$$= A + B e^{-kt/m}$$

The *particular integral* is $s = mgt/k$, so that the general solution of the equation is

$$s = A + B e^{-kt/m} + \frac{mgt}{k}$$

Remember $V_T = mg/k$, so write the equation as

$$s = A + B e^{-gt/V_T} + V_T t$$

Since $s = 0$ when $t = 0$, $A + B = 0$, so $A = -B$.

Also

$$v = \frac{ds}{dt}$$

$$= -\left(\frac{Bg}{V_T}\right)e^{-gt/V_T} + V_T$$

Now $v = 0$ when $t = 0$, so $Bg/V_T = V_T \Rightarrow B = V_T^2/g$

This gives the solution, as before

$$s = V_T t - \frac{V_T^2}{g}(1 - e^{-gt/V_T})$$

Deciding on positive direction

It is easy to get confused with the signs when writing down Newton's Second Law. Decide which is the direction in which you are measuring *positive displacement* and take this as your positive direction for *all variables*. A positive value of v means the body is moving in this direction. In the equation of motion, resistive forces will be in a direction opposite to that in which the body is moving. Thus, when the resistive force is proportional to velocity, write it as $-kv$ where k is a positive constant. If in fact the velocity is negative, the value of $-kv$ will be positive, still in the opposite direction to the motion. Your equation is correct whichever direction the body is actually moving in.

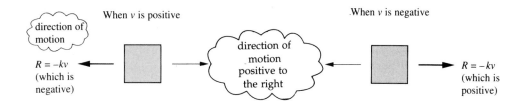

This does not work when the resistive force is of the form $-kv^2$. You need different equations depending on which way the body is travelling. The two cases have to be considered separately. The next example demonstrates this.

EXAMPLE

A heavy ball is retarded by air resistance which is modelled by kv^2, where v m s^{-1} is the speed of the ball, and k is a positive constant. When falling under gravity in air, its terminal velocity is V m s^{-1}.

(i) Write an expression for the total force on the ball when it is descending and hence derive an expression for V^2 in terms of k, g and the mass of the ball.

(ii) The ball is projected vertically upwards with an initial speed u m s^{-1}. Derive a differential equation, involving V, but not k, covering its ascent.

(iii) Show that, when $u = 50$ and $V = 100$, the ball reaches a height of almost 114 m.

Solution

(i) Assume that the mass of the ball is m and the resistance is kv^2. When descending, the net downward force is $mg - kv^2$.

The terminal velocity V is obtained when the force is zero, so $mg = kV^2$.

$$\Rightarrow \quad V^2 = \frac{mg}{k}$$

(ii) When ascending, both the weight and the air resistance are in the same direction. Take the positive direction as *upwards*, so the forces are negative and the initial velocity is positive.

Since part (iii) involves *displacement*, write acceleration in the form $v \dfrac{dv}{ds}$.

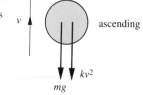

upwards positive

ascending

The equation of motion is then

$$- mg - kv^2 = mv \frac{dv}{ds}$$

$$-\frac{mg}{k} - v^2 = \frac{m}{k} v \frac{dv}{ds}$$

Substituting $\dfrac{mg}{k} = V^2$ gives

$$- V^2 - v^2 = \left(\frac{V^2}{g}\right) v \frac{dv}{ds}$$

$$v \frac{dv}{ds} = -\frac{g(V^2 + v^2)}{V^2}$$

(iii) Separating variables and integrating:

$$-\int v \frac{dv}{(V^2 + v^2)} = \frac{g}{V^2} \int ds$$

At the start $s = 0$, $v = U$. At the top of the ascent $s = h$, $v = 0$. Applying these limits to the integration:

$$- [\tfrac{1}{2}\ln(V^2 + v^2)]_U^0 = \frac{g}{V^2} [s]_0^h$$

$$\tfrac{1}{2}\ln\left(\frac{V^2 + U^2}{V^2}\right) = \frac{gh}{V^2}$$

$$h = \frac{V^2}{2g}\ln\left(1 + \frac{U^2}{V^2}\right)$$

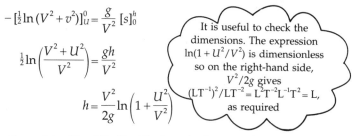

It is useful to check the dimensions. The expression $\ln(1 + U^2/V^2)$ is dimensionless so on the right-hand side, $V^2/2g$ gives $(LT^{-1})^2/LT^{-2} = L^2T^{-2}L^{-1}T^2 = L$, as required

Substituting the values $V = 100$, $U = 50$ given in the question:

$$h = \frac{(100)^2}{2g} \ln 1.25 = 113.8$$

The ball reaches a height of 113.8 m.

For Discussion

Can a body descending under air resistance start off with a speed *greater* than the terminal velocity? How? What happens subsequently?

When a ball is thrown up in the air, does it take longer to go up or come down? Bear in mind energy is dissipated by the air resistance and consider the velocity at corresponding points on the way up and the way down.

The first thorough treatment of the motion of bodies in a resisting medium was given by Newton who devoted the second of the three books of his master work Philosophiae Naturalis Principia Mathematica (first edition 1687) to this topic. He considered motion where the fluid resistance is proportional to the velocity, the square of the velocity and a combination of the two. He also investigated the physical causes of fluid resistance.

Terminal velocity under resistive force was an important part of R. A. Millikan's classic experiment in 1896 to measure the charge on an electron (the smallest quantum of charge). He observed tiny charged droplets of oil rising or falling in an electric field and was able to measure the terminal velocity very accurately. The field causes an electric force on the charged particle in addition to the gravitational force and viscous force of the air. By observing, over many hours, the terminal velocity of the same droplet using different values of the electric field, Millikan was able to eliminate various other unknown quantities (e.g. the droplet's mass) and calculate the charge on the droplet. Such measurements on hundreds of droplets showed that the charge was always an integer multiple of the smallest charge: the charge on an electron.

Exercise 3B

1. The resistance to a cork of mass $10\,\text{g}$ falling in air with velocity $v\,\text{m s}^{-1}$ can be modelled as $kv^2\,\text{N}$, where k is constant.
   (i) Write down the equation of motion and hence deduce an expression for the terminal velocity in terms of k.
   (ii) Find the value of k, given that the terminal velocity is $15\,\text{m s}^{-1}$.
   (iii) At what speed is the resistive force equal to half the weight of the cork?

2. In an experiment with Millikan's oil-drop apparatus, the resistive force on a spherical oil droplet of radius $r\,\text{m}$ with a velocity $v\,\text{m s}^{-1}$ can be modelled as $3.1 \times 10^{-4}\,rv\,\text{N}$. The density of the oil is $800\,\text{kg m}^{-3}$.
   (i) Find the mass of an oil drop of radius 1 micron ($10^{-6}\,\text{m}$).
   (ii) Write down the equation of motion for such an oil drop falling under gravity. Hence determine the terminal velocity.
   (iii) A droplet moving at its terminal velocity is observed to take 20 seconds to fall $5\,\text{mm}$. What is its radius? (Take $g = 9.8\,\text{m s}^{-2}$.)

3. Moira pushes herself off from the bank of a frozen pond with an initial speed U and then slides across the ice until she stops. The horizontal force is a resistance of kv

newtons, where v is her velocity and k is a constant. Moira's mass is $40\,\text{kg}$.
   (i) Give Newton's Second Law for the motion, writing acceleration as $\dfrac{dv}{dt}$.
   (ii) Find Moira's speed after $1\,\text{s}$, in terms of U and k.
   (iii) Show that her speed reduces by the same proportion each second.
   (iv) Given that U is $5\,\text{m s}^{-1}$ and the resistance is $5v\,\text{N}$, find the distance she has travelled after $5\,\text{s}$.
   (v) Show that she will not go further than $40\,\text{m}$.

4. A car of mass $m\,\text{kg}$, moving at $25\,\text{m s}^{-1}$ on a level road, runs out of petrol and tries to coast to a lay-by $0.6\,\text{km}$ further on. It is acted on by a resistive force of magnitude $m(v^2 + 1)/200\,\text{N}$, where $v\,\text{m s}^{-1}$ is the car's speed. Show that the car reaches the lay-by.

5. A sphere of radius $R\,\text{m}$ dropped in water is resisted by a viscous force $6\pi R\eta v\,\text{N}$, where $v\,\text{m s}^{-1}$ is the velocity of the sphere and η is the coefficient of viscosity. Also according to Archimedes principle, a body immersed in a fluid receives an upthrust equal to the weight of fluid displaced by the volume

Exercise 3B continued

of the body. Experiments are performed by releasing glass marbles in the water. ρ_g denotes the density of glass and ρ_w the density of water.

(i)   Find the weight of water which the sphere displaces and hence write the equation of motion for the sphere as it descends.

(ii) Show that the terminal velocity V_T has the value

$$\left(\frac{2R^2 g}{9\eta}\right)(\rho_g - \rho_w)g$$

and that the equation of motion may be written

$$V_T - v = \left(\frac{2R^2 \rho_g}{9\eta}\right)\frac{dv}{dt}$$

(iii) Taking ρ_g as 3000 kg m^{-3}, ρ_w as 1000 kg m^{-3}, the value of η as 1 in SI units, and g as 10 m s^{-2}, find the terminal velocity of a 1 cm diameter marble. Deduce the terminal velocity for a 2 cm diameter marble.

(iv) How long does it take for the 1 cm marble to reach 99% of its terminal velocity?

6. A skier is descending a slope 30° to the horizontal against resistive forces proportional to her speed. The terminal velocity under these conditions is 20 m s^{-1}. She has managed to reach this velocity when she arrives at the bottom and begins to ascend at an angle of 2°. She continues against the same type of resistive forces (without pushing with her sticks) until she stops.

(i)   If the resistive forces have magnitude kv, and the skier has mass m, derive the terminal velocity for the descent, and hence express k in terms of g and m.

(ii) Once the skier starts to ascend, what is the total force down the slope? Express this in terms of g, m and her velocity v.

(iii) Write down the equation of motion of the skier up the slope and show that she comes to a halt just over 11 seconds after beginning to ascend.

(iv) How far up the slope does she travel?

7. A ball bearing released from rest in a fluid falls under gravity. When its distance below the point of release is x, its speed v satisfies the equation

$$-\frac{gx}{V} = v + V \ln\left(1 - \frac{v}{V}\right)$$

where V is a constant.

(i)   Find $\dfrac{dv}{dx}$.

(ii) By writing acceleration as $v\dfrac{dv}{dx}$, show that the motion can be explained by assuming a resistive force proportional to velocity (assuming no buoyancy effects). Show that V is the terminal velocity under these conditions.

(iii) Show that the velocity after time t is given by

$$v = V(1 - e^{-t/t_0})$$

where t_0 is the time which the ball bearing would take to reach velocity V if falling under gravity with no resistance.

8. A ball of mass m is thrown vertically upwards with an initial speed U. The air resistance, proportional to the square of the speed of the ball, is such that in free fall under gravity, the ball would reach a terminal speed V.

(i)   Show that when the particle is moving upwards with speed v, the retardation is of magnitude $g\left(1 + \dfrac{v^2}{V^2}\right)$.

(ii) Show that the ball reaches a maximum height of

$$\frac{V^2}{2g}\ln\left(1 + \frac{U^2}{V^2}\right).$$

(iii) Find the speed with which it returns to its starting point.

(iv) Find an expression for the speed of the ball as a function of time during its ascent.

[MEI 1990 adapted]

9. A body of mass m kg is dropped vertically into a deep pool of liquid. Once in the liquid, it is subject to gravity, an upward buoyancy force of $\frac{6}{5}$ times its weight, and a resistive force of $2mv^2$ N opposite to its direction of travel when it is travelling at v m s^{-1}.

(i) Show that the body stops sinking less than $\frac{\pi}{4}$ seconds after it enters the pool.

(ii) Suppose now that the body enters the liquid with speed 1 m s^{-1}. Show that the

body descends to a depth of $\frac{1}{4}\ln 2$ m and that it returns to the surface with speed $1/\sqrt{2}$ m s^{-1}.

(iii) Show further that it returns at a time

$$\frac{\pi}{8} + \frac{1}{4}\ln\frac{(\sqrt{2}+1)}{(\sqrt{2}-1)}$$

Take the value of g to be 10 m s^{-2}.

[O&C 1993]

Gravitational force

Gravitation is one of the fundamental forces of the universe, responsible for the motions of the planets, satellites and comets and indeed for the large-scale structure of the universe. All bodies attract each other by a gravitational force. It is very tiny for pairs of everyday objects, but large and important for objects of an astronomical size, such as the sun, the moon and the planets. Two particles attract each other with a force proportional to the product of their masses, m_1m_2, and inversely proportional to the square of the distance, r, between them:

$$\text{force} = \frac{Gm_1m_2}{r^2}$$

The constant G is known as the gravitational constant. For uniform spherical bodies, the force is along the line joining their centres and the distance is measured between their centres.

At the Earth's *surface*, an object of mass m is attracted by a force $\dfrac{GMm}{R^2}$, where M is the mass of the Earth and R is its radius. Applying Newton's Second Law gives

$$\frac{GMm}{R^2} = ma$$

So a, the acceleration of a body at the Earth's surface, is $\dfrac{GM}{R^2}$. This is g, the acceleration due to gravity. The Earth is so large that g can be assumed to be constant near to the surface. However, for objects such as meteorites or returning spacecraft, the continuous change in the gravitational force due to the changing distance from the centre of the Earth must be taken into account when the motion is analysed.

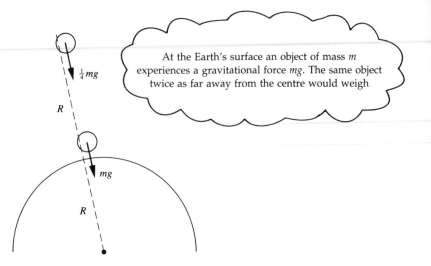

At the Earth's surface an object of mass m experiences a gravitational force mg. The same object twice as far away from the centre would weigh

EXAMPLE

(i) Denoting the Earth's radius by R metres, show that the gravitational force on a body of mass m kg above the Earth's surface and a distance s metres from the Earth's centre $(s > R)$ is $\dfrac{mgR^2}{s^2}$.

(ii) A projectile is fired vertically from the Earth's surface with an initial velocity of u m s^{-1} and reaches a maximum height of h m. Neglecting air resistance, derive from Newton's Second Law an expression giving u^2 in terms of R, g and h.

(iii) For what launch speed would the projectile just reach a height equal to the radius of the Earth, 6400 km?

(iv) What is the minimum launch speed if the projectile is never to return?

Solution

(i) At the Earth's surface, the force on a body of mass m is $\dfrac{GMm}{R^2}$ newtons. So

$$\frac{GMm}{R^2} = \text{mass} \times \text{acceleration} = mg$$

$$\Rightarrow \quad GM = R^2 g$$

Above the Earth's surface at a distance s metres from the centre, the force on a body of mass m is

$$\frac{GMm}{s^2} = \frac{R^2 gm}{s^2} \quad \text{(substituting } GM = R^2 g)$$

$$= \frac{mgR^2}{s^2}$$

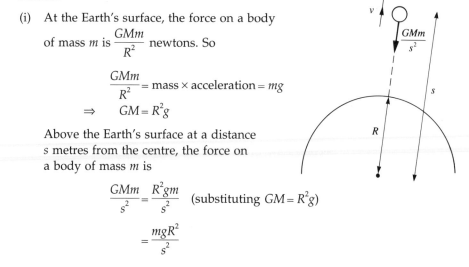

(ii) Take the positive direction (increasing s) upwards. The force is acting downwards, hence the negative sign when applying the equation of motion:

$$-\frac{mgR^2}{s^2} = mv\frac{dv}{ds}$$

The $v \dfrac{dv}{ds}$ form for acceleration is used, since the problem involves distances and velocities.

Dividing by m and separating the variables give

$$v \, dv = -\frac{R^2 g}{s^2} \, ds$$

Remember, s is the distance from the centre of the earth; the motion begins when $s = R$ and finishes when $s = R + h$, where h is the maximum height. The velocity v has initial value u and is zero at the maximum height. Using these as the limits of integration

$$\int_u^0 v \, dv = -R^2 g \int_R^{R+h} \frac{1}{s^2} \, ds$$

$$\Rightarrow \quad -\tfrac{1}{2}u^2 = R^2 g \left[\frac{1}{(R+h)} - \frac{1}{R} \right]$$

$$\Rightarrow \quad \tfrac{1}{2}u^2 = R^2 g \left[\frac{1}{R} - \frac{1}{(R+h)} \right]$$

(iii) When h is equal to R, the above formula gives

$$\tfrac{1}{2}u^2 = R^2 g \left[\frac{1}{R} - \frac{1}{(R+R)} \right]$$

$$= R^2 g \left(\frac{1}{2R} \right)$$

$$= \tfrac{1}{2}Rg$$

$$\Rightarrow \quad u = \sqrt{(Rg)}$$

$$= \sqrt{(6.4 \times 10^6 \times 9.8)} \quad (R = 6400 \text{ km} = 6400 \times 10^3 \text{ m})$$

$$= 7920$$

The launch speed is approximately 7.9 km s^{-1}.

(iv) The equation at the end of the part (ii) relates the launch speed, u, to the height reached, h:

$$\tfrac{1}{2}u^2 = R^2 g \left[\frac{1}{R} - \frac{1}{(R+h)} \right]$$

$$= Rg - \frac{R^2 g}{(R+h)}$$

If h is large enough, the second term on the right-hand side becomes negligibly small. You can see that in the limit $\tfrac{1}{2}u^2 = Rg$. For an initial velocity given by this equation, h becomes infinitely large and the projectile never returns. The value of u is then

$$u = \sqrt{(2Rg)} = 11\,200$$

This is 11.2 km s^{-1}. For obvious reasons this value of u is known as the *escape velocity*.

An alternative solution using energy methods is on page 65.

Investigation

Escape velocity

Assume that all of a group of planets, including the Earth, can be modelled as spheres of equal densities.

(i)   Show that the escape velocity from the surface of a planet is proportional to the radius. [Remember: $g = GM/R^2$]

(ii) What would be the escape velocity from the surface of a planet with a radius $\frac{1}{1000}$th that of the Earth's?

(iii) Many asteroids (minor planets between the orbits of Mars and Jupiter) are only a few kilometres in radius. Could you hit a tennis ball into space from the surface of an asteroid? Could you jump off into space?

Variation of *g* on the Earth's surface

The apparent weight of a body varies slightly at different parts of the Earth's surface. One reason is that the Earth is rotating. Unless you are at one of the poles, you are rotating in a circle round the Earth's axis and a small part of the gravitational force supplies the central acceleration for circular motion. You would otherwise be thrown into space. This has nothing to do with gravity but does affect what we measure as *g*.

In addition the Earth is not spherical but bulges at the equator. The equatorial radius is about 1 part in 300 greater than the polar radius. Thus someone at the equator is further from the centre than someone at a pole.

These effects combine to make *g* about 0.5% less at the equator than at the poles.

For Discussion

How much faster does a pendulum swing at the poles?

HISTORICAL NOTE

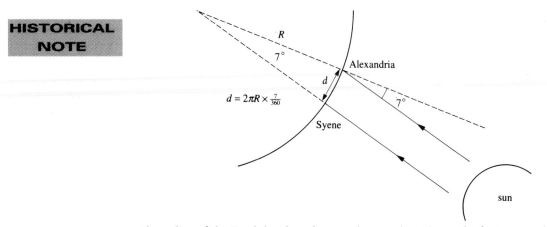

$$d = 2\pi R \times \tfrac{7}{360}$$

The radius of the Earth has been known since ancient times; the first reasonable estimate was by Eratosthenes (born in 276 BC). He was curator of the library at Alexandria and measured the elevation of the sun at noon on midsummer day to be about 7°. Due south at Syene (now Aswan), at the same time, on the same day, the sun was known to be overhead. So a 7° arc along the circumference of the earth corresponded to the known distance between Alexandria and Syene. Thus the circumference of the Earth and hence its radius could be estimated.

The value of g can be measured directly, so if the gravitational constant G can be determined, the mass of the Earth can be deduced using g=GM/R². Henry Cavendish, a brilliant but reclusive British physicist, performed a classic experiment to measure G in 1798. A rod with lead weights at each end was suspended on a fine fibre. When large weights are brought near the suspended ones, the tiny gravitational attraction causes a minute twist of the rod. This can be amplified and measured by the movement of a beam of light reflected from a mirror attached to the fibre.

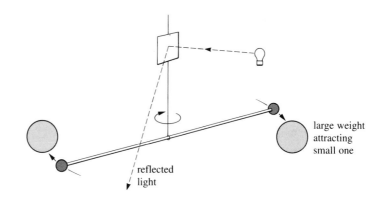

large weight
attracting
small one

reflected
light

Exercise 3C

1. The gravitational acceleration on a body distance r from the centre of a uniform sphere of mass M has magnitude GM/r^2, where G is the gravitational constant.
   (i) What are the dimensions of G?
   (ii) Assuming the Earth can be modelled as a perfect sphere of mass M and radius R, write down an expression for a_h, the acceleration due to gravity at a distance h above the Earth's surface. What symbol is normally used for a_0?
   (iii) Show that

$$a_h = g\left(1 + \frac{h}{R}\right)^{-2}$$

   (iv) Hence show that when h is small compared to the radius of the Earth, so that $(h/R)^2$ is negligible,

$$a_h = g\left(1 - \frac{2h}{R}\right)$$

   (v) Given that the radius of the Earth is about 6400 Km, by what percentage does gravitational acceleration differ from g at a height of 50 km?

2. The gravitational acceleration on a body distance x from the centre of the Earth has magnitude k/x^2, where k is a constant. An artificial satellite is in a circular orbit of radius r about the centre of the Earth.
   (i) Show that $k = gR^2$, where R is the radius of the Earth and g has its usual significance.
   (ii) Write down the gravitational acceleration on the satellite in terms of R, r and g. Hence show that the time T taken for the satellite to orbit the Earth is given by

$$T = \frac{2\pi}{R}\sqrt{\left(\frac{r^3}{g}\right)}$$

(iii) Show that a satellite in circular orbit whose period is 24 hours (a geosynchronous satellite) will be about 36 000 km above the Earth's surface. (Take g as 9.8 m s^{-2} and the radius of the Earth as 6400 km.)

3. A meteor of mass m is attracted towards the moon by a gravitational force mk/r^2, where k is a constant, and r is the distance of the meteor from the centre of the moon. The meteor initially has negligible velocity and is at a distance R above the moon's surface, where R is the radius of the moon. It then falls directly towards the moon.

(i) Starting from Newton's Second Law, find the velocity of the meteor when it smashes into the surface, in terms of k and R.

(ii) The radius of the moon is 1750 km and the gravitational acceleration at the surface is 1.6 m s^{-2}. Using this information, find the value of k (in SI units) and hence find the actual final velocity of the meteor.

4. A body of unit mass *inside* a uniform sphere distance r from its centre O experiences a gravitational attraction towards the centre of GM_r/r^2, where G is the gravitational constant and M_r is the mass of material inside the sphere of radius r. (In other words it is as if the body were on the surface of a sphere of radius r, all the matter further from the centre than r will have no net gravitational effect on the body.)

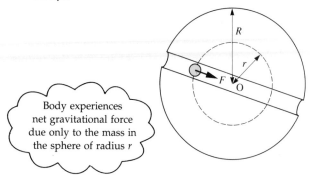

Body experiences net gravitational force due only to the mass in the sphere of radius r

Suppose a straight tube could be drilled right through the Earth, modelled as a uniform sphere of radius R and total mass M. A ball is dropped into the tube at the surface of the Earth.

(i) Work out the mass of a sphere of the Earth of radius r, in terms of R, M and r.

(ii) Write down an expression for the gravitational force on the ball when at a distance r from the centre.

(iii) Hence show that the ball will oscillate with simple harmonic motion with an amplitude equal to the radius of the earth. Determine the period.

5. A missile is projected vertically, with speed u, from the Earth's surface with the aim of reaching the moon. The missile travels in a straight line from the Earth to the moon (whose motion is ignored). It is assumed that the only forces acting on the missile during its journey are the gravitational attractions of the Earth and moon. M_e and M_m denote the masses of the Earth and moon, R the radius of the Earth, and d the distance between the centres of the Earth and moon. The gravitational attraction between two bodies masses m_1 and m_2 may be taken as Gm_1m_2/r^2, where r is the distance between their centres and G is the gravitational constant.

(i) Write down an expression for the acceleration of the missile when it has reached a point a distance x from the Earth's centre (i.e. height $x - R$ above the Earth's surface).

(ii) Show that the acceleration is zero when

$$x = \frac{d}{1 + \sqrt{\left(\dfrac{M_m}{M_e}\right)}}$$

(iii) Integrate your expression for the acceleration to derive an expression for the velocity of the particle as a function of the distance x.

(iv) Explain how you would calculate the minimum launch speed required to reach the moon (you need not actually derive this).

Work, energy and impulse with variable forces

The concepts of work, energy and impulse are very valuable in the context of variable forces. In particular, the principle of conservation of mechanical energy, which has been previously used to solve problems without having to calculate acceleration explicitly, often enables you to solve variable-force problems which would not be easily dealt with by integration.

You have already partly covered this topic. *Mechanics 2* gave the definition of the impulse of a variable force and *Mechanics 3* showed how to calculate the work against the tension when stretching an elastic string. This section reviews the definitions of work, energy and impulse when variable forces are involved and applies them to the resistive and gravitational forces you have already met in this chapter.

Work done by a variable force

When a body moves a short distance δs along a line solely under the action of a parallel force F, you know that the force has done *an element of work* $\delta W = F \, \delta s$. When the force is *varying*, the total work is the sum of all these elements, that is $\displaystyle\int F \, ds$.

N O T E

You may not be satisfied with this and similar informal arguments. A more rigorous treatment is as follows.

Assuming the force increases from F to $F + \delta F$ as the body moves; then

$$F \, \delta s < \delta W < (F + \delta F)\delta s$$

$$\Rightarrow \quad F < \frac{\delta W}{\delta s} < F + \delta F$$

As $\delta s \to 0$, $\delta F \to 0$ and $\dfrac{\delta W}{\delta s} \to \dfrac{dW}{ds}$, hence

$$\frac{dW}{ds} = F$$

By integration, the total work done over the period of the motion is

$$W = \int F \, ds$$

In *Mechanics 3*, $\displaystyle\int F \, ds$ was used to find the work done in extending an elastic string, of stiffness k, by an amount s. When extended by x, the tension in the string is, by Hooke's Law, kx. Hence the total work is

$\displaystyle\int_0^s kx \, dx = \tfrac{1}{2}ks^2$, which is the elastic energy in a stretched spring.

Work done = increase in kinetic energy

You have seen this important result, which follows from Newton's Second Law:

$$F = ma = mv \frac{dv}{ds}$$

Separating the variables and taking u and v as the starting and finishing velocities respectively, you get

$$\int F \, ds = \int mv \, dv$$

$$= \tfrac{1}{2}mv^2 - \tfrac{1}{2}mu^2$$

work done by force = increase in kinetic energy

EXAMPLE

A particle of mass 1 kg moves along a line with a velocity v m s^{-1} under the influence of a resistive force of magnitude kv^2 N, where k is a constant. Initially, the velocity of the particle is 10 m s^{-1} and the force continues to act until the particle has slowed down to 5 m s^{-1}.

(i) Use kinetic energy considerations to write down the work done by the resistive force.

(ii) Solve the equation of motion of the particle and express the displacement x in terms of v. Hence show that the particle travels a distance $s = \dfrac{1}{k}\ln 2$ m while the force acts.

(iii) From part (ii) express v and hence F in terms of x. Hence confirm by integration the result obtained in part (i).

Solution

(i)
$$\text{Work done} = \text{final KE} - \text{initial KE}$$

$$= \tfrac{1}{2} \times 5^2 - \tfrac{1}{2} \times 10^2 \text{ J}$$

$$= -37.5 \text{ J}$$

This is negative because the force is in the opposite direction to the displacement.

(ii) Applying $F = ma$ gives

$$-kv^2 = v \frac{dv}{dx} \quad (m = 1, \; a = v\frac{dv}{dx}, \text{ where } x \text{ is displacement})$$

Separating the variables gives

$$-k \, dx = \frac{dv}{v}$$

$$-k \int dx = \int \frac{dv}{v}$$

$$\Rightarrow \qquad -kx = \ln v + c$$

When $x = 0$, $v = 10$, hence $c = -\ln 10$. So

$$-kx = \ln v - \ln 10$$

$$= -\ln \frac{10}{v}$$

$$\Rightarrow \quad x = \frac{1}{k} \ln \frac{10}{v} \qquad \text{①}$$

When $v = 5$ (final velocity)

$$s = \frac{1}{k} \ln 2$$

(iii) Making v the subject in ① gives $v = 10\,e^{-kx}$. Therefore

$$F = -kv^2$$

$$= -100k\,e^{-2kx}$$

Work done is $\displaystyle\int F\,dx$:

$$\int F\,dx = \int_0^s -100k\,e^{-2kx}\,dx$$

$$= 50 \int_0^s -2k\,e^{-2kx}\,dx$$

$$= 50[e^{-2kx}]_0^s$$

Now $s = \dfrac{1}{k} \ln 2$, so applying the upper limit gives

$$e^{-2ks} = e^{-2\ln 2}$$

$$= e^{\ln \frac{1}{4}}$$

$$\Rightarrow \quad e^{-2ks} = \tfrac{1}{4}$$

Thus the work done $= 50(\tfrac{1}{4} - 1)$

$$= -37.5$$

This is $-37.5\,\text{J}$, as before.

Force–distance graph

Note that if you plot force against distance (figure 3.2), the work done $\displaystyle\int F\,dx$ is the area under the graph.

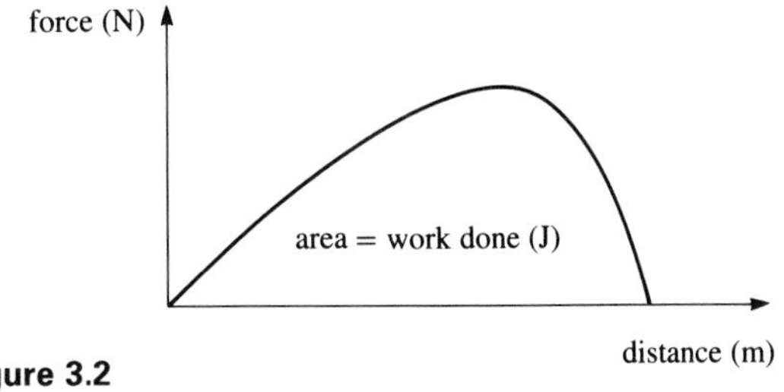

force (N)

area = work done (J)

distance (m)

Figure 3.2

Investigation

Simple harmonic motion by the energy method

In *Mechanics 3*, you investigated the spring–mass oscillator: a particle of mass m suspended on a spring oscillating vertically about its equilibrium position. You can derive the same results by the energy method.

A particle of mass m is suspended from a spring of natural length l and stiffness k. Without referring to the equilibrium position, write down an expression for the total energy when the string is extended by x and moving with velocity v. The total energy is constant throughout the motion.

Differentiate the energy equation and show that the mass oscillates with simple harmonic motion. Deduce the equilibrium position *from* the simple harmonic motion equation.

Gravitational potential energy

For a body moving near the Earth's surface, the work done by the gravitational force is often expressed as the change in the *potential energy* of the body. When a body of mass m rises a distance h, the work done *against* the gravitational force mg is mgh and you know this is the increase in potential energy of the body.

When the gravitational force *changes*, such as in the case of a missile launched from the Earth's surface into space, the work done is obtained by integration as follows.

The gravitational force on a body of mass m at a distance r_1 from a body of mass M is

$$F = -\frac{GMm}{r_1^2}$$

The positive direction will be that of increasing r, hence the minus sign. Suppose a body of mass m is pulled away from a distance r_1 to a distance r_2 (figure 3.3), what work is done *against* the gravitational force?

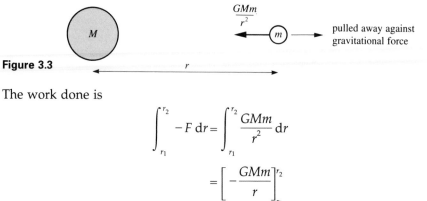

Figure 3.3

The work done is

$$\int_{r_1}^{r_2} -F \, dr = \int_{r_1}^{r_2} \frac{GMm}{r^2} \, dr$$

$$= \left[-\frac{GMm}{r} \right]_{r_1}^{r_2}$$

$$= \frac{GMm}{r_1} - \frac{GMm}{r_2}$$

This is positive $(r_2 > r_1)$, since increasing the separation between the bodies involves positive work against the gravitational force.

Just as in the case of constant gravitation, the increase in *potential energy* is defined as the work done against the force. With this type of motion, the zero level of potential energy is normally taken at infinity. Then the gravitational potential energy of a body of mass m at a distance r from a body of mass M is defined as the work done in bringing them together from infinity, namely $-\dfrac{GMm}{r}$.

The work done against gravitation in moving a spacecraft towards or away from a planet is thus the increase in potential energy, which according to the formula depends only on the initial and final values of r. And since

$$\text{work done against force} = \text{loss in kinetic energy}$$

it follows once more that

$$\text{gain in potential energy} = \text{loss in kinetic energy}$$

The following example shows the use of gravitational potential energy on a problem solved in a previous example by using the equation of motion.

EXAMPLE

A ballistic missile, fired vertically from the Earth's surface with an initial velocity of u just reaches a height h. Using energy methods, derive an expression giving u^2 in terms of R (the radius of the Earth) g and h. Assume the gravitational acceleration at a distance r from the centre of the Earth is $\dfrac{GM}{r^2}$, where $GM = gR^2$.

Solution

At launch, the kinetic energy of the missile is $\frac{1}{2}mu^2$ and its potential energy is $-\dfrac{GMm}{R}$. At the missile's highest point, its velocity is zero and since it is a distance $(R + h)$ from the centre of the Earth, the potential energy is $-\dfrac{GMm}{(R + h)}$.

So using the principle of conservation of energy:

$$\text{loss in kinetic energy} = \text{gain in potential energy}$$

$$\tfrac{1}{2}mu^2 = GMm\left[-\frac{1}{(R + h)} + \frac{1}{R}\right]$$

$$\Rightarrow \quad u^2 = 2gR^2\left[\frac{1}{R} - \frac{1}{(R + h)}\right] \quad \text{since } GM = gR^2$$

This is the result obtained previously on page 57.

Conservative forces

It is shown above that the work done against the gravitational force depends only on the initial and final position of the body. This means that when the body returns to its starting point, no net work has been done, so the kinetic energy will be the same as before. That is, the force conserves the total mechanical energy. Forces which have this property are known as conservative, *a term you met in* Mechanics 2. *It can be shown that conservation of mechanical energy is equivalent to the fact that the work done is dependent only on the initial and final positions.*

Potential energy is associated only with conservative forces. *The gravitational force, the electrical force between charged particles, the tension in a spring (obeying Hooke's Law)—these are all examples of conservative forces. Frictional forces and the forces involved in non-elastic collisions are not conservative. They are known as* dissipative *as they result in a reduction of the total mechanical energy.*

Power with variable forces

As you know from *Mechanics 2*, power is defined as the rate at which work is done. The definition is independent of whether or not the force is varying.

$$P = \frac{dW}{dt}$$

$$\Rightarrow \quad W = \int P \, dt$$

where the integration is performed over the total time period. This implies that the area under the power–time graph (figure 3.4) is equal to the work done.

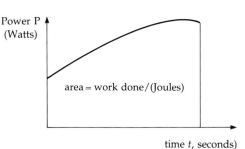

Figure 3.4

Power = force × velocity

You have seen that $F = dW/ds$, where W is the work done. Hence by the chain rule

$$\frac{dW}{dt}\frac{dt}{ds} = F$$

$$\Rightarrow \quad \frac{dW}{dt} = F\frac{ds}{dt}$$

Therefore, $\qquad\qquad P = Fv$

where v is the velocity of the body on which F acts.

This is a familiar result. Note that *constant power does not imply constant force.* When a car engine exerts a thrust of F N and the car is travelling with a speed v, the rate of working, i.e. power, is Fv J s^{-1}. If the power is constant, and v varies then F must vary.

EXAMPLE

When working at a constant power of 2.5 kW against a resistance proportional to the square of its speed, the maximum speed a vehicle can attain on a level road is 50 m s⁻¹. If the vehicle accelerates from rest under the same conditions, how far does it travel before it attains half the maximum speed? The mass of the vehicle is 1500 kg.

Solution

The power is 2500 W = force × velocity, so the engine's driving force has magnitude 2500/velocity.

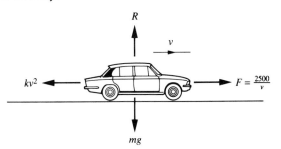

When the speed is v m s⁻¹ and resistance is kv^2 N, Newton's Second Law gives

$$\frac{2500}{v} - kv^2 = 1500v \frac{dv}{dx} \qquad \text{①}$$

The maximum speed of 50 m s⁻¹ occurs when the acceleration is zero. So

$$\frac{2500}{50} = k(50)^2$$

$$\Rightarrow \quad k = \frac{1}{50}$$

Substituting this value of k in ① gives

$$\frac{2500}{v} - \frac{v^2}{50} = 1500v \frac{dv}{dx}$$

$$\Rightarrow \quad 125\,000 - v^3 = 75\,000v^2 \frac{dv}{dx}$$

This can be solved by separating the variables and performing integration. The limits of v are 0 to $\frac{1}{2}V$ (i.e. 25) as x goes from 0 to s. So

$$\int_0^s dx = \int_0^{25} \frac{75\,000v^2}{(125\,000 - v^3)} dv$$

$$\Rightarrow \quad s = (-75\,000/3)[\ln(125\,000 - v^3)]_0^{25}$$

$$= -25\,000 \ln \frac{(125\,000 - 25^3)}{125\,000}$$

$$= -25\,000 \ln \tfrac{7}{8}$$

$$= 25\,000 \ln \tfrac{8}{7}$$

$$= 3338$$

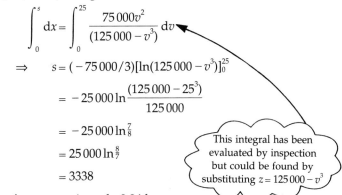

This integral has been evaluated by inspection but could be found by substituting $z = 125\,000 - v^3$

The distance is therefore approximately 3.34 km.

Impulse of a variable force

For a constant force, the *impulse* is force × time for which it acts.

When the force varies, the impulse over a small time interval δt is $F\,\delta t$, and so the total impulse over a period is defined as $\int F\,dt$ (see figure 3.5).

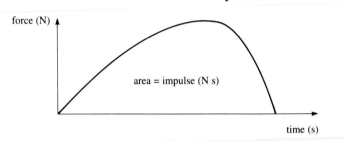

force (N)

area = impulse (N s)

time (s)

Figure 3.5

Over a period T in which the velocity changes from U to V

$$\int_0^T F\,dt = \int_U^V m\frac{dv}{dt}\,dt \quad \left(\text{since } F = m\frac{dv}{dt}\right)$$

$$\int_0^T F\,dt = \int_U^V m\,dv$$

$$= mV - mU$$

when m is constant.

This is the result which you saw in *Mechanics 2*, that the *impulse of a force is equal to the change in momentum*. This applies even when the force is varying.

Any collision involves variable forces. When two snooker balls, A and B, collide, they are in contact for a very short time. During that time the force between them is not constant. It is zero just as they touch, builds up to a maximum while they deform slightly and then goes down to zero again as the balls rebound. But by Newton's Third Law the force of A on B is equal and opposite to that of B on A *at every moment*. So if the total impulse on A is $\int F\,dt$, that on B is

$$\int -F\,dt = -\int F\,dt$$

The sum of the impulses on A and on B is zero and so the total momentum change is zero. The principle of conservation of momentum applies even though forces are variable.

Exercise 3D

1. A particle of mass $3\,\text{kg}$ initially at rest is subject to a force whose magnitude increases linearly with time from zero to $2\,\text{N}$ over a period of 1 second. What is the kinetic energy of the particle at the end of this period?

2. An object of mass $10\,\text{kg}$ is acted on by a force whose magnitude varies according to the distance from the starting point O, as shown in the graph. The force acts in a constant direction.

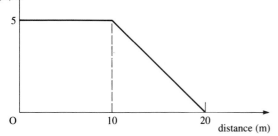

    (i)   What work has been done by the force when the object reaches the point $20\,\text{m}$ from O?

    (ii) If the object starts from rest, what is its final speed?

   (iii) What is its final momentum?

   (iv) What is the total impulse of the force over the period?

3. An object of mass $10\,\text{kg}$ begins at rest at O (see diagram), and is acted on by a force $F = 5 - \dfrac{s^2}{80}$, where s is the distance from O.

    (i)   What is the speed when $s = 20$?

    (ii) What is the total impulse of the force?

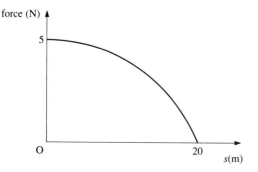

4. A pendulum consists of a weight on the end of a non-extensible string of length l and negligible mass. The weight is released from a point when the string is taut and horizontal, so that it swings down.

    (i)   Use the energy method to find the velocity of the weight at the lowest point.

    (ii) Hence show that the tension in the string at the lowest point is three times what it is when the weight is suspended at rest.

5. The gravitational potential energy of a body of mass m at a distance r from the earth's centre can be taken as $-R^2mg/r$, where R is the radius of the earth and g is the gravitational acceleration at the surface.

    (i)   Give an expression for the total kinetic and potential energy of a missile of mass m fired with velocity u from the Earth's surface.

    (ii) Use conservation of mechanical energy (neglecting forces other than gravitation) to calculate the value of u for the missile to escape from the earth completely. (Take g as $10\,\text{m s}^{-2}$ and the Earth's radius as $6400\,\text{km}$.)

6. A particle of mass m, moving in the sun's gravitational field, at a distance x from the centre of the sun, experiences a force Gmx^{-2} (where G is a constant) directly towards the sun.

    (i)   Show that, if at some time $x = h$ and the particle is travelling directly away from the sun with speed V, then x cannot become arbitrarily large unless $V^2 \geqslant 2Gh^{-1}$.

    (ii) A particle is initially motionless a great distance from the sun (of radius R). If, at some later time, it is at a distance h from the centre of the sun, how long after that will it take to fall into the sun?

7. A car of mass $800\,\text{kg}$ moves along a straight horizontal road with its engine working at a constant rate of $40\,\text{kw}$. It starts from rest and after t seconds its speed is $v\,\text{m s}^{-1}$. The resistance during this period is equal to $25\,v\,\text{N}$.

(i) Show that

$$32v \frac{dv}{dt} = 1600 - v^2$$

(ii) Deduce that the car cannot go faster than 40 m s^{-1}.

(iii) Express t in terms of v and find the time taken for the car to reach a speed of 35 m s^{-1} from rest.

8. The resistance to a car is proportional to the square of its speed. The car can cruise at a constant speed of 120 km h^{-1} on the level exerting a power of P W, and it can coast (i.e. roll without power) down a hill of 1 in 20 at the same speed.

(i) What would be its acceleration, at a power of P W, on the level at 60 km h^{-1}?

(ii) If it maintains this power while accelerating from 60 km h^{-1} to 90 km h^{-1} on the level, find how far it would go during this acceleration.
(Take g as 9.8 m s^{-2}.)

9. A catapult as shown projects a particle on a smooth horizontal surface. A light elastic string of natural length $8a$ and modulus of elasticity $2mg$ is attached to two points P and Q at a distance $8a$ apart on a horizontal table. A stone of mass m is attached to the mid-point, at a point D, drawn back a distance $3a$ to a point C and then released.

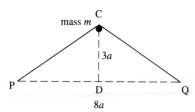

(i) What energy is stored in the elastic string when the mass is at point C?

(ii) Use conservation of energy to find the velocity of the stone when it reaches D.

(iii) The stone leaves the string when it reaches D, and is then acted on by a resistance of magnitude mkv, where v is the speed of the particle and k is a constant. Show that it comes to rest at a distance $\sqrt{(ga)}/k$ from D.

10. A train starts from a station. The tractive force exerted by the engine is at first constant and equal to F. However, after the speed attains the value u, the engine works at a constant rate P, where $P = Fu$. The mass of the engine and the train together is M. Forces opposing motion may be neglected.

(i) Show that the engine will attain a speed v, with $v \geqslant u$, after a time

$$\frac{M}{2P} (u^2 + v^2).$$

(ii) Show that in this time it will have travelled a distance

$$\frac{M}{6P} (2v^3 + u^3)$$

[O&C 1993]

11. A motor car of mass m accelerates along a level straight road. Over a period, the engine works at a rate which can be modelled by $pv + q$, where v is the velocity of the car and p and q are positive constants. Its speed at the beginning of the period is q/p and at the end of the period is $3q/p$. Resistance can be neglected.

(i) What is the thrust on the engine when the car is moving with velocity v?

(ii) Write down the equation of motion. Show that the variables can be separated to give

$$\int \frac{p}{m} \, dt = \int \frac{pv}{(pv + q)} \, dv$$

(iii) Show that the length of the period is

$$(2 - \ln 2) \frac{mq}{p^2}.$$

12. In a race, runners run 5 km in a straight line to a fixed point and then turn and run back to the starting point. A steady wind of 3 m s^{-1} is blowing from the start to the turning point. At a steady racing pace, a certain runner expends energy at a constant rate of 300 W. Two resistive forces act. One is of constant magnitude 50 N. The other, arising from air resistance, is of magnitude $2w$ N, where $w \text{ m s}^{-1}$ is the

runner's speed relative to the air. (Ignore effects due to acceleration and deceleration at the start and end.)

(i) Give a careful argument to derive formulae from which the runner's steady speed in each half of the race may be found.

(ii) Calculate, to the nearest second, the time the runner will take for the whole race.

The runner may use alternative tactics, expending the same total energy during the race as a whole, but applying different constant powers, x_1 W in the outward trip, and x_2 W on the return trip.

(iii) Prove that with the wind as above, if the outward and return speeds are v_1 m s^{-1} and v_2 m s^{-1} respectively, then $(v_1 + v_2)$ is independent of the choices for x_1 and x_2.

(iv) Hence show that these alternative tactics allow the runner to run the whole race approximately 15 s faster.

[O&C 1990]

Motion under variable force in two dimensions

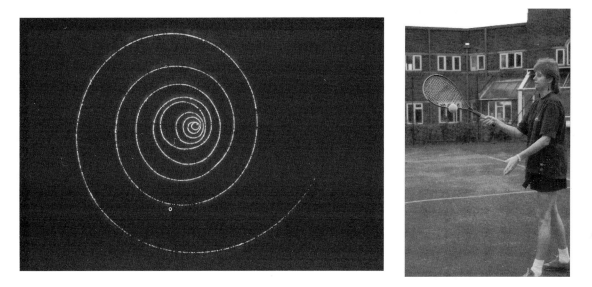

The photographs show two situations where forces are creating acceleration in a direction other than that of the velocity. The velocity thus changes direction leading to motion in more than one dimension.

You have already analysed, in *Mechanics 1*, the motion of a projectile under gravity, at least in the case where air resistance is ignored. There are no horizontal forces, so the horizontal speed is constant. The only vertical force is gravity, so the vertical motion can be analysed using the constant acceleration formula. The solution uses the fact that Newton's Second Law can be applied to the resolved parts of force and acceleration in *any* direction.

When the forces are varying, the same principle applies. Chapter 1 gave some examples where a force is given as a *function of time*:

$$\mathbf{F} = X_t\mathbf{i} + Y_t\mathbf{j} = m\ddot{\mathbf{r}}$$

where X_t and Y_t are the components of the force in the x and y directions at time t. As you have seen, the velocity $\dot{\mathbf{r}}$ and position $\mathbf{r}$ of a body under the influence of such a force can be obtained by integrating each component separately.

This section considers cases of two-dimensional motion where the magnitude and direction of forces are dependent on the *velocity* of an object (e.g. resisted motion) or on its *position* (e.g. gravitation or other *force fields*).

For Discussion

What forces act on the objects in the two photographs? How, if at all, do they depend on the position or velocity of the objects?

Resisted motion in two dimensions

Consider a body moving in two dimensions under a resistive force whose magnitude is directly proportional to the speed. This is the case with viscous flow, as you saw earlier in the chapter.

The resistive force $\mathbf{R}$ is in the opposite direction to the velocity $\mathbf{v}$ and its magnitude is proportional to $|\mathbf{v}|$, so

$$\mathbf{R} = -k\mathbf{v} \quad \text{for } k > 0$$

$$= -kv_x\mathbf{i} - kv_y\mathbf{j}$$

where v_x and v_y are, respectively, the x and y components of $\mathbf{v}$.

Assuming there are no other forces, applying $m\dfrac{d\mathbf{v}}{dt} = \mathbf{R}$ gives

$$m\frac{d\mathbf{v}}{dt} = -kv_x\mathbf{i} - kv_y\mathbf{j}$$

$$\Rightarrow \quad m\frac{dv_x}{dt}\mathbf{i} + m\frac{dv_y}{dt}\mathbf{j} = -kv_x\mathbf{i} - kv_y\mathbf{j}$$

Equating the x and y components separately gives

$$m\frac{dv_x}{dt} = -kv_x$$

$$m\frac{dv_y}{dt} = -kv_y$$

The x component of acceleration depends only on the x component of velocity, and similarly with the y component. So each of these equations can be solved exactly as shown in the one-dimensional resisted motion problems earlier in the chapter. The example takes you through this.

A missile of unit mass is projected horizontally with speed u under water. Apart from gravity, it experiences a resistive force of k times the velocity. Take the unit vector $\mathbf{i}$ along the direction of projection and $\mathbf{j}$ vertically upwards.

(i)   Show the equation of motion can be written as

$$\dot{\mathbf{v}} = -k\mathbf{v} - g\mathbf{j}$$

Deduce the equations for v_x and v_y separately, where v_x and v_y are, respectively, the x and y components of $\mathbf{v}$.

(ii) Solve the equations and express v_x and v_y in terms of time t.

(iii) Write the vector equation for $\mathbf{v}$ in terms of time t and hence show that the direction of the acceleration is constant.

(iv) Assuming the missile is at the origin at time $t = 0$, show that its position at time t is

$$\mathbf{r} = \frac{u}{k}(1 - e^{-kt})\mathbf{i} - \left(\frac{gt}{k} - \frac{g}{k^2}(1 - e^{-kt})\right)\mathbf{j}$$

Sketch the path of the missile.

Solution

(i)   The gravitational force is directed downwards and thus is $-g\mathbf{j}$. The resistive force is against the direction of motion and of magnitude $k\mathbf{v}$, hence is $-k\mathbf{v}$.

Applying $m\dfrac{\mathrm{d}\mathbf{v}}{\mathrm{d}t} = \mathbf{F}$ gives

$$\frac{\mathrm{d}\mathbf{v}}{\mathrm{d}t} = -k\mathbf{v} - g\mathbf{j}$$

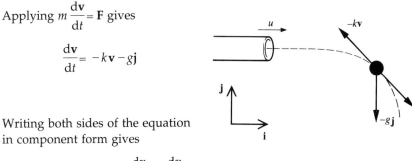

Writing both sides of the equation in component form gives

$$\frac{\mathrm{d}v_x}{\mathrm{d}t}\mathbf{i} + \frac{\mathrm{d}v_y}{\mathrm{d}t}\mathbf{j} = -kv_x\mathbf{i} - (g + kv_y)\mathbf{j}$$

Hence

$$\frac{\mathrm{d}v_x}{\mathrm{d}t} = -kv_x \qquad\qquad ①$$

and

$$\frac{\mathrm{d}v_y}{\mathrm{d}t} = -g - kv_y \qquad\qquad ②$$

(ii) *Horizontal direction*: separate the variables in ①, and integrate:

$$\int \frac{\mathrm{d}v_x}{v_x} = \int -k\,\mathrm{d}t$$

$$\Rightarrow \qquad \ln v_x = -kt + c$$

The value of v_x is u when $t = 0$, so $c = \ln u$. Therefore

$$\ln v_x = -kt + \ln u$$

$$\ln\left(\frac{v_x}{u}\right) = -kt$$

$$v_x = u\,e^{-kt}$$

Vertical direction: separate the variables in ②, and integrate:

$$\int \frac{dv_y}{(kv_y + g)} = \int -dt$$

$$\Rightarrow \quad \frac{1}{k}\ln(kv_y + g) + c' = -t$$

When $t = 0$, $v_y = 0$, so $\ln g + c'k = 0$, giving $kc' = -\ln g$. Hence

$$\ln\left(\frac{kv_y + g}{g}\right) = -kt$$

$$\Rightarrow \quad \frac{kv_y}{g} + 1 = e^{-kt}$$

$$\Rightarrow \quad v_y = \frac{g}{k}(e^{-kt} - 1) \quad \text{(note } v_y \text{ is negative)}$$

(iii) Putting the components of velocity together gives

$$\mathbf{v} = u\,e^{-kt}\mathbf{i} + \frac{g}{k}(e^{-kt} - 1)\mathbf{j}$$

Differentiate with respect to t to find acceleration:

$$\frac{d\mathbf{v}}{dt} = -ku\,e^{-kt}\mathbf{i} - g\,e^{-kt}\mathbf{j}$$

$$= -e^{-kt}(uk\mathbf{i} + g\mathbf{j})$$

This is parallel to $-uk\mathbf{i} - g\mathbf{j}$, which has a constant direction.

(iv)
$$\mathbf{v} = \frac{d\mathbf{r}}{dt} = u\,e^{-kt}\mathbf{i} + \frac{g}{k}(e^{-kt} - 1)\mathbf{j}$$

Integrating each component gives

$$\mathbf{r} = -\frac{u}{k}e^{-kt}\mathbf{i} - \frac{g}{k^2}e^{-kt}\mathbf{j} - \frac{gt}{k}\mathbf{j} + \mathbf{c}$$

When $t = 0$, the missile is at the origin. This gives

$$\mathbf{c} = \frac{u}{k}\mathbf{i} + \frac{g}{k^2}\mathbf{j}$$

Thus

$$\mathbf{r} = \frac{u}{k}(1 - e^{-kt})\mathbf{i} - \frac{gt}{k}\mathbf{j} + \frac{g}{k^2}(1 - e^{-kt})\mathbf{j}$$

$$= \frac{u}{k}(1 - e^{-kt})\mathbf{i} - \left(\frac{gt}{k} - \frac{g}{k^2}(1 - e^{-kt})\right)\mathbf{j}$$

In the horizontal direction the speed dies away and the missile never exceeds a distance of u/k. Vertically the missile approaches a terminal speed of g/k, as you could deduce from the equation of vertical motion

$$\frac{dv_y}{dt} = -g - kv_y \quad \text{(remember } v_y \text{ is negative).}$$

The path followed is thus as shown.

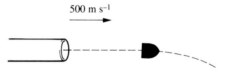

Investigation

Two-dimensional resisted motion under v^2 law

Two-dimensional motion with air resistance modelled as a force proportional to the square of velocity gives equations which cannot be readily solved analytically.

In this investigation, try to model the case of a projectile fired through the air under gravity and against a resistive force proportional to the square of the velocity. Write the equation of motion and separate into equations for the x and y components. Why is it that these cannot be solved in a similar way to the previous example?

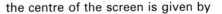

If you have studied Euler's numerical method (*Mechanics 4: Differential Equations*) try to see how you would apply this to working out the initial motion of a bullet of mass 0.01 kg fired horizontally with a velocity of 500 m s^{-1} under a resistive force of $5 \times 10^{-5} v^2$ N.

For Discussion

A bullet is fired horizontally over level ground and at the identical moment another bullet is dropped from the same height as the gun. A naive mathematician says they both hit the ground at the same time. Which actually hits the ground first and why?

Exercise 3E

1. In a cathode-ray tube, an electron of mass m travelling horizontally with speed u is deflected by a uniform vertical electric force of magnitude eE, where e is the charge on the electron and E is the strength of the electric field. This force acts only while the electron is passing between two plates, a horizontal distance a. After this it travels in a straight line to the screen, which is a horizontal distance b from the end of the plates. Assuming all other forces are negligible, show that the deflection d from the centre of the screen is given by

$$d = (\tfrac{1}{2}a + b)\frac{eEa}{mu^2}$$

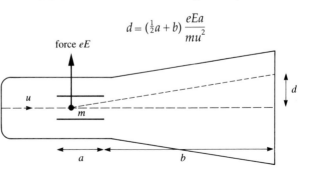

2. A particle has position vector
   $\mathbf{r} = a\cos\omega t\mathbf{i} + a\cos^2\omega t\mathbf{j}$, where a is in metres, t
   is in seconds and $\omega = \pi/20\text{ s}^{-1}$.
   (i)   Find the period T of the motion.
   (ii) Sketch the path of the particle, showing
         its position every 5 s for $0 \leqslant t \leqslant T$.
   (iii) Find expressions for the velocity and
         acceleration vectors of the particle as
         functions of time t.
   (iv) Show that the speed $v\text{ m s}^{-1}$ of the
         particle is given by

   $$v^2 = (a\omega\sin\omega t)^2(5 - 4\sin^2\omega t).$$

   (v)   Find the times in the first 20 s when the
         speed is a maximum and determine
         these maximum values in terms of a.

   [MEI 1993]

3. A particle of mass m moves in a vertical
   plane against a resistive force of magnitude
   mk times its speed, where k is a positive
   constant. At time t the particle is at the
   point (x, y) with respect to the horizontal
   axis Ox and the vertically upward axis Oy.
   At time $t = 0$ the particle is at O and moving
   so that its horizontal and vertical
   components of velocity are (U, V).
   (i)   Show that the vertical component of
         velocity v at time t whilst the particle is
         ascending satisfies the equation

   $$\frac{dv}{dt} = -(kv + g)$$

   (ii) Solve this equation to find the time
         taken by the particle to achieve its
         greatest height above Ox.
   (iii) Show that x satisfies the equation

   $$\frac{dx}{dt} = U - kx$$

   and find the horizontal distance the
   particle has travelled when it reaches
   its greatest height.

4. A projectile is fired horizontally with a
   speed u. In the horizontal direction there is
   a deceleration of magnitude kv_x (where v_x is
   the horizontal velocity). In the vertical
   direction the speed is such that the resistive
   force may be neglected and the
   acceleration is simply g. Find the horizontal
   (x) and downward vertical (y) distances
   travelled after time t, and hence show that
   the equation of the trajectory is

   $$x = \frac{u}{k}(1 - e^{-kv(2y/g)})$$

5. As part of some design tests for a wire-
   guided missile, an unpowered projectile is
   to be fired to find its maximum range when
   trailing light wires. The projectile has mass
   m and is fired with velocity $\mathbf{V}$. It trails
   behind it the guidance wires which link it
   to the firing site. These wires exert a force
   of $-n^2m\mathbf{r}$ on the projectile, where $\mathbf{r}$ is the
   position vector of the projectile relative to
   the firing site and n is a constant.
   (i)   Neglecting air resistance, show that, at
         time t after firing,

   $$\mathbf{r} = \frac{1}{n}(\sin nt)\mathbf{V} + \frac{1}{n^2}(1 - \cos nt)\mathbf{g}$$

   where $\mathbf{g}$ is the acceleration due to
   gravity.
   (ii) Find the maximum range of the
         projectile over a horizontal plane and
         the angle of elevation required to
         produce it.

   [MEI 1985]

6. A projectile of mass m is fired with velocity
   $\mathbf{u}$ and is subject to a gravitational field $\mathbf{g}$
   and a resistive force which opposes
   motion. The resistive force has a magnitude
   proportional to the speed of the projectile
   and is such that in free fall the projectile
   would reach a terminal speed of V.
   (i)   Find the constant of proportionality
         associated with the resistive force.
   (ii) Write down a vector differential
         equation for the motion of the
         projectile and hence show that its
         position vector relative to the point of
         firing at a time t after firing is given by

   $$\mathbf{r} = \frac{V}{g}\left[\left(\mathbf{u} - \frac{V}{g}\mathbf{g}\right)(1 - e^{-gt/V}) + \mathbf{g}t\right]$$

   [MEI 1988]

Force dependent on position: fields of force

At some time you have probably plotted the 'lines of force' round a magnet using iron filings or small compass needles (figure 3.6). The line through each point gives the direction of the magnetic *field* at that point, i.e. the direction of the magnetic force which is causing the compass needle to align.

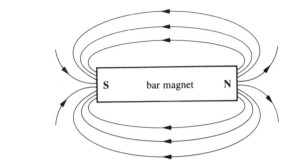

Figure 3.6

The notion of a *vector field* is used when every point of space is associated with some vector representing, for example, the force acting at that particular point. Another example is the gravitational field around a planet (figure 3.7). A particle of mass m placed at any point P will be attracted towards the planet by a force GMm/r^2 N, where G is the gravitational constant and M the mass of the planet. The acceleration of the particle of mass m will therefore be directed towards the planet and of magnitude GM/r^2. Thus an acceleration vector can be drawn at every point of space, pointing towards the planet. The magnitude GM/r^2 is called the field *strength* or *intensity*.

direction of
acceleration field
round planet

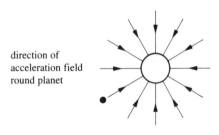

Figure 3.7

Fields of force or acceleration are most commonly associated with gravitational, electric or magnetic fields (the so-called 'action at a distance' forces where there is no physical contact between the interacting objects). Analysis of two- and three-dimensional physical situations often involve other types of vector field—for example, the *velocity* of air flowing over an aircraft wing.

EXAMPLE

A charged particle of mass m is moving under gravity near a vertical, long, straight, charged wire. The effect of the charged wire is to repel the particle with a force perpendicular to the wire. When the particle is at a distance d from the wire, the magnitude of this electric force is k/d N, where k is a constant. Assuming the wire coincides with the y axis and the particle is in the xy plane, write down the vector representing the total force **F** on the particle when it is at a point (x, y).

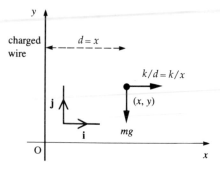

Solution

Using unit vectors as shown, the gravitational force is mg in the $-\mathbf{j}$ direction and the electric force is k/x in the $\mathbf{i}$ direction. So

$$\mathbf{F}(x, y) = \frac{k}{x}\mathbf{i} - mg\mathbf{j}$$

The force **F** varies at different points in space. It is a function of position, which is why it is written $\mathbf{F}(x, y)$, although in this particular problem the value of **F** depends only on x. It may equally be written $\mathbf{F}(\mathbf{r})$, showing **F** is a function of **r**, the position vector of each point.

EXAMPLE

A star of mass M is at the origin of the xy plane. What is the gravitational acceleration vector at a point $P(x, y)$?

Solution

The acceleration vector is directed along PO and its magnitude is

$$\frac{GM}{r^2} = \frac{GM}{x^2 + y^2}$$

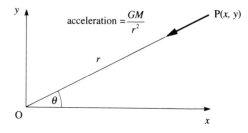

Resolving in the x and y directions gives

$$F_x = -\frac{GM}{x^2 + y^2}\cos\theta$$

$$= -\frac{GM}{x^2 + y^2} \times \frac{x}{\sqrt{(x^2 + y^2)}} = -\frac{GMx}{(x^2 + y^2)^{\frac{3}{2}}}$$

Similarly

$$F_y = -\frac{GMy}{(x^2 + y^2)^{\frac{3}{2}}}$$

$$\Rightarrow \quad \mathbf{F} = -\frac{GMx}{(x^2 + y^2)^{\frac{3}{2}}}\mathbf{i} - \frac{GMy}{(x^2 + y^2)^{\frac{3}{2}}}\mathbf{j}$$

$$= -\frac{GM}{(x^2 + y^2)^{\frac{3}{2}}}(x\mathbf{i} + y\mathbf{j})$$

NOTE *$r^2 = x^2 + y^2$, which suggests a simplification of the function for F. You will see in Chapter 4 that 'central force' problems like this are easier to describe and solve using polar co-ordinates rather than xy co-ordinates.*

Work done by a force field on a moving particle

Energy methods give a means of investigating some aspects of motion in a force field. As a particle moves along some path through a force field, work is done on it by the force field and the total work done is equal to the gain in kinetic energy. How do you calculate this work when a force is given as a function of position?

With reference to figure 3.8, look at work done as the particle moves from some point $P(x, y)$ on the path to $Q(x + \delta x, y + \delta y)$. The displacement is $\delta \mathbf{s} = \delta x \mathbf{i} + \delta y \mathbf{j}$. If this is so small that any change in **F** can be ignored, the work done, as explained in Chapter 1, is

$$\mathbf{F} . \delta \mathbf{s} = F_x \delta x + F_y \delta y$$

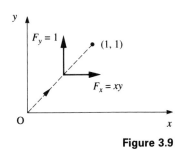

The total work done is the sum of these infinitesimal contributions over the complete path:

Figure 3.8

$$\int \mathbf{F} . \mathrm{d}\mathbf{s} = \int F_x \, \mathrm{d}x + \int F_y \, \mathrm{d}y$$

Before examining what this integral means, look at a simple example. Suppose a force field has the form

$$\mathbf{F}(x, y) = xy\mathbf{i} + \mathbf{j}$$

Imagine a particle moves in a straight line from the origin O (figure 3.9) to the point $(1, 1)$. The work done by the force field is

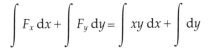

$$\int F_x \, \mathrm{d}x + \int F_y \, \mathrm{d}y = \int xy \, \mathrm{d}x + \int \mathrm{d}y$$

Figure 3.9

The equation of the path is the line $y = x$, and the limits of both y and x are 0 to 1. Substituting $y = x$ in the first integral and the appropriate limits in each integral gives

$$\int_0^1 x^2 \, \mathrm{d}x + \int_0^1 \mathrm{d}y = [\tfrac{1}{3}x^3]_0^1 + [y]_0^1$$

$$= \tfrac{4}{3}$$

The integral $\displaystyle\int \mathbf{F} \cdot d\mathbf{s}$ is known as a *line* integral because it is evaluated along a specified path. Once you know the relationship between y and x on this path, you can convert each of the integrals $\displaystyle\int F_x(x,y)\, dx$ and $\displaystyle\int F_y(x,y)\, dy$ to one involving just a single variable by substituting for x or y as appropriate.

In general, the work done by the field when a particle moves between two points will depend on the path taken. Using the force field above, assume the particle now moves along the x axis to $(1,0)$ and then parallel to the y axis to $(1,1)$, as shown in figure 3.10.

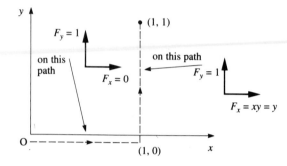

Figure 3.10

While moving along the x axis, y is zero, so $\displaystyle\int xy\, dx$ and $\displaystyle\int dy$ are both zero. (Physically this is because F_x is zero along the x axis and F_y does not do any work when the motion is along the x axis.)

During the second part of the motion, x is constant, so the limits of x are equal and $\displaystyle\int F_x\, dx = 0$. The limits of y are 0 to 1, so

$$\int F_y\, dy = \int_0^1 dy = 1$$

Hence the work done by the force field on the particle along this path is 1 N. This is less than the work done in the direct path.

N O T E

Constrained paths in a force field
In problems like those just discussed, you may well wonder 'Why does the particle move along the given path when the given force field would not cause it to do that?' There must be other forces on the particle. One practical case is where a charged particle is moving in combined electric and magnetic fields. The electrical force is a function of its position and forms the force field which is doing the work. But a magnetic field causes a force on a charged particle which is always at right angles to its motion. *Thus it constrains the motion of the particle without doing any work. You can think of the particle constrained by moving inside a smooth tube, where the reaction of the tube (the magnetic force) does no work.*

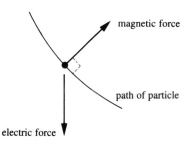

magnetic force

path of particle

electric force

EXAMPLE

A force is defined as

$$\mathbf{F}(x, y) = 2xy\mathbf{i} + x^2\mathbf{j}$$

(i) Evaluate the work done by this field when a particle moves in a straight line from the origin to $(2, 4)$.

(ii) Do the same over some other path between these two points and check that this particular field gives the same result.

Solution

$$\text{Work done} = \int \mathbf{F} . \mathbf{ds} = \int 2xy \, dx + \int x^2 \, dy$$

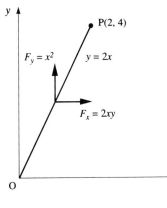

(i) This particle moves along the line $y = 2x$ from $(0, 0)$ to $(2, 4)$.
Since $y = 2x$ and the limits of x are 0 to 2, the first integral is

$$\int_0^2 2xy \, dx = \int_0^2 4x^2 \, dx$$

$$= [\tfrac{4}{3}x^3]_0^2 = \tfrac{32}{3}$$

In the second integral, substitute $dy = 2dx$ to give

$$\int_0^2 x^2 \, dy = \int_0^2 2x^2 \, dx$$

$$= [\tfrac{2}{3}x^3]_0^2 = \tfrac{16}{3}$$

You will get the same result if you substitute $x = \tfrac{1}{2}y$ in the original integral $\int x^2 \, dy$

The total work done is $\tfrac{32}{3} + \tfrac{16}{3} = 16$ units.

(ii) Suppose it moves along the curve $y = x^2$, which passes through $(0, 0)$ and $(2, 4)$.

Again, work done

$$= \int \mathbf{F} . \mathbf{ds} = \int 2xy \, dx + \int x^2 \, dy$$

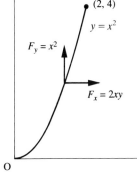

Substitute $y = x^2$ in the first integral and $x^2 = y$ in the second. The limits of x are $(0,2)$ and of y are $(0,4)$. These give

$$\int_0^2 2x^3 \, dx + \int_0^4 y \, dy = [\tfrac{1}{2}x^4]_0^2 + [\tfrac{1}{2}y^2]_0^4$$

$$= 16$$

This is the same result as before. If you check *any* path you will find you obtain the same result.

NOTE

Conservative force fields
A conservative force was defined earlier as one in which the work done depends only on the initial and final positions. Similarly, in two- and three-dimensional motion a conservative force field is one in which the work done in moving between two points is independent of the path followed. Gravitational, electric and magnetic fields are all conservative. As you saw in the one-dimensional case, this means that a potential energy function can be defined, and that energy methods can be applied to problems of motion in such fields.

Path given by parametric equation
The path taken by a particle may be given in parametric form. The line integral may then be worked out in terms of the parameter.

EXAMPLE

Find the work done in moving a particle from the point $(2,0)$ in an anticlockwise direction round a circle of radius 2, and whose centre is the origin in the force field

$$\mathbf{F} = (2x - y)\mathbf{i} + (x + y)\mathbf{j}$$

What is the result if the particle is moved in the opposite direction?

Solution

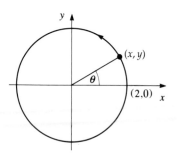

Work done
$$\int \mathbf{F} \cdot d\mathbf{s} = \int (2x - y) \, dx + \int (x + y) \, dy$$

The equation of the circle is $x^2 + y^2 = 4$. But it is easier to work in the parametric form $x = 2\cos \theta$, $y = 2\sin \theta$. Then the anticlockwise motion can be described by θ increasing with time.

Differentiating x and y with respect to θ gives

$$dx = -2\sin \theta \, d\theta \qquad dy = 2\cos \theta \, d\theta$$

The path starts on the x axis, where $\theta = 0$, and goes right round until $\theta = 2\pi$.

$$\int \mathbf{F} \cdot d\mathbf{s} = \int (2x - y)\, dx + \int (x + y)\, dy$$

$$= \int_0^{2\pi} (4\cos\theta - 2\sin\theta) \times (-2\sin\theta)\, d\theta + \int_0^{2\pi} (2\cos\theta + 2\sin\theta) \times (2\cos\theta)\, d\theta$$

$$= \int_0^{2\pi} (4 - 4\sin\theta\cos\theta)\, d\theta$$

$$= [4\theta - 2\sin^2\theta]_0^{2\pi} = 8\pi$$

It makes no difference where the particle starts if it is going round a closed loop, as in this example. However, the answer will have the *opposite sign* if the particle goes the other way round (integrating between limits of 2π to zero). In this case, the answer of -8π would mean that the field did negative work on the particle: it was tending to move against the force.

Exercise 3F

1. A particle is moved through a force field $\mathbf{F} = 3xy\mathbf{i} - y^2\mathbf{j}$ N, over a path given in parametric form by $x = t$, $y = 2t^2$ between the origin and the point $(1, 2)$.

   (i) Show that the work done by the field is

   $$\int (3xy\, dx - y^2\, dy)$$

   where the integral is evaluated along the path.

   (ii) Convert this to a definite integral in the single variable t and hence show that the work done by the field is $-\frac{7}{6}$ J.

2. A particle moves in the xy plane under the influence of a force field defined by

   $$\mathbf{F} = (x^2 - y^2)\mathbf{i} + 2xy\mathbf{j}$$

   Find the work done by the force in moving the particle from the origin $(0, 0)$ to the point $(1, 1)$ by each of the three paths

   (i) the straight lines from $(0, 0)$ to $(0, 1)$ and then to $(1, 1)$;

   (ii) the straight line from $(0, 0)$ to $(1, 1)$;

   (iii) along the parabola $y = x^2$.

   Sketch the path in each case. Show further that the work done in moving the particle between any two points on the circle

   $x^2 + y^2 - 2x = 0$ by a path round its circumference is constant.

   [MEI 1985]

3. A particle is moving in the xy plane under the influence of a force field defined by

   $$\mathbf{F} = \frac{(x\mathbf{i} + y\mathbf{j})}{(x^2 + y^2)}$$

   Find the work done on the particle in moving it from the point $(1, 1)$ to the point $(2, 4)$ by each of the following paths:

   (i) the straight line from $(1, 1)$ to $(2, 1)$ followed by the straight line from $(2, 1)$ to $(2, 4)$;

   (ii) the straight line direct from $(1, 1)$ to $(2, 4)$.

   [MEI 1990]

4. A particle moves in a force field

   $$\mathbf{F} = (3x - 4y)\mathbf{i} + (4x + 2y)\mathbf{j}$$

   Find the work done *on* the particle in moving it once round the ellipse

   $$\left(\frac{x}{5}\right)^2 + \left(\frac{y}{7}\right)^2 = 1$$

   in an anticlockwise direction starting from the point $(5, 0)$.

Exercise 3F continued

State, with reasons, whether the starting point or the direction of motion make any difference to the work done on the particle.

[MEI 1987]

5. A charged particle of mass m moves under gravity in a vertical plane where Ox is a horizontal axis and Oy an upward vertical axis. It is also under the influence of an electric force due to a charged wire along Oy. *The total* force on the particle when it is at the position (x, y) where $x > 0$, is

$$\mathbf{F} = \frac{-k}{x}\mathbf{i} - mg\mathbf{j}$$

where k is a constant. The particle moves along a path with equation $y = f(x)$ from (x_1, y_1) to (x_2, y_2).

(i) Show that, *whatever* the path, the work done by the field is

$$k \ln\left(\frac{x_1}{x_2}\right) - mg(y_2 - y_1)$$

(ii) Use this result to explain why the gravitational potential energy of the particle when at (x, y) may be defined as mgy.

(iii) Define an electric potential energy in this situation.

KEY POINTS

- When a particle is moving along a line under a variable force F, Newton's Second Law gives a differential equation. It is generally solved by writing acceleration as

  $$\frac{dv}{dt} \qquad \text{when } F \text{ is given as function of time } t$$

  $$v\frac{dv}{ds} \qquad \text{when } F \text{ is given as function of displacement } s$$

  $$\frac{dv}{dt} \text{ or } v\frac{dv}{ds} \qquad \text{when } F \text{ is given as function of velocity } v$$

- Where forces are conservative, variable force problems may be handled using the principle of conservation of energy.

- A *resistive* force is opposite to the direction of motion, and its magnitude often depends on the speed: typically it is proportional to v or v^2.

- The *terminal velocity* of a falling body is reached when the resistive force is balanced by the force of gravity.

- The gravitational force between two particles is proportional to the product of their masses and inversely proportional to the square of the distance between them: $F = \dfrac{Gm_1m_2}{r^2}$

- In vector form, Newton's Second Law is $\mathbf{F} = \mathbf{ma}$. Two-dimensional problems may be solved by regarding this as two equations for the $\mathbf{i}$ and $\mathbf{j}$ directions.

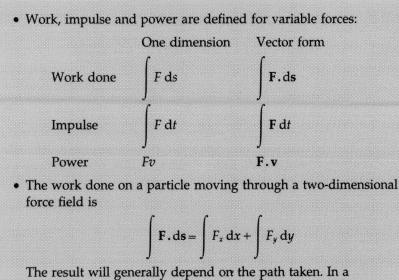

- Work, impulse and power are defined for variable forces:

	One dimension	Vector form
Work done	$\displaystyle\int F\,\mathrm{d}s$	$\displaystyle\int \mathbf{F}\cdot\mathbf{ds}$
Impulse	$\displaystyle\int F\,\mathrm{d}t$	$\displaystyle\int \mathbf{F}\,\mathrm{d}t$
Power	Fv	$\mathbf{F}\cdot\mathbf{v}$

- The work done on a particle moving through a two-dimensional force field is

$$\int \mathbf{F}\cdot\mathbf{ds} = \int F_x\,\mathrm{d}x + \int F_y\,\mathrm{d}y$$

The result will generally depend on the path taken. In a conservative field, the work done is independent of the path.

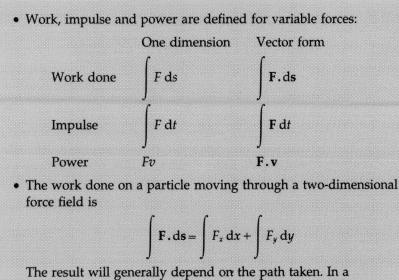

Motion described in polar co-ordinates

What make planets go round the sun? At the time of Kepler, some people answered this problem by saying there were angels behind them beating their wings and pushing the planets around an orbit. As you will see the answer is not far from the truth. The only difference is that the angels sit in a different direction and their wings push inwards.

Richard Feynman

Planets orbit round the sun because of the gravitational attraction which the sun exerts on them. A special case of this type of motion is that of a particle moving in a circle where, as you know, the acceleration, and therefore the force, is directed towards the centre of the circle. Mechanics problems involving such *central* forces are often more easily dealt with using polar co-ordinates rather than the Cartesian co-ordinates you are more used to. This chapter deals with such problems.

The first section is a short introduction to polar co-ordinates. This gives sufficient detail for your needs in mechanics, but the topic is covered more fully in *Pure Mathematics 5* and you may find the extra background there useful.

Definition of polar co-ordinates

So far when analysing motion in two dimensions, you have used an *xy* co-ordinate system (also known as a rectangular or Cartesian frame of reference). The position of a point is described using two co-ordinates which give its distance from two rectangular axes (figure 4.1)

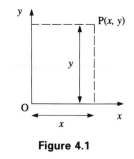

Figure 4.1

But there are other ways of specifying position. If, using radar, you give the position of a ship or aircraft, you are likely to give its distance and bearing (direction) from the radar station. This is effectively a polar co-ordinate system.

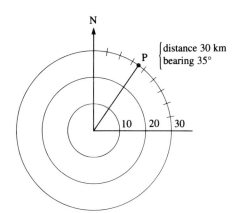

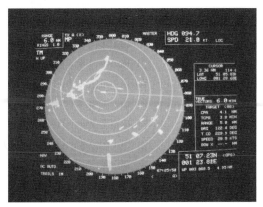

distance 30 km, bearing 35°)

Polar co-ordinates r and θ

When using polar co-ordinates in mathematics the position of a point P is given as (r, θ), where r is the distance OP of P from the origin O, and θ is the angle OP make with the x axis (figure 4.2).

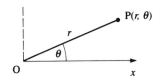

Figure 4.2

The angle is measured in an anticlockwise sense from the x axis, so a negative value of θ corresponds to a clockwise rotation. Also r may be negative, in which case the distance is in the opposite direction: along PO produced. This means that there are many ways of giving the co-ordinates of the same point. Figure 4.3 shows three alternative sets of co-ordinates for each of four points A, B, C, D. It is conventional to choose the form with $r \geqslant 0$ and $-\pi < \theta \leqslant \pi$. This is the first form given in each case.

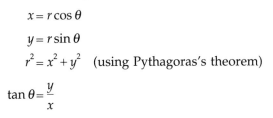

Figure 4.3

The relationships between the (x, y) and (r, θ) co-ordinates of a point can be deduced from figure 4.4.

$$x = r\cos\theta$$
$$y = r\sin\theta$$
$$r^2 = x^2 + y^2 \quad \text{(using Pythagoras's theorem)}$$
$$\tan\theta = \frac{y}{x}$$

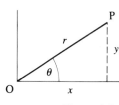

Figure 4.4

Motion described in polar co-ordinates

If you use $x = r\cos\theta$, $y = r\sin\theta$ to work out the x and y co-ordinates of the point C in figure 4.3 from the various polar forms given there, you will find you get $(-2\sqrt{2}, -2\sqrt{2})$ every time, showing that the different forms are equivalent.

N O T E

The origin in polar co-ordinates is often called the pole *and the x axis the* initial line. *There is no strict need to draw a y axis when drawing polar curves (although it often helps).*

The polar equations of curves

Just as an equation involving x and y ($y = x^2$, for example) leads to a curve in the xy plane, so an equation involving r and θ leads to a curve relative to the polar axes. Take, for example, $r = 2\cos\theta$.

If you were plotting this by hand, you would take a succession of values of θ, work out the corresponding values of r and then plot the points. Some are shown in figure 4.5, but you would obviously have to plot many more before realising that this curve is, in fact, a circle.

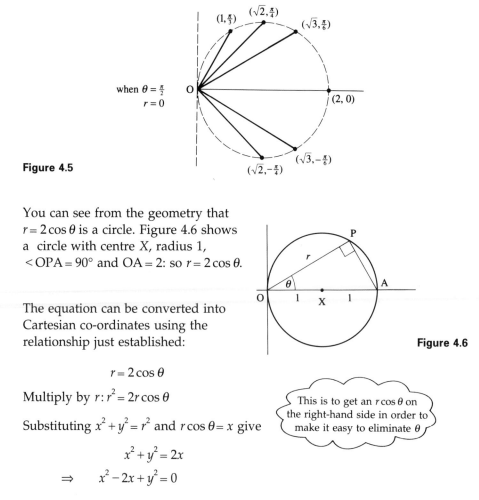

Figure 4.5

You can see from the geometry that $r = 2\cos\theta$ is a circle. Figure 4.6 shows a circle with centre X, radius 1, $<$OPA $= 90°$ and OA $= 2$: so $r = 2\cos\theta$.

The equation can be converted into Cartesian co-ordinates using the relationship just established:

Figure 4.6

$$r = 2\cos\theta$$

Multiply by r: $r^2 = 2r\cos\theta$

Substituting $x^2 + y^2 = r^2$ and $r\cos\theta = x$ give

This is to get an $r\cos\theta$ on the right-hand side in order to make it easy to eliminate θ

$$x^2 + y^2 = 2x$$
$$\Rightarrow \quad x^2 - 2x + y^2 = 0$$

Completing the square in x gives

$$(x-1)^2 + y^2 = 1$$

showing that the equation represents a circle centre $(1,0)$ and radius 1.

Notice the symmetry about the x axis, which is always the case with a curve where r depends only on $\cos\theta$. This is because $\cos(\theta) = \cos(-\theta)$, so if the point (r, θ) is on the curve, then so is the point $(r, -\theta)$.

The curves θ=constant and r=constant
These are the simplest equations in polar co-ordinates. Two examples are

$$\theta = \frac{\pi}{3} \quad \left(\text{line through the origin, angle } \frac{\pi}{3}\right)$$

$$r = 2 \quad \text{(circle, centre at origin, radius 2)}$$

This part included
if negative values
of r allowed

EXAMPLE

Show that the polar equation $r = d\sec(\theta - \alpha)$, where d and α are constants, represents a straight line
(i) by converting to xy co-ordinates;
(ii) by interpreting the polar equation geometrically.

Solution

(i)

$$r = d\sec(\theta - \alpha)$$

$$\Rightarrow \quad r\cos(\theta - \alpha) = d$$

$$\Rightarrow \quad r\cos\theta\cos\alpha + r\sin\theta\sin\alpha = d$$

Convert to xy co-ordinates by substituting $r\cos\theta = x$, $r\sin\theta = y$:

$$x\cos\alpha + y\sin\alpha = d$$

which is a straight line.

(ii) AB is a straight line. Draw OM, the perpendicular from O. Denote the angle it makes with the initial line by α. Then any point P on this line will have co-ordinates (r, θ). Denote the length OM by d.

You can see from the geometry that

$$d = r\cos(\theta - \alpha)$$

$$\Rightarrow \quad r = d\sec(\theta - \alpha)$$

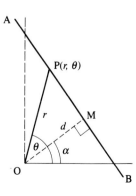

Some common polar curves

The graphs in figure 4.7 are familiar to anyone who has studied polar co-ordinates. Their equations are simple in polars but would be complicated in xy co-ordinates.

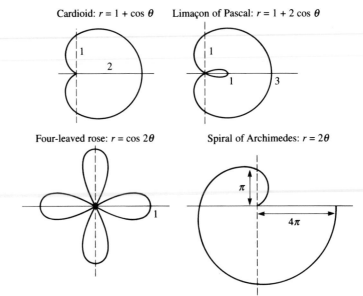

Figure 4.7

Vectors in polar co-ordinates

Imagine P is a comet which remains in the same plane but moves around a fixed sun (which is taken as the origin O). At any moment, there are various vectors associated with P, such as its position vector ($\overrightarrow{OP}$), its velocity, its acceleration and the force acting on it (figure 4.8).

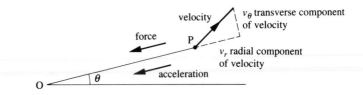

Figure 4.8

Up to now, you have generally described vectors in terms of their x and y components, thus $\mathbf{v} = v_x\mathbf{i} + v_y\mathbf{j}$. But many problems are easier to deal with if you take components of vectors not in the x and y directions but in *radial* and *transverse* directions. Radial means the direction out along the 'radius' OP. Transverse means at right angles to this, taken in the direction of increasing θ.

In figure 4.8, both the force and acceleration are directed towards the sun and therefore have only radial components. The value of these components will be negative, since the vectors are in the direction PO while the positive radial direction is OP. The velocity of P has both radial and transverse components, shown as v_r and v_θ in figure 4.8.

You have seen an example in *Mechanics 3*, where the velocity and acceleration of a particle moving round a circle were resolved into radial and transverse components (figure 4.9).

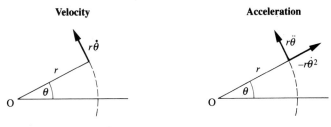

Figure 4.9

In this case, although the velocity and acceleration are changing direction, the acceleration is always along the radius and the velocity is always perpendicular to the radius: that is, in the transverse direction when the origin is taken at the centre of the circle. This is why polar co-ordinates are useful for such a problem; the components of the velocity and acceleration assume a simple form when taken in the radial and transverse directions.

The unit vectors $\hat{\mathbf{r}}$ and $\hat{\boldsymbol{\theta}}$

In figure 4.10, P has polar co-ordinates (r, θ) with respect to an origin O and an initial line which is the x axis. As you know, when using Cartesian co-ordinates, it is useful to work in terms of unit vectors $\mathbf{i}$ and $\mathbf{j}$ in the x and y directions. Similarly in polar co-ordinates, it is useful to define unit vectors $\hat{\mathbf{r}}$ and $\hat{\boldsymbol{\theta}}$ in the radial and transverse directions respectively.

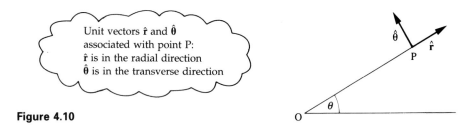

Unit vectors $\hat{\mathbf{r}}$ and $\hat{\boldsymbol{\theta}}$
associated with point P:
$\hat{\mathbf{r}}$ is in the radial direction
$\hat{\boldsymbol{\theta}}$ is in the transverse direction

Figure 4.10

Note that the position vector of P relative to O can always be written in terms of the unit vector $\hat{\mathbf{r}}$ as $\overrightarrow{OP} = \mathbf{r} = r\hat{\mathbf{r}}$.

EXAMPLE

A particle P moves around a circle, centre O and radius a, with constant angular velocity ω. Write down (i) its position vector, (ii) its velocity and (iii) its acceleration, all in terms of unit vectors $\hat{\mathbf{r}}$ and $\hat{\boldsymbol{\theta}}$ at P defined relative to a polar co-ordinate system with its origin at O.

Solution

The diagram shows the unit vectors $\hat{\mathbf{r}}$ and $\hat{\boldsymbol{\theta}}$ associated with P.

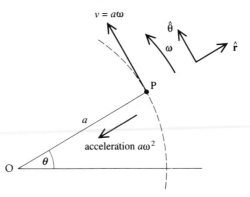

(i)   The position vector of P relative to O is then $a\hat{\mathbf{r}}$.

(ii) The speed of P is $a\omega$ in the direction shown, i.e. the transverse direction. The unit vector in this direction is $\hat{\boldsymbol{\theta}}$. Hence, the velocity of P is $a\omega\hat{\boldsymbol{\theta}}$.

(iii) The acceleration is directed towards the centre and has magnitude $a\omega^2$. The unit vector in this direction is $-\hat{\mathbf{r}}$. Hence the acceleration is $-a\omega^2\hat{\mathbf{r}}$.

Variation of $\hat{\mathbf{r}}$ and $\hat{\boldsymbol{\theta}}$ with position

Unlike $\mathbf{i}$ and $\mathbf{j}$, the directions of the unit vectors $\hat{\mathbf{r}}$ and $\hat{\boldsymbol{\theta}}$ vary according to position. You can see this from figure 4.11.

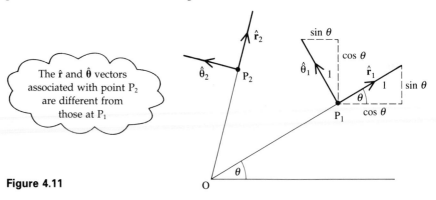

The $\hat{\mathbf{r}}$ and $\hat{\boldsymbol{\theta}}$ vectors associated with point P_2 are different from those at P_1

Figure 4.11

Resolving in the x and y directions gives

$$\hat{\mathbf{r}} = \cos\theta\mathbf{i} + \sin\theta\mathbf{j}$$

$$\hat{\boldsymbol{\theta}} = -\sin\theta\mathbf{i} + \cos\theta\mathbf{j}$$

and both these vary as θ varies.

This means that a *constant* vector has different radial and transverse components when associated with different positions of P. In figure 4.12 a particle moves from Q (position vector **i**) to R (position vector **i** + **j**) with constant velocity 2**j**. At Q the velocity is entirely in the transverse direction, and its magnitude is 2. At R, radial and transverse components are each √2. So the velocity at Q is 2**θ̂** and at R is √2**θ̂** + √2**r̂**. The velocity is the same at the two places; it is **r̂** and **θ̂** which have changed as θ has changed from zero to 45°.

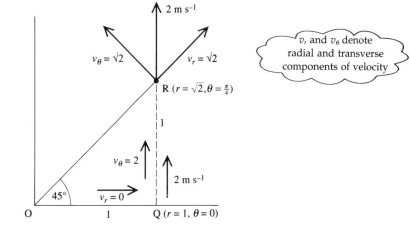

v_r and v_θ denote radial and transverse components of velocity

Figure 4.12

Exercise 4A

1. Write down the polar equations relative to the origin O of
   (i)   a circle radius a, centre O;
   (ii) a straight line with one end at O and making an angle α with the initial line;
   (iii) a straight line parallel to the initial line and at a distance c from it;
   (iv) a straight line through the point with polar co-ordinates $(d, 0)$ and making an angle β with the initial line.

2. (i)   Sketch the parabola $y^2 = 4(x+1)$.
   (ii) Show that its polar equation is $r(1 - \cos\theta) = 2$.
   (iii) Mark on your sketch the point on the curve where $\dfrac{dr}{d\theta} = 0$. Interpret this result.

3. A curve has the equation $r^2 = \cos 2\theta$.
   (i)   For what values of θ ($-\pi < \theta \leqslant \pi$) does the curve exist?

   (ii) Why do you know that the curve is symmetrical about the initial line?
   (iii) Taking only positive values of r, plot points of the curve for values of θ between $\frac{-\pi}{4}$ and $\frac{\pi}{4}$ and hence sketch this part of the curve.
   (iv) Use symmetry to complete the sketch.

4. A point P has polar co-ordinates (r, θ), relative to an origin and an initial line which is taken as the x axis.
   (i)   Write down the unit vectors **r̂** and **θ̂** in terms of **i** and **j** and θ.
   (ii) Use the scalar product formula to confirm that $\hat{r}.\hat{r} = 1$, $\hat{\theta}.\hat{\theta} = 1$, $\hat{r}.\hat{\theta} = 0$ and interpret this result.
   (iii) Draw sketches to show **r̂** and **θ̂** at the following positions of P, in polar co-ordinates: $(1, 0)$, $(1, \frac{\pi}{4})$, $(2, \frac{\pi}{4})$.

5. A particle P has position vector $3\mathbf{i} + 4\mathbf{j}$ relative to an origin O and is moving with a speed 3 in the direction of $\mathbf{j}$. Unit vectors $\hat{\mathbf{r}}$ and $\hat{\boldsymbol{\theta}}$ are defined at P relative to a polar co-ordinate system whose origin is O and whose initial line is the x axis.
   (i)   Write down the velocity of P in terms of $\mathbf{i}$ and $\mathbf{j}$.
   (ii) Write down the position vector of P in terms of $\hat{\mathbf{r}}$.
   (iii) Write down the velocity of P in terms of $\hat{\mathbf{r}}$ and $\hat{\boldsymbol{\theta}}$.
   (iv) Write down $\hat{\mathbf{r}}$ and $\hat{\boldsymbol{\theta}}$ in terms of $\mathbf{i}$ and $\mathbf{j}$.

6. The equation of a curve in polar co-ordinates is $r = 4\cos\theta$. A particle P, initially at the point $r = 4, \theta = 0$, moves round the curve in the direction of increasing θ, with constant angular velocity $\frac{\pi}{8}\text{rad s}^{-1}$ relative to O.
   (i)   Sketch the curve.
   (ii) What is the value of θ after 2 seconds? What is it after t seconds?
   (iii) How far is P from the origin after t seconds?

7. A particle P starts at a point with polar co-ordinates $(1, \frac{\pi}{2})$ and moves parallel to the initial line, in a direction with θ decreasing with a constant speed of $2\,\text{m s}^{-1}$. At any moment, unit vectors $\hat{\mathbf{r}}$ and $\hat{\boldsymbol{\theta}}$ are associated with the position of P. Find, in terms of $\hat{\mathbf{r}}$ and $\hat{\boldsymbol{\theta}}$:
   (i)   the initial velocity of P,
   (ii) the velocity when $\theta = \frac{\pi}{4}$,
   (iii) the approximate velocity when P has moved a great distance from O.

Velocity and acceleration in polar co-ordinates

You were reminded in Chapter 1 that, when the position of a moving body is given in terms of its x and y co-ordinates, you can find the components of velocity and acceleration simply by differentiation. For example, when the position of a body at time t is given by

$$\mathbf{r} = \cos t\mathbf{i} + \sin t\mathbf{j}$$

its velocity is

$$\dot{\mathbf{r}} = -\sin t\mathbf{i} + \cos t\mathbf{j}$$

and its acceleration is

$$\ddot{\mathbf{r}} = -\cos t\mathbf{i} - \sin t\mathbf{j}$$

However, when a position is given in polar co-ordinates, the velocity and acceleration cannot be found quite so simply. This is because unlike $\mathbf{i}$ and $\mathbf{j}$ the vectors $\hat{\mathbf{r}}$ and $\hat{\boldsymbol{\theta}}$ are themselves changing as the position of the body changes. In this section some important formulae are derived which enable you to find the velocity and acceleration of a body whose motion is specified in polar co-ordinates.

Rate of change of $\hat{\mathbf{r}}$ and $\hat{\boldsymbol{\theta}}$

As P moves, its components of position in polar co-ordinates (r, θ) vary and so each component can be regarded as a function of time. As the position changes, the directions of $\hat{\mathbf{r}}$ and $\hat{\boldsymbol{\theta}}$ also change. It is necessary to find the rate of change of these unit vectors with respect to time.

With reference to figure 4.13

$$\hat{\mathbf{r}} = \cos\theta\mathbf{i} + \sin\theta\mathbf{j}$$

Differentiating with respect to θ gives

$$\frac{d\hat{\mathbf{r}}}{d\theta} = -\sin\theta\mathbf{i} + \cos\theta\mathbf{j}$$

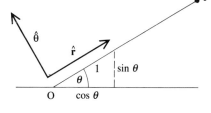

To find $\dfrac{d\hat{\mathbf{r}}}{dt}$, you must use the chain rule:

Figure 4.13

$$\frac{d\hat{\mathbf{r}}}{dt} = \frac{d\hat{\mathbf{r}}}{d\theta} \times \frac{d\theta}{dt}$$

$$= (-\sin\theta\mathbf{i} + \cos\theta\mathbf{j})\frac{d\theta}{dt}$$

$$= \dot{\theta}\hat{\boldsymbol{\theta}}$$

since $\hat{\boldsymbol{\theta}} = -\sin\theta\mathbf{i} + \cos\theta\mathbf{j}$, as shown previously.

The rate of change of $\hat{\mathbf{r}}$ is in the direction of $\hat{\boldsymbol{\theta}}$ and has magnitude $\dot{\theta}$.

Similarly
$$\hat{\boldsymbol{\theta}} = -\sin\theta\mathbf{i} + \cos\theta\mathbf{j}$$
$$\frac{d\hat{\boldsymbol{\theta}}}{dt} = (-\cos\theta\mathbf{i} - \sin\theta\mathbf{j})\frac{d\theta}{dt}$$
$$= -\dot{\theta}\hat{\mathbf{r}}$$

The rate of change of $\hat{\boldsymbol{\theta}}$ is in the *opposite* direction to $\hat{\mathbf{r}}$ and also has magnitude $\dot{\theta}$.

Geometrical interpretation

Figure 4.14 may help to give you a geometrical interpretation of the first result. As P moves from Q to Q' in a small time δt, the direction of the unit vector $\hat{\mathbf{r}}$ changes from OA to OA'.

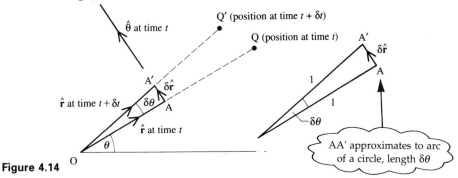

Figure 4.14

AA' represents the change in $\hat{\mathbf{r}}$, i.e. $\delta\hat{\mathbf{r}}$. The angle OAA' is $\dfrac{\pi}{2} - \dfrac{\delta\theta}{2}$: this

approaches a right angle as $\delta\theta$ tends to zero. In other words, an infinitesimal change in $\hat{\mathbf{r}}$ *is in the direction* $\hat{\boldsymbol{\theta}}$. The length of $\delta\hat{\mathbf{r}}$, since $\hat{\mathbf{r}}$ is a unit vector, is approximately $\delta\theta$. So

$$\delta\hat{\mathbf{r}} \approx \delta\theta\,\hat{\boldsymbol{\theta}} \quad \Rightarrow \quad \frac{\delta\hat{\mathbf{r}}}{dt} \approx \frac{\delta\theta}{\delta t}\,\hat{\boldsymbol{\theta}}$$

In the limit as $\delta t \to 0$, $\dfrac{d\hat{\mathbf{r}}}{dt} = \dot{\theta}\,\hat{\boldsymbol{\theta}}$

Note that changes in r (movements along OQ) have no effect on $\hat{\mathbf{r}}$ or $\hat{\boldsymbol{\theta}}$, so that neither $\dfrac{d\hat{\mathbf{r}}}{dt}$ nor $\dfrac{d\hat{\boldsymbol{\theta}}}{dt}$ depends on $\dot{r}$, only on $\dot{\theta}$.

For Discussion

Interpret similarly the meaning of $\dfrac{d\hat{\boldsymbol{\theta}}}{dt} = -\dot{\theta}\,\hat{\mathbf{r}}$

Velocity in polar co-ordinates

As pointed out earlier, the position vector of P can be expressed in terms of $\hat{\mathbf{r}}$:

$$\mathbf{r} = r\hat{\mathbf{r}}$$

To find the velocity of P, you must differentiate $\mathbf{r}$ with respect to t. Because both r and $\hat{\mathbf{r}}$ depend on t, this requires the use of the product rule:

$$\mathbf{v} = \frac{dr}{dt}\,\hat{\mathbf{r}} + r\,\frac{d\hat{\mathbf{r}}}{dt}$$

$$= \dot{r}\hat{\mathbf{r}} + r\dot{\theta}\hat{\boldsymbol{\theta}} \quad \left(\text{since } \frac{d\hat{\mathbf{r}}}{dt} = \dot{\theta}\hat{\boldsymbol{\theta}} \text{ from above}\right)$$

This gives two important results:

- The radial component of velocity is $\dot{r}$.
- The transverse component of velocity is $r\dot{\theta}$.

You have met the transverse component of velocity in connection with circular motion: $r\dot{\theta} = r\omega$, where ω is the angular velocity.

Finding the speed from components of velocity
Since $\hat{\mathbf{r}}$ and $\hat{\boldsymbol{\theta}}$ are unit vectors at right angles, the modulus of the velocity vector (i.e. the speed) is obtained from the radial and transverse components by Pythagoras's theorem (figure 4.15):

$$v^2 = \dot{r}^2 + (r\dot{\theta})^2$$

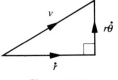

Figure 4.15

NOTE

Differentiating scalar × vector

In differentiating $r\hat{\mathbf{r}}$ it has been assumed that the product rule for differentiation can be applied to the case of a scalar times a vector. To confirm this, suppose

$$\mathbf{p} = u\mathbf{q}$$

where $\mathbf{p}$ and $\mathbf{q}$ are vectors, u is a scalar and they are all functions of t.

Then, expressing $\mathbf{q}$ in Cartesian component form:

$$\mathbf{p} = u(q_1\mathbf{i} + q_2\mathbf{j})$$
$$= uq_1\mathbf{i} + uq_2\mathbf{j}$$

where q_1 and q_2 are functions of t.

It is shown in Chapter 1 that a vector can be differentiated by dealing with each Cartesian component separately. Applying the product rule to each gives

$$\dot{\mathbf{p}} = (\dot{u}q_1 + u\dot{q}_1)\mathbf{i} + (\dot{u}q_2 + u\dot{q}_2)\mathbf{j}$$
$$= \dot{u}(q_1\mathbf{i} + q_2\mathbf{j}) + u(\dot{q}_1\mathbf{i} + \dot{q}_2\mathbf{j})$$
$$\Rightarrow \quad \dot{\mathbf{p}} = \dot{u}\mathbf{q} + u\dot{\mathbf{q}}$$

Hence the product rule is valid.

EXAMPLE

A small spider is crawling outwards along one of the radial rails of a playground roundabout, with a speed of 10 cm s^{-1}. The spider does not realise that the roundabout has started to rotate (although it feels a bit wobbly). It clings on and maintains the same speed relative to the rail. When the spider is 20 cm from the centre of the roundabout, it is rotating once every 4 seconds.

(i)   What is the transverse component of the spider's velocity at this instant?
(ii) What is the spider's speed relative to the earth at this instant?

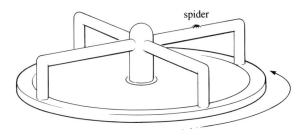

Solution

(i)   The roundabout is rotating once every 4 s, so the angular velocity is

$$\dot{\theta} = \frac{2\pi}{4} = \frac{\pi}{2} \text{ rad s}^{-1}$$

The transverse component

$$r\dot\theta = \frac{20\pi}{2} = 10\pi$$

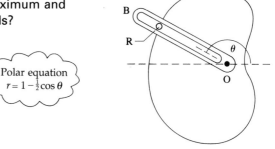

The transverse speed is thus 31.4 cm s^{-1}.

(ii) The radial speed $\dot r = 10$ cm s^{-1}.

The total speed v is given by

$$v^2 = (r\dot\theta)^2 + \dot r^2$$
$$= (10)^2 + (10\pi)^2$$
$$= 1087.0 \quad \text{(to 5 significant figures)}$$

So $\qquad v = 33$ (to 2 significant figures)

Spider's speed is approximately 33 cm s^{-1}.

EXAMPLE

The diagram shows a fixed piece of metal (a cam) whose shape is given by the polar equation, relative to origin O, $r = 1 - \frac{1}{2}\cos\theta$ (a limaçon). A roller R is moved by a slotted arm OB which rotates about O with constant angular velocity ω. R is constrained (by a spring not shown) to follow the edge of the metal shape.

(i) What are the radial and transverse components of the velocity of R as a function of θ?

(ii) What are its maximum and minimum speeds?

Polar equation
$r = 1 - \frac{1}{2}\cos\theta$

Solution

(i) To work out the components of velocity, you need to find $\dot r$ and $\dot\theta$.

The angular velocity $\dot\theta = \omega$ is constant.

The roller follows the path $r = 1 - \frac{1}{2}\cos\theta$, so $\dot r$ can be expressed in terms of θ by differentiating. Using the chain rule to differentiate with respect to time t gives

$$\dot r = \frac{dr}{d\theta} \times \frac{d\theta}{dt}$$
$$= (\tfrac{1}{2}\sin\theta)\dot\theta$$

So the radial velocity $\dot r = \frac{1}{2}\omega\sin\theta$.

The transverse velocity $\qquad r\dot\theta = r\omega$

$$= (1 - \tfrac{1}{2}\cos\theta)\omega$$

Substituting for r from original equation since the answer is required in terms of θ

(ii) The total speed v is given by
$$v^2 = \dot{r}^2 + r\dot{\theta}^2 = (\tfrac{1}{2}\omega\sin\theta)^2 + (1 - \tfrac{1}{2}\cos\theta)^2\omega^2$$
$$= \omega^2(\tfrac{1}{4}\sin^2\theta + 1 - \cos\theta + \tfrac{1}{4}\cos^2\theta)$$
$$= \omega^2(\tfrac{5}{4} - \cos\theta) \quad (\text{since } \sin^2\theta + \cos^2\theta = 1)$$

Maximum v is when $\cos\theta = -1$, $v = \omega\sqrt{(\tfrac{9}{4})} = \dfrac{3\omega}{2}$

Minimum v is when $\cos\theta = 1$, $v = \omega\sqrt{(\tfrac{1}{4})} = \dfrac{\omega}{2}$

Acceleration in polar co-ordinates

To find the acceleration in polar co-ordinates, differentiate the velocity using the product rule (the second term on the right-hand side involves a triple product):

$$\mathbf{v} = \dot{r}\hat{\mathbf{r}} + r\dot{\theta}\hat{\boldsymbol{\theta}}$$

$$\frac{d\mathbf{v}}{dt} = \left(\dot{r}\frac{d\hat{\mathbf{r}}}{dt} + \ddot{r}\hat{\mathbf{r}}\right) + \left(r\dot{\theta}\frac{d\hat{\boldsymbol{\theta}}}{dt} + \dot{r}\dot{\theta}\hat{\boldsymbol{\theta}} + r\ddot{\theta}\hat{\boldsymbol{\theta}}\right)$$

Substituting $\dfrac{d\hat{\mathbf{r}}}{dt} = \dot{\theta}\hat{\boldsymbol{\theta}}$ and $\dfrac{d\hat{\boldsymbol{\theta}}}{dt} = -\dot{\theta}\hat{\mathbf{r}}$ (these results were obtained above) gives

$$\frac{d\mathbf{v}}{dt} = \dot{r}\dot{\theta}\hat{\boldsymbol{\theta}} + \ddot{r}\hat{\mathbf{r}} - r\dot{\theta}\dot{\theta}\mathbf{r} + \dot{r}\dot{\theta}\hat{\boldsymbol{\theta}} + r\ddot{\theta}\hat{\boldsymbol{\theta}}$$

$$= (\ddot{r} - r\dot{\theta}^2)\hat{\mathbf{r}} + (r\ddot{\theta} + 2\dot{r}\dot{\theta})\hat{\boldsymbol{\theta}}$$

Hence the following important results:

- The radial component of acceleration is $\ddot{r} - r\dot{\theta}^2$.
- The transverse component is $r\ddot{\theta} + 2\dot{r}\dot{\theta}$.

Activity

Check by differentiating that the transverse component can be written

$$\frac{1}{r}\frac{d}{dt}\left(r^2\frac{d\theta}{dt}\right)$$

As you will see, this form is very useful.

Special case: circular motion

When a body is moving so that r is constant, say $r = a$, it is moving in a circle (figure 4.16). Both $\ddot{r}$ and $\dot{r}$ are zero. Then the radial component is

$$\ddot{r} - r\dot{\theta}^2 = -a\dot{\theta}^2$$

and the transverse component is

$$r\ddot{\theta} + 2\dot{r}\dot{\theta} = a\ddot{\theta}.$$

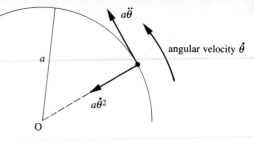

Both of these are familiar from your knowledge of circular motion in *Mechanics 3*. The radial component is, of course, negative because the acceleration is directed to the centre of the circle.

Figure 4.16

EXAMPLE

A particle is moving in a circle in such a way that the velocity and acceleration are at right angles. Show that it must be moving with constant angular velocity.

Solution

Circular motion implies $r =$ constant: $\dot{r} = \ddot{r} = 0$

Velocity
$$\mathbf{v} = \dot{r}\hat{\mathbf{r}} + r\dot{\theta}\hat{\boldsymbol{\theta}}$$
$$= r\dot{\theta}\hat{\boldsymbol{\theta}}$$

Acceleration
$$\mathbf{a} = (\ddot{r} - r\dot{\theta}^2)\hat{\mathbf{r}} + (r\ddot{\theta} + 2\dot{r}\dot{\theta})\hat{\boldsymbol{\theta}}$$
$$= -r\dot{\theta}^2\hat{\mathbf{r}} + r\ddot{\theta}\hat{\boldsymbol{\theta}}$$

Velocity and acceleration are at right angles, that is, $\mathbf{a}.\mathbf{v} = 0$. So
$$(-r\dot{\theta}^2\hat{\mathbf{r}} + r\ddot{\theta}\hat{\boldsymbol{\theta}}).(r\dot{\theta}\hat{\boldsymbol{\theta}}) = 0$$

Multiplying out (scalar products obey the usual algebra of multiplication) gives
$$-r^2\dot{\theta}^3\hat{\mathbf{r}}.\hat{\boldsymbol{\theta}} + r^2\ddot{\theta}\dot{\theta}\hat{\boldsymbol{\theta}}.\hat{\boldsymbol{\theta}} = 0$$

But $\hat{\mathbf{r}}.\hat{\boldsymbol{\theta}} = 0$ and $\hat{\boldsymbol{\theta}}.\hat{\boldsymbol{\theta}} = 1$ (since they are unit vectors at right angles). Hence $r^2\ddot{\theta}\dot{\theta} = 0$.

But $r \neq 0$ and $\dot{\theta} \neq 0$, so $\ddot{\theta} = 0 \implies \dot{\theta}$ is constant, by integration.

Special case: angular velocity constant
When the angular velocity $\dot{\theta}$ has a constant value ω, then $\ddot{\theta} = 0$. The radial component of acceleration is $\ddot{r} - r\omega^2$ and the transverse component reduces to $2\dot{r}\omega$.

This models the situation where a body is moving along the radius of a wheel which is rotating with constant angular velocity. As the next example shows, a transverse force is necessary in order to maintain the radial motion.

EXAMPLE

A spider of mass m is on a radial rail of a roundabout rotating at a constant angular velocity once every 4 seconds.
(i) Show that when the spider is at rest relative to the rail 20 cm from the centre, it experiences a force (in addition to its weight) of magnitude $F = m\dfrac{\pi^2}{20}$ N, directed towards the centre of the roundabout.
(ii) When the spider begins to crawl outwards along the rail it has to exert a *transverse* force of the same magnitude as F in order to stop being thrown off sideways. How fast is it moving along the rail?

Solution

(i) The roundabout rotates once every 4 s, so its angular velocity

$$\omega = \dot{\theta} = \frac{2\pi}{4} = \frac{\pi}{2} \text{ rad s}^{-1}$$

When the spider is at rest relative to the roundabout, its acceleration towards the centre is

$$r\omega^2 = \frac{1}{5}\left(\frac{\pi}{2}\right)^2 = \frac{\pi^2}{20} \quad (20 \text{ cm is } \tfrac{1}{5} \text{ m})$$

Applying $F = ma$ gives

$$F = m\frac{\pi^2}{20} \text{ N}.$$

(ii) When the spider's speed along the rail is $\dot{r}$, the transverse acceleration is

$$2\dot{r}\dot{\theta} = 2\dot{r}\left(\frac{\pi}{2}\right)$$

as $\dot{\theta} = \omega = \frac{\pi}{2}$ and $\ddot{\theta} = 0$.

Applying Newton's Second Law in the transverse direction, the force exerted on the spider in this direction is

$$2m\dot{r}\left(\frac{\pi}{2}\right) = F = m\frac{\pi^2}{20} \quad \text{(from part (i))}$$

Hence $\dot{r} = \dfrac{\pi}{20}$.

The unit here is m s^{-1}, so the spider's speed relative to the rail is

$$100 \times \frac{\pi}{20} = 15.7 \text{ cm s}^{-1}.$$

Note that the transverse force exerted *by* the rail *on* the spider as it moves out is in the direction of rotation. When the spider runs *towards* the centre, the transverse force is in the opposite direction.

Investigation

If you have an opportunity, try walking quickly along a radial path on a rotating merry-go-round. But be careful, the transverse acceleration is unexpected and you may be thrown off balance.

Investigation continued

Using a reasonable estimate of the rotation speed of a large roundabout, work out the sideways force on your feet as you walk quickly from the circumference to the centre. Which way will you tend to fall over?

If you do stay upright, what is your path in space? What is its equation in polar co-ordinates? (As a substitute, mark a radial path on a potter's wheel and try to follow it with a pencil as the wheel rotates)

Further discussion of the transverse component

The expression for the transverse component of acceleration $r\ddot{\theta} + 2\dot{r}\dot{\theta}$ is not at all obvious. You have seen the first term *radius × angular acceleration* when studying circular motion; you would expect this to contribute to the transverse acceleration. But where does $2\dot{r}\dot{\theta}$ come from?

To try to get an intuitive understanding, examine the situation of a particle moving along a radial spoke of a rotating wheel, rather like the spider in the preceding example. Assume that the particle's speed *along the spoke* is a constant u m s^{-1} and that the wheel has a constant angular velocity of ω rad s^{-1}.

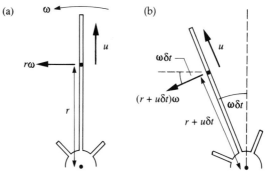

Figure 4.17

At a certain time t, the situation is as shown in figure 4.17a, with the particle distance r from the centre. At this moment, the transverse direction is to the left in figure 4.17a and the speed of the particle in this direction is $r\omega$. To find the acceleration in this direction, we are going to find the total component of velocity of the particle *in this direction* a moment later.

Figure 4.17b shows the position a small time δt later when the wheel has turned through the angle $\omega\,\delta t$. The particle has moved an extra distance $u\delta t$ along the spoke. The radial motion now has a component of velocity $u\sin(\omega\delta t)$ to the left. The total component of velocity to the left is now

$$u\sin(\omega\delta t) + (r + u\delta t)\omega\cos(\omega\delta t) \approx u\omega\delta t + (r + u\delta t)\omega$$

$$= r\omega + 2u\omega\delta t$$

Since angle $\omega\delta t$ is small, $\sin(\omega\delta t) \approx \omega\delta t$ and $\cos(\omega\delta t) \approx 1$

The *increase* in velocity *in this direction* is thus approximately $2u\omega\delta t$, so the average rate of change of velocity over the period δt is $2u\omega$. As $\delta t \rightarrow 0$, the approximation becomes exact and the instantaneous acceleration in the radial direction is $2u\dot{\theta}$.

Looking at the equation above you can see that half of the $2u\dot{\theta}$ arises because the *radial motion is changing its direction*, the other half because the *transverse speed itself increases* as the particle moves further from the centre.

Motion of a particle on a polar curve

You sometimes need to find the velocity and acceleration of a particle constrained to move on a curve whose polar equation you know. In such cases you normally differentiate the polar equation with respect to *time* and use the chain rule to express $\dot{r}$ and $\ddot{r}$ in terms of $\dot{\theta}$ and $\ddot{\theta}$. The next example demonstrates this.

EXAMPLE

A particle R moves in a circular path with polar equation $r = 2a\cos\theta$ referred to an origin O on the circumference. The angular velocity ω about O, is constant.

(i)   Using these co-ordinates, express $\dot{r}$ and $\ddot{r}$ in terms of θ and ω.
(ii) Find the radial and transverse components of acceleration in terms of θ.
(iii) Hence show that the acceleration of R has a constant magnitude $4a\omega^2$.

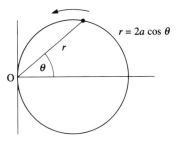

$r = 2a\cos\theta$

Solution

(i)   The equation of the path is $r = 2a\cos\theta$. Differentiate with respect to time t using the chain rule:

$$\frac{dr}{dt} = \frac{dr}{d\theta}\frac{d\theta}{dt}$$

Now $\dfrac{dr}{d\theta} = -2a\sin\theta$ and $\dfrac{d\theta}{dt}$ is the angular velocity ω. Hence

$$\dot{r} = -2a\omega\sin\theta$$

Differentiating again with respect to t gives

$$\ddot{r} = -2a\omega\cos\theta\,\dot{\theta} = -2a\omega^2\cos\theta$$

(ii) The radial component of acceleration, a_r, is

$$\ddot{r} - r\dot{\theta}^2 = -2a\omega^2\cos\theta - 2a\cos\theta\,\omega^2 \quad \text{(substituting for } \ddot{r}, r \text{ and } \dot{\theta} \text{ from part (i))}$$

$$= -4a\omega^2\cos\theta$$

The transverse component, a_θ, is

$$r\ddot{\theta} + 2\dot{r}\dot{\theta} = 2(-2a\sin\theta)\omega \quad (\ddot{\theta} = 0)$$

$$= -4a\omega^2\sin\theta$$

(iii) The magnitude of the total acceleration is

$$\sqrt{(a_r^2 + a_\theta^2)} = \sqrt{[(4a\omega^2\cos\theta)^2 + (4a\omega^2\sin\theta)^2]}$$

$$= 4a\omega^2\sqrt{(\cos^2\theta + \sin^2\theta)}$$

$$= 4a\omega^2$$

N O T E   *Coriolis acceleration*

Because the earth is rotating from west to east, an aircraft aiming south from the North Pole to a point on the equator is getting further from the earth's axis of rotation. Thus as the aircraft moves south, the speed of rotation of the ground immediately underneath it increases. The aircraft needs to accelerate in an easterly direction to avoid moving to the right of its chosen course. This is similar to your having to undergo a transverse acceleration to maintain a radial motion when walking across a roundabout.

Taking off from the equator and travelling north has a similar effect. At the equator, the aircraft will share the earth's easterly rotation speed there. As the aircraft travels north, the rotation speed of the ground underneath decreases. Once again the aircraft will veer to its right, relative to the ground. In the southern hemisphere, the tendency is to veer to the left (think out why this should be so).

This phenomenon is known as the Coriolis effect, and has a significant effect on weather and tidal systems. It is the reason why, in the northern hemisphere, winds blow around storm centres in an anticlockwise direction (looking from above); in the southern hemisphere the winds are clockwise.

Exercise 4B

1. A particle is moving round the circle whose equation in polar co-ordinates is $r = a$. Its angular velocity after t seconds is $\omega = 2t$ rad s^{-1}. Use the formulae for polar components of velocity and acceleration to confirm

(i)   the radial component of velocity is zero;

(ii) the transverse component of velocity is $2at$;

(iii) the radial component of acceleration is $-4at^2$;

(iv) the transverse component of acceleration is $2a$.

2. A particle moves with constant angular velocity ω with respect to the origin, along the following curves. In each case find (in terms of θ and ω), the radial and transverse components of velocity, and hence the speed of the particle when at the point (r, θ).

(i) $r = e^\theta$ (ii) $r = 3\cos\theta$

(iii) $r = 1 + \cos\theta$ (iv) $r(1 + \cos\theta) = 2$

3. A particle moves with constant angular velocity ω with respect to the origin, along the following curves. In each case find (in terms of θ and ω), the radial and transverse components of acceleration when at the point (r, θ).

(i) $r = 2\theta$ (ii) $r = e^\theta$

(iii) $r = 3\sin\theta$ (iv) $r = 1 - \sin\theta$

4. A merry-go-round is rotating at a uniform angular speed of 1 rad s^{-1}. A supervisor tries to run at 3 m s^{-1} along a radial path through the horses, from the edge to the centre. As long as he remains on his feet, what is the magnitude of the transverse acceleration he undergoes? Compare this with the acceleration of gravity.

5. L is the fixed position of a light. A moth M moves in a plane containing L. At time t the moth is at the point with polar co-ordinates (r, θ) relative to L and a fixed line in the plane. The moth flies so that its velocity always makes a constant angle α, $(0 < \alpha < \frac{\pi}{2})$ with ML as shown in the diagram. When $t = 0$, the moth is at the point with polar co-ordinates $(R, 0)$.

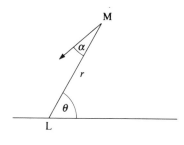

By considering the radial and transverse components of the velocity of M,

(i) show that $\dfrac{dr}{d\theta} = -r\cot\alpha$;

(ii) find the polar equation of the path of M.

[London 1993]

6. A particle P is moving in a plane with constant angular speed ω about a fixed origin O. At all times the force on it is at right angles to the line OP.
   (i) Write down the equation of motion for the radial direction.
   (ii) Show that this is satisfied by

   $$r = A\,e^{\omega t} + B\,e^{-\omega t}$$

   where A and B are arbitrary constants.
   (iii) At time $t = 0$, OP = 2 and the particle is moving at right angles to OP. Find the particular solution for r.
   (iv) Draw a polar sketch graph to illustrate the motion.

7. A particle P moves in a plane along the spiral with equation $r = a\,e^{2\theta}$, where a is constant.

(i) Show that the radial component of acceleration is

$$a_r = a\,e^{2\theta}\left[2\,\frac{d^2\theta}{dt^2} + 3\left(\frac{d\theta}{dt}\right)^2\right]$$

(ii) Find the transverse component in terms of a, $\dot\theta$ and $\ddot\theta$.

(iii) Given that at all times the two components are equal, find the relationship between $\dot\theta$ and $\ddot\theta$, and show it is satisfied by $\dot\theta = C\,e^{\theta}$ where C is an arbitrary constant.

8. A particle P moves on the circle with equation $r = 2a\sin\theta$, where (r, θ) are polar co-ordinates relative to a fixed origin O and an initial line OA. When P is at the point with co-ordinates (r, θ), it has speed $\lambda\,\mathrm{cosec}^2\,\theta$, where λ is constant. Prove that at time t

$$\frac{d\theta}{dt} = \frac{\lambda}{2a\sin^2\theta} \quad \text{and} \quad \frac{dr}{dt} = \lambda\,\frac{\cos\theta}{\sin^2\theta}$$

Show that the transverse component of the acceleration is zero. Hence find in terms of a, λ and r, the magnitude of the acceleration of P.

[AEB 1992]

Central forces

Newton's demonstration that the observed elliptical orbit of a planet around the sun rose from an inverse square law of gravitation was one of the greatest triumphs for the application of mathematics to the real world. This is an example which can be modelled as a *central force* problem (figure 4.18) where the *only* significant force on a body (in this case a planet) is always directed towards or away from a fixed origin (in this case the sun).

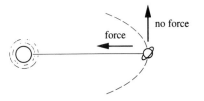

Figure 4.18

Central force problems, the topic of this section, are important in physics and, as you will see, there are two general principles which assist the mathematical analysis.

- Central force orbits are in a plane
- $r^2\dot{\theta}$ is constant.

Central force orbits are in a plane

The path of a body moving solely under the influence of a central force is in a plane. You can see this from figure 4.19, in which P is a planet and S is the sun, and **v** is the velocity of the planet at some moment. The line SP and the direction of **v** define a plane (the plane of the page of this diagram). Since the force is along PS, there is no component of force or velocity perpendicular to the page, so the planet must remain in the plane at all time. This is true of *any* central force motion.

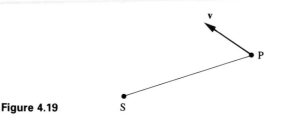

Figure 4.19

N O T E

The planets in the solar system all move approximately in the same plane, presumably a result of whatever process formed the planets originally. Pluto deviates the greatest amount, with an orbit inclined at 17° to the Earth's; Mercury is next, at 7°.

Constant value of $r^2\dot{\theta}$

Where there is only a central force, the transverse acceleration must be zero, since there is no transverse force. You saw earlier that the transverse acceleration can be written as $\dfrac{1}{r}\dfrac{d}{dt}(r^2\dot{\theta})$. Therefore $\dfrac{1}{r}\dfrac{d}{dt}(r^2\dot{\theta}) = 0$ which gives by integration

$$r^2\dot{\theta} = \text{constant} = h$$

This is an important result which is very helpful in analysing the motion of bodies acted on only by a central force. It is conventional to denote the value of $r^2\dot{\theta}$ by h.

Transverse velocity
Another way to write $r^2\dot{\theta}$ is

$$r^2\dot{\theta} = r(r\dot{\theta}) = rv_\theta$$

that is, r times the *transverse component* of velocity.

Figure 4.20 represents a comet in a highly eccentric orbit about the sun at O. As the comet gets nearer to the sun, r gets smaller and since rv_θ is constant, the transverse component of the velocity must become proportionately greater. The transverse velocity will be greatest when the comet is at Q (in figure 4.20 the velocity is wholly transverse at that point).

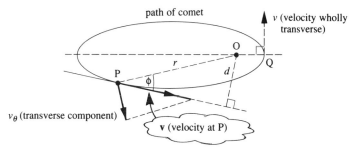

Figure 4.20

There is another way of looking at rv_θ. When the comet is at P

$$rv_\theta = rv \sin \phi = vd$$

This is the speed of the comet multiplied by the closest distance it would pass the sun *if the comet were to continue in a straight line*. This form is useful in problems where an orbiting particle starts at a great distance away so that it is more or less moving in a straight line to begin with.

Angular momentum

Let m denote the mass of an orbiting body. Then $mr^2\dot\theta$ can be written as $r \times m(r\dot\theta)$, which is the *angular momentum* (or *moment of momentum*) of the body about O. You will meet this concept in *Mechanics 6*. The fact that $r^2\dot\theta$ is *constant* for motion under a central force is equivalent to the *law of conservation of angular momentum*.

N O T E

Kepler's Second Law

The constancy of $r^2\dot\theta$ has a geometrical significance which led to the realisation that a force towards the sun was responsible for the motion of the planets. Look at the area swept out by OP as the particle P moves from A(r, θ) to a nearby position A'(r + δr, θ + δθ) in a small time δt.

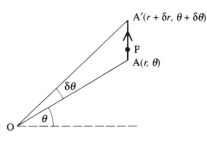

The area OAA' is found using the standard result that the area of a triangle ABC is $\frac{1}{2}ab\sin C$. In this case a=r, b=r+δr, and C=δθ.

$$\tfrac{1}{2}r(r+\delta r)\sin\delta\theta \approx \tfrac{1}{2}r^2\delta\theta \quad \text{(as } \delta\theta \to 0, \sin\delta\theta \to \delta\theta \text{ and } \delta r\delta\theta \to 0)$$

Thus an area approximately equal to $\frac{1}{2}r^2\delta\theta$ is swept out in time δt. In the limit as $\delta\theta \to 0$, $\frac{1}{2}r^2\dfrac{d\theta}{dt}$ is the rate at which area is covered. But $r^2\dfrac{d\theta}{dt}$ is constant throughout the motion. This means that OP sweeps out area at a constant rate. This is often stated in the form of Kepler's Second Law: the line joining the sun to a given planet sweeps out equal areas in equal times.

This law applies to motion under any *central force, not just the inverse square law of gravitation.*

The following example shows how the knowledge of the orbit under a central force enables the force to be calculated, using the fact that $r^2\dot\theta$ is constant.

EXAMPLE

A particle moves along the curve given by the polar equation $r = e^{\theta}$ when under the action of a central force directed towards the pole. When $\theta = 0$, its speed is u.

(i) Express $\dot r$ in terms of r and $\dot\theta$ and hence show that the transverse and radial speeds are always equal. Work out their value when $\theta = 0$.

(ii) Write down the value of $r^2\dot\theta$ when $\theta = 0$.

(iii) Hence express $\dot r$ and then $\ddot r$ in terms of r and u.

(iv) Find the radial acceleration and deduce that the force is inversely proportional to the cube of the distance of the particle from the pole.

Solution

(i) To obtain the radial velocity, $v_r = \dot r$, differentiate $r = e^{\theta}$ using the chain rule,

$$\frac{dr}{dt} = \frac{dr}{d\theta}\frac{d\theta}{dt}$$

$$\Rightarrow \quad \dot r = e^{\theta}\dot\theta$$

The transverse velocity $v_{\theta} = r\dot\theta = e^{\theta}\dot\theta$. So $v_r = v_{\theta} = e^{\theta}\dot\theta$.

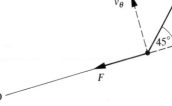

Note: this means that the particle always travels at 45° to the radius.

The initial speed is u, so $u^2 = v_r^2 + v_{\theta}^2 = 2v_r^2$ (since $v_r = v_{\theta}$)

$$\Rightarrow \quad v_r = \frac{u}{\sqrt 2}$$

$$\Rightarrow \quad v_r = v_{\theta} = \frac{u}{\sqrt 2} \text{ initially}$$

(ii) When $\theta = 0$, $r = e^0 = 1$ and $v_\theta = \dfrac{u}{\sqrt{2}}$ from part (i), so

$$r^2\dot\theta = rv_\theta = \frac{u}{\sqrt{2}}$$

Since the particle is under a central force, $r^2\dot\theta = \text{constant} = \dfrac{u}{\sqrt{2}}$ at all points in its path.

(iii) $\dot r = e^\theta\dot\theta = r\dot\theta$ from part (i) and $r^2\dot\theta = \dfrac{u}{\sqrt{2}}$ from part (ii)

$$\dot r = \frac{u}{r\sqrt{2}}$$

Differentiating with respect to t again gives

$$\ddot r = -\frac{u}{r^2\sqrt{2}}\dot r$$

$$= -\frac{u^2}{2r^3} \quad \left(\text{substituting } \dot r = \frac{u}{r\sqrt{2}}\right)$$

(iv) The radial acceleration is

$$\ddot r - r\dot\theta^2 = -\frac{u^2}{2r^3} - r\left(\frac{u}{r^2\sqrt{2}}\right)^2$$

$$= -\frac{u^2}{2r^3} - \frac{u^2}{2r^3}$$

$$= -\frac{u^2}{r^3}$$

Applying $F = ma$ in the radial direction, the force is of magnitude $\dfrac{mu^2}{r^3}$ and so it is inversely proportional to the cube of the distance.

Orbits under an inverse square law of attraction

This section examines motion under an attractive central force whose magnitude varies with the inverse square of the distance.

An important example which can be modelled in this way pervades much of this chapter: the motion of an astronomical body (e.g. the Earth) in orbit round a much larger one (the sun), which can be assumed to be stationary. The force is, of course, gravity. You saw in Chapter 1 that the attractive force between a planet (mass m) and the sun (mass M) has magnitude GMm/r^2, where G is the gravitational constant and r is the distance between the bodies (figure 4.21).

This force may be written in a vector form as either

$$\mathbf{F} = -\left(\frac{GMm}{r^3}\right)\mathbf{r} \quad \text{or} \quad \mathbf{F} = -\left(\frac{GMm}{r^2}\right)\hat{\mathbf{r}}$$

These show that the force is directed along PO (because of the negative sign) and its magnitude is the magnitude of $\mathbf{r}$ (i.e. r) times GMm/r^3, giving GMm/r^2, as required (figure 4.21).

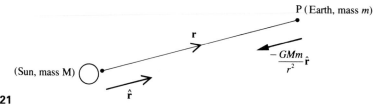

Figure 4.21

Equations of ellipse, parabola and hyperbola

An inverse-square-law central force results in an orbit which is a *conic*: a circle, ellipse, parabola or hyperbola. If you are unfamiliar with the polar equations of these curves, you should read Appendix 1 (pages 173–177). The polar equation of the conic is

$$r = \frac{l}{(1 + e\cos\theta)}$$

where l and e are constants. The type of curve depends on the value of the eccentricity, e:

$e = 0$ circle

$e < 1$ ellipse

$e = 1$ parabola

$e > 1$ hyperbola

Historically, the planetary orbits were found to be ellipses with the sun at a focus, before the existence and nature of the central force was known. If you know the path is an ellipse and you know the force on the planet is always directed towards the sun, you can show that the force obeys an inverse square law. Question 9 in Exercise 4C takes you through this.

It is more difficult to find the type of orbit, given the inverse square law of force. This is the subject of the next section.

Determining the orbit from the inverse square law

Starting from the assumption of an inverse square law of force on a planet, we now show that the planet's orbit must be a conic curve.

The principle of the calculation is as follows:
(i)   Write down the equation of motion $F = ma$, using the radial component of acceleration.
(ii) Since this equation will have three variables (r, θ and t), eliminate t using the fact that $r^2\dot{\theta} = $ constant (there is no transverse force).

(iii) Solve the differential equation to obtain r in terms of θ, i.e. the equation of the path.

However, the detail is a little tricky because in order to produce a differential equation that is easy to solve, you make a substitution $r = 1/u$ at stage (ii). It is worth noting how this works below because this substitution $r = 1/u$ can be useful for other central force problems. The details are as follows:

(i)   Apply $F = ma$ in the *radial* direction. The force k/r^2 is given per *unit mass*, so the mass cancels out. The equation of motion is thus

$$-\frac{k}{r^2} = \ddot{r} - r\dot{\theta}^2 \qquad \textcircled{1}$$

(ii) Since there is no transverse force, $r^2\dot{\theta} = h$, a constant. Substitute $r = 1/u$ (see comment above) and so

$$\frac{\dot{\theta}}{u^2} = h$$

Differentiate r with respect to t, using the chain rule:

$$\frac{dr}{dt} = \frac{dr}{du} \times \frac{du}{dt} = \frac{dr}{du} \times \frac{du}{d\theta} \times \frac{d\theta}{dt}$$

Hence

$$\dot{r} = -\frac{1}{u^2}\dot{u} = -\frac{1}{u^2}\frac{du}{d\theta}\dot{\theta}$$

Substituting $\dot{\theta} = u^2 h$ gives

$$\dot{r} = -h\frac{du}{d\theta}$$

Differentiating with respect to t again:

$$\ddot{r} = -h\frac{d^2u}{d\theta^2}\dot{\theta}$$

$$= -h^2u^2\frac{d^2u}{d\theta^2} \quad \text{(again putting } \dot{\theta} = u^2h\text{)}$$

Substituting for $\ddot{r}$ and $\dot{\theta}$ in $\textcircled{1}$ eliminates *time* from the equation to give

$$-ku^2 = -h^2u^2\frac{d^2u}{d\theta^2} - h^2u^3$$

Divide by h^2u^2 and rearrange:

$$\frac{d^2u}{d\theta^2} + u = \frac{k}{h^2}$$

(iii) You can see the value of the substitution $r = 1/u$; the expression for the radial acceleration simplifies because of the factor u^2h^2. The result is a differential equation familiar from *Mechanics 3*: it is similar to the standard simple harmonic motion equation with $\omega = 1$, referred to a displaced origin. You can immediately write down the general solution:

$$u = A \cos (\theta + \varepsilon) + \frac{k}{h^2}$$

where A and ε are constants depending on the initial conditions.

$$\Rightarrow \quad \frac{1}{r} = A \cos (\theta + \varepsilon) + \frac{k}{h^2}$$

This is the polar equation of the orbit.

The initial line (x axis) can be chosen in any direction. Choose it so that $\varepsilon = 0$:

$$\frac{1}{r} = A \cos \theta + \frac{k}{h^2} \qquad \qquad ②$$

To see that this is the equation of a conic, write it as

$$r = \frac{1}{A \cos \theta + \dfrac{k}{h^2}}$$

$$= \frac{\dfrac{h^2}{k}}{\dfrac{Ah^2}{k} \cos \theta + 1}$$

You can match it against the polar equation of a conic.

$$r = \frac{l}{(1 + e \cos \theta)}$$

where $l = \dfrac{h^2}{k}$ and the eccentricity $e = A \dfrac{h^2}{k}$. The type of curve depends on the value of e (see page 112).

This completes the proof that the orbit under an inverse square law of force is a conic.

Notice that with the equation in this form, the initial line of the polar co-ordinates is the axis of the curve: this was the result of choosing $\varepsilon = 0$ in the foregoing proof. This means that, whichever the type of orbit, when $\theta = 0$ *the velocity is purely transverse.* Suppose you know the value, v_0, of the velocity at that point when $r = r_0$ (see figure 4.22). Then the eccentricity e can be determined as follows.

Since the velocity is transverse

$$h = r_0 v_0$$

Substituting $r = r_0$ and $\theta = 0$ in the equation of the orbit:

$$r_0 = \frac{\dfrac{h^2}{k}}{e + 1} = \frac{h^2}{k(e + 1)} \qquad (\cos 0° = 1)$$

$$\Rightarrow \quad e + 1 = \frac{h^2}{kr_0}$$

$$= \frac{r_0 v_0^2}{k} \quad \text{(since } h = r_0 v_0\text{)}$$

$$\Rightarrow \quad e = \frac{r_0 v_0^2}{k} - 1$$

So the eccentricity, e, is given by $\frac{r_0 v_0^2}{k} - 1$. Thus the type of curve is

determined by the value of $\frac{r_0 v_0^2}{k}$:

when $\frac{r_0 v_0^2}{k} < 2$ then $e < 1$, giving an ellipse

when $\frac{r_0 v_0^2}{k} = 2$ then $e = 1$, giving a parabola

when $\frac{r_0 v_0^2}{k} > 2$ then $e > 1$, giving a hyperbola

When $\frac{r_0 v_0^2}{k} = 1$, e is zero and the orbit is a circle of radius r_0. You can see this

using your knowledge of circular motion:

$$\frac{r_0 v_0^2}{k} = 1$$

$$\Rightarrow \quad \frac{v_0^2}{r_0} = \frac{k}{r_0^2} \quad \text{(the force per unit mass towards the sun)}$$

But v_0^2/r_0 is the acceleration of a body moving with velocity v_0 in a circle radius r_0. The velocity is just right for circular motion given the central force, so the planet remains in a circular orbit.

What if e in the above expression turns out to be negative, i.e. $r_0 v_0^2/k < 1$? The polar equation is still an ellipse but one in which the pole is the *further* focus (figure 4.22).

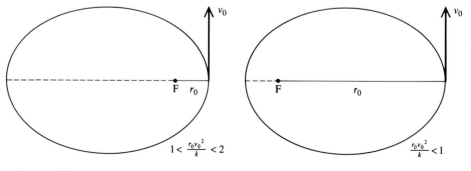

Figure 4.22

For Discussion

The initial conditions of orbital motion

The value of r_0 and v_0 when $\theta = 0$ form the *initial conditions* of the original differential equation of motion. The following discussion may help you to see how different types of orbit are obtained.

Imagine you are a giant, able to start a solar system by flinging planets, comets etc. into orbit about a sun. Even for a giant, the sun might be rather hot, so you position yourself a good distance r_0 from it.

You are going to throw a planet exactly at right angles to the line between yourself and the sun. Then the speed at which you throw it will determine the orbit. By mistake you just let the first one go with no velocity; it simply drops into the sun. You throw the next five progressively faster and get a different type of orbit in each case as shown. What curves are represented by these orbits? What range of initial speeds do they represent?

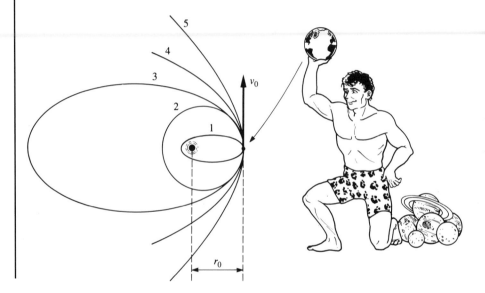

**HISTORICAL
NOTE**

The nature of planetary orbits has been of interest since ancient times. Even after Copernicus in the 16th century, when it began to be generally accepted that the planets went round the sun rather than the Earth, the observations could not be made to fit any of the theories of the time. Such theories had always been based on complicated compositions of circles. In 1609, the German mathematician Kepler published his first two laws, which were formulated from the examination of a large mass of observational data. Referring specifically to the planet Mars, the first law was 'the planet moves in an ellipse with the sun as focus'. Kepler had chosen Mars because of the available accurate data, and because Mars had the least circular orbit of the planets known at that time (excluding Mercury, which was too near the sun for sufficient measurements to have been made). Kepler himself was the first to coin the word 'focus' from the Latin for 'fireplace' because the sun was there. You have already seen his second law about equal areas. The third law, published nine years later, says 'the square of the time a planet takes to go round the sun is proportional to the cube of its mean distance'. (See the investigation on page 115.)

Kepler had no idea why the orbits were ellipses, or indeed why planets went round the sun at all. He still had the general view at the time that if something moves, it must be pushed from behind. He died in 1630, 12 years before the birth of Newton, who was to use Kepler's results in developing the theory of gravity.

Investigation

The following table gives data about the planets in the solar system. It also includes Halley's comet, which has been observed every 75 years or so since 240 BC. Comets orbit the sun like planets, but their orbits are highly elliptical (i.e. have large eccentricities) compared with the planets. Some appearances of Halley's comet have been spectacular, like that of 1066 recorded in the Bayeaux tapestry, but its last visit in 1986 was a disappointment to the naked-eye observer.

	Semi-major axis a (10^6 km)	Period (years)	Eccentricity e
Mercury	57.9	0.241	0.206
Venus	108	0.615	0.006 80
Earth	150	1.00	0.0167
Mars	228	1.88	0.0934
Jupiter	778	11.9	0.0484
Saturn	1430	29.5	0.0557
Uranus	2870	84.1	0.0472
Neptune	4500	165	0.008 57
Pluto	5980	249	0.249
Halley's comet		76	0.967

The first column gives the semi-major axis, a: this is the mean of the minimum and maximum distances, respectively $a(1-e)$ and $a(1+e)$. The second column gives the time for the complete orbit round the sun, given in terms of Earth years (365.24 days). The third column is the eccentricity of the elliptical orbit. Other relevant data are: the gravitational constant $G = 6.67 \times 10^{-11}$ N m^2 kg^{-2}; the mass of the sun $M = 1.99 \times 10^{30}$ kg.

Using the notation from page 112 where the orbit was derived from the inverse square law:

force per unit mass on planet is $\dfrac{k}{r^2}$, where $k = GM$

$h = r^2\dot{\theta}$, constant for a given orbit

$\dfrac{h^2}{k} = l = a(1-e^2)$, where l is the semi-latus rectum

(i) Calculate the minimum (perihelion) and maximum (aphelion) distances from the sun for each object. Show that the orbit of Pluto passes inside that of Neptune. How near and far does Halley's comet go, compared with the planets?

(ii) Calculate the speed of Mercury and of Pluto at the extremes (i.e. the ends of the major axis) of their orbits.

Investigation continued

(iii) Note that most of the orbits have small eccentricities, i.e. are nearly circular. Model those with smallest eccentricities, Venus, Neptune and Earth, as if their orbits *were* circular with radii given by the first column. In each case, work out the acceleration towards the centre and hence, using methods for circular motion, find the periods of rotation, checking against the values in the second column.

(iv) Kepler's Third Law states that the square of the orbital period T of a planet is proportional to the cube of the semi-major axis a of the orbit. In other words: $T^2 = wa^3$ for some constant w. Draw an appropriate graph to verify this relationship, and find w from it. Hence find the semi-major axis of Halley's comet.

Note: the value of w is $4\pi^2/GM$. Exercise 4C, Question 3 asks you to show Kepler's Third Law for the particular case of circular orbits.

Energy in central force systems

Energy methods are as useful in application to central force systems as they are in so many other parts of dynamics. Consider once more a comet in orbit round the sun. It has kinetic energy which changes as its speed increases and decreases. This change is balanced by the work done by the gravitational force on the comet. But as you saw in Chapter 3, this is easy to calculate via the change in potential energy. For a comet of mass m distance r from the sun of mass M, the potential energy is $-GMm/r$. This is defined with the zero level of energy at infinity.

Thus for a comet in motion about the sun

$$\text{kinetic energy} + \text{potential energy} = \text{constant} = E \text{ (say)}$$

$$\tfrac{1}{2}mv^2 - \frac{GMm}{r} = E$$

As you can see from the next example, this energy equation together with the zero transverse acceleration equation, $r^2\dot\theta = $ constant, can be used very effectively to make deductions about the orbit.

NOTE

The above reference to the potential energy under an inverse square law has ignored the fact that the previous treatment in Chapter 3 referred to motion only in one dimension, i.e. directly towards or away from the central body, the sun in this case. In fact, the potential energy formula GMm/r is true for motion in two (or three) dimensions. Because the force is always pointing towards the centre, work is only done for the radial component of any move. No work is done for movements along the transverse direction. So the work done against gravitation when a planet moves from a distance r_1 to r_2 is

$$GMm\left(\frac{1}{r_1} - \frac{1}{r_2}\right)$$

no matter what path it takes.

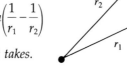

Imagine dividing the path from A to B into small radial and transverse steps (see diagram). Work is done against the gravitational force only in the radial steps, since the force is radial. So the total work against the force is

$$\int F\,\mathrm{d}r = GMm \int_{r_1}^{r_2} \frac{1}{r^2}\,\mathrm{d}r = GMm\left(\frac{1}{r_1} - \frac{1}{r_2}\right)$$

EXAMPLE

An Earth satellite of mass m is launched from a space station which is orbiting at a distance R from the Earth's centre. The Earth's gravitational attraction on the satellite is of the form mk/r^2, where r is the distance from the centre of the Earth at any moment and k is a constant. The satellite is launched with a speed (relative to the Earth) of $\sqrt{(k/R)}$ at an angle of $60°$ with the line from the Earth to the space station (see diagram).

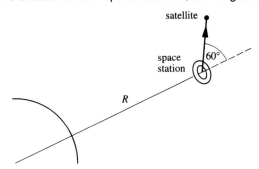

(i) Write down expressions for the initial kinetic energy and gravitational potential energy (due to the Earth) of the satellite.

(ii) Evaluate the constant $h = r^2\dot{\theta}$ for the satellite orbit about the Earth, in terms of k and R.

(iii) Denote by v and s the speed and distance of the satellite at a point where it is moving purely transversely (i.e. at right angles to the line between the Earth and the satellite). Using the fact that h is constant, show $v^2 = \dfrac{3kR}{4s^2}$.

(iv) Write down an expression for the total energy at this point, and hence derive another equation connecting v and s.

(v) Eliminate v from these equations and show that one value of s is $3R/2$. Write down the other value and deduce the length of the major axis of the elliptical orbit of the satellite.

Solution

(i) At launch, the kinetic energy is

$$\tfrac{1}{2}mu^2 = \tfrac{1}{2}m\,\frac{k}{R}$$

and the gravitational potential energy is $-\dfrac{km}{R}$.

The total energy is

$$\tfrac{1}{2}m\,\frac{k}{R} - \frac{km}{R} = -\tfrac{1}{2}m\,\frac{k}{R}$$

(ii) $r^2\dot{\theta} = h$ (a constant)

At launch, $r = R$ and $r\dot{\theta}$ = transverse speed = $\sqrt{\dfrac{k}{R}} \cos 30°$. So

$$h = R\sqrt{\left(\dfrac{k}{R}\right)}\cos 30°$$

$$= \tfrac{1}{2}\sqrt{3kR}$$

(iii) When the velocity is purely transverse

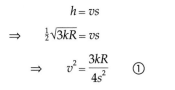

$$h = vs$$

$$\Rightarrow \quad \tfrac{1}{2}\sqrt{3kR} = vs$$

$$\Rightarrow \quad v^2 = \dfrac{3kR}{4s^2} \qquad \text{①}$$

(iv) The total energy E at this point is $\tfrac{1}{2}mv^2 - \dfrac{mk}{s}$. This is the same as the initial

energy $-\tfrac{1}{2}m\dfrac{k}{R}$ from part (i) so

$$\tfrac{1}{2}mv^2 - \dfrac{mk}{s} = -\tfrac{1}{2}\dfrac{mk}{R}$$

$$\Rightarrow \quad v^2 = \dfrac{2k}{s} - \dfrac{k}{R}$$

(v) Substituting for v^2 from ① gives

$$\dfrac{3kR}{4s^2} = \dfrac{2k}{s} - \dfrac{k}{R}$$

$$3R^2 = 8Rs - 4s^2$$

$$3R^2 - 8Rs + 4s^2 = 0$$

$$(3R - 2s)(R - 2s) = 0$$

$$\Rightarrow \quad s = \dfrac{3R}{2} \quad \text{or} \quad \dfrac{R}{2}$$

The length of the major axis is the
sum of the two extreme distances, namely
$\dfrac{3R}{2} + \dfrac{R}{2} = 2R$.

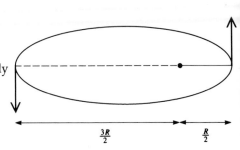

Relationship between total energy and major axis of ellipse
The last part of the previous example is a case of a more general result. Assume the orbit of a planet is an ellipse, and that at one end of the major axis (A or A') the planet's velocity is v and the distance from the sun is r. Taking the central force as mk/r^2, the total energy is

$$E = \tfrac{1}{2}mv^2 - m\frac{k}{r} \quad \text{(kinetic energy + potential energy)}$$

Also, since the velocity at this point is wholly transverse, then $h = vr$. Thus

$$E = \tfrac{1}{2}m\left(\frac{h^2}{r^2}\right) - m\left(\frac{k}{r}\right)$$

$$\Rightarrow \quad 2r^2E + 2kmr - mh^2 = 0$$

This is a quadratic equation in r, the two roots r_1 and r_2 giving the values of r at each end of the major axis. Now the sum of the roots of a quadratic $ax^2 + bx + c = 0$ is $-b/a$. In this case

$$r_1 + r_2 = -\frac{km}{E}$$

But $r_1 + r_2 = 2a$, the major axis of the ellipse. So $E = -\dfrac{km}{2a}$.

Thus there is a simple relationship between the total energy and the major axis of the ellipse.

For Discussion

Relationship between energy and type of orbit under an inverse square law force

You saw previously that a body under an attractive central force $-k/r^2$ per unit mass moves in an ellipse, hyperbola or parabola depending on the value of $r_0v_0^2/k$, where v_0 is the speed of the body when at its closest point r_0 to the origin.

Now the total energy E = kinetic energy + potential energy

$$= \tfrac{1}{2}mv_0^2 - \frac{km}{r_0}$$

$$= \frac{mk}{2r_0}\left(\frac{r_0v_0^2}{k} - 2\right)$$

Thus E is negative, zero or positive according to the sign of $\dfrac{r_0v_0^2}{k} - 2$.

Can you see a connection between the total energy and the type of curve, and justify why this should be the case? Remember that the zero level of gravitational energy was defined to be at infinity.

The orbit when the central force is proportional to distance

One type of central force is best analysed without using polar co-ordinates. This is where the force is proportional to distance and is attractive:

$$\mathbf{F} = -kr\hat{\mathbf{r}} = -k\mathbf{r}$$

The orbit turns out to be an ellipse, but this time the centre of the force is not at a focus but at the geometrical centre of the ellipse. Perhaps uniquely among central force problems, this one is best handled using xy co-ordinates (because the component of force in the x direction depends purely on the x co-ordinate, and similarly for y). The example demonstrates this using the case of a conical pendulum.

EXAMPLE

A pendulum consists of a particle of mass m on the end of a long string of length l which is attached to a point Q. The particle is set swinging round so that the angle ϕ between the string and the vertical is always so small that ϕ^2 is negligible although ϕ is not necessarily constant. In these circumstances, the motion of the particle can be modelled as if it remains in the same horizontal plane throughout. Take O as the point where this plane cuts the vertical through Q.

(i)   Show that the particle experiences a force of magnitude kr towards O, where r is the distance OP and $k = \dfrac{mg}{l}$.

(ii) By taking xy axes in the plane of the motion, with an origin at O, show that the equation of motion can be written

$$\ddot{x} = -\omega^2 x \qquad \ddot{y} = -\omega^2 y$$

for some constant ω.

(iii) By choosing the axes so that $x = 0$ and $\dot{y} = 0$ when $t = 0$, show that the path of the particle is an ellipse.

Solution

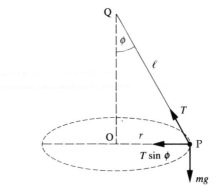

(i) Resolving vertically, $T\cos\phi = mg$ as there is negligible vertical motion. When ϕ is small, $\cos\phi \approx 1 - \dfrac{\phi^2}{2} \approx 1$, since ϕ^2 is negligible. Hence $T = mg$: the variation in the tension is negligible.

The horizontal component of T is always towards O and has magnitude

$$F = T\sin\phi = \frac{mgr}{l}$$

Thus the motion may be modelled by an attractive central force proportional to r: $F = kr$, where $k = \dfrac{mg}{l}$.

(ii) Take xy co-ordinates in the plane of motion, with an origin at O, as in the diagram. Then applying $ma = F$ in the x and y directions gives

$$m\ddot{x} = -\left(\frac{mg}{l}\right)r\cos\theta = -\left(\frac{mg}{l}\right)x$$

$$m\ddot{y} = -\left(\frac{mg}{l}\right)r\sin\theta = -\left(\frac{mg}{l}\right)y$$

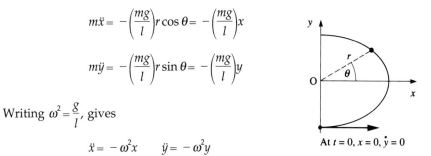

At $t = 0$, $x = 0$, $\dot{y} = 0$

Writing $\omega^2 = \dfrac{g}{l}$, gives

$$\ddot{x} = -\omega^2 x \qquad \ddot{y} = -\omega^2 y$$

This is the simple harmonic equation, with the well-known general solution

$$x = A\sin\omega t + B\cos\omega t \qquad y = C\sin\omega t + D\cos\omega t$$

Since $x = 0$ when $t = 0$, then $B = 0$.

Also when $t = 0$, $\dot{y} = \omega C\cos\omega t - \omega D\sin\omega t = 0$, so $C = 0$. Hence

$$x = A\sin\omega t \qquad y = D\cos\omega t$$

Eliminating t gives

$$\frac{x^2}{A^2} + \frac{y^2}{D^2} = 1$$

This is the equation of an ellipse with major axis A and minor axis D. The values of the constants could be determined if you knew more details of the initial conditions.

For Discussion

You may be dubious about the modelling assumptions made in the foregoing example. Imagine the bob of a 2 m conical pendulum swinging in an ellipse about its equilibrium position with minor and major axes of 20 cm and 60 cm respectively. The bob does not quite remain in a horizontal plane because it is higher when 60 cm from the centre than it is when 20 cm from it. But how much higher? Do you think it is valid to neglect the deviation from a horizontal plane?

M5

For Discussion continued

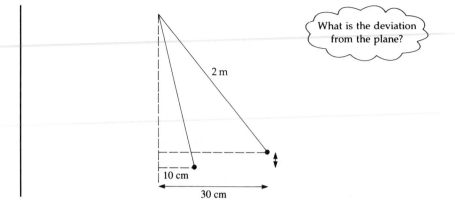

What is the deviation from the plane?

2 m

10 cm

30 cm

Exercise 4C

1. A satellite of mass m is in a circular orbit about a planet and moving with speed V. Show that its total energy is $-\frac{1}{2}mV^2$. (The gravitational potential energy of a satellite distance r from a planet under a central force k/r^2 per unit mass is $-km/r$.)

2. The ratio of the angular velocity ω of the planet Mercury at perihelion (nearest the sun) to that at aphelion (furthest) is 9:4. Show that the eccentricity e of its orbit is 0.2. (Use the fact that $r^2\omega$ is constant and that the perihelion and aphelion distances are respectively $a(1-e)$ and $a(1+e)$, where a is the semi-major axis.)

3. Suppose a solar system is modelled as a set of planets rotating in circular orbits about a fixed sun under an inverse square law of attraction k/r^2. Show that the square of the period of rotation is proportional to the cube of the radius of the orbit (Kepler's Third Law).

4. A satellite is launched with a speed V relative to the earth at a distance R from the earth in the direction shown.

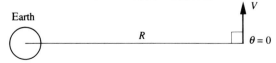

It goes into orbit with a polar equation relative to the earth of

$$r = \frac{h^2}{k(1 + e\cos\theta)}$$

where $h = r^2\dot\theta$ (constant for the orbit) and k and e are constants.

(i) Show $e = \dfrac{RV^2}{k} - 1$.

(ii) For what value of V (in terms of R and k) is the orbit a circle?

(iii) If V is 20% more than its value in part (ii), what is the furthest distance the satellite would travel from earth?

5. A particle P, moving in a plane, has co-ordinates (r, θ) referred to a fixed pole O and an initial line in the plane. Given that the particle moves under a force directed towards O, show that at time t, $r^2\dot\theta = h$, where h is a constant.

The particle moves round the circle whose polar equation is $r = 2a\cos\theta$, where $-\frac{\pi}{2}\leqslant\theta\leqslant\frac{\pi}{2}$ and a is a positive constant. Show that

$$\dot r = -\frac{2ah\sin\theta}{r^2}$$

Show also that the magnitude of the force of attraction is proportional to r^{-5}

[London 1987]

6. A particle P of mass m moves in a plane under the action of a force directed towards a fixed origin O in the plane. Initially it is at a point X distance a from O and moving at (see diagram). As the particle moves, its velocity always makes an angle of 45° with the radius vector PO.

Taking r as the distance OP and θ as the angle POX, show

(i) $r^2\dot{\theta}$ is constant $(=h)$;

(ii) $\dot{r} = -r\dot{\theta}$;

(iii) and hence show $\dot{r} = -\dfrac{h}{r}$ and $\ddot{r} = -\dfrac{h^2}{r^3}$.

(iv) Given that the central force has a magnitude $\dfrac{mk}{r^3}$ and is directed towards O, express h in terms of k.

(v) Use the initial conditions to show that $k = a^2 u^2$.

7. A planet of mass m moves round the sun in an elliptical orbit described by the polar equation

$$r = \frac{l}{1 + e\cos\theta}$$

where e and l are constants. Also $r^2\dot{\theta} = h$, a constant.

(i) Show that the radial component of velocity is $\dfrac{eh\sin\theta}{l}$, and write down the transverse component in terms of h and r.

(ii) Derive an expression for the speed of the planet.

(iii) Hence show that the kinetic energy of the planet is equal to

$$\frac{mh^2}{lr} - \frac{mh^2}{2l^2}(1 - e^2)$$

8. A particle P moves in a fixed plane and its position in that plane is specified by polar co-ordinates (r, θ), where r metres is the distance of P from an origin O and θ is the angle which OP makes with a fixed initial line OA. Initially P is held at a point on OA

which is 1 m from O, and is projected at right angles to OA with a speed of $2\,\mathrm{m\,s^{-1}}$. Throughout its motion, P experiences an acceleration towards O of magnitude $\frac{8}{9}(r-1)\,\mathrm{m\,s^{-1}}$, where $r \geqslant 1$.

By considering the transverse motion of P, show that $\dot{\theta} = \dfrac{2}{r^2}$ and that the radial equation of motion of P then becomes

$$\frac{\mathrm{d}^2 r}{\mathrm{d}t^2} = \frac{4}{r^3} - \frac{8}{9}(r-1)$$

Integrate this expression to show that

$$\left(\frac{\mathrm{d}r}{\mathrm{d}t}\right)^2 = 4 - \frac{4}{r^2} - \frac{8}{9}(r-1)^2$$

and hence, or otherwise, show that $r \leqslant 3$ in the subsequent motion.

[AEB 1987]

9. Assume that the orbit of a planet is an ellipse with the sun at one focus, and that the force on the planet is directed towards the sun. Write the polar equation of the ellipse, relative to an origin at the sun, in the form

$$\frac{1}{r} = \frac{(1 + e\cos\theta)}{l}$$

(i) By differentiating the equation, find $\dot{r}$ in terms of $\dot{\theta}$.

(ii) Using the fact that $r^2\dot{\theta} = h$, a constant for motion under a central force, write $\dot{r}$ in terms of l, h, e and θ. Hence show that

$$\ddot{r} = \left(\frac{eh^2}{lr^2}\right)\cos\theta$$

(iii) Use the values of $\ddot{r}$ and $\dot{\theta}$ to work out the radial component of acceleration.

(iv) Hence show that the radial force has the form

$$F_r = -\frac{mk}{r^2}$$

where m is the mass of the planet and $k = h^2/l$.

10. An investigation of atomic scattering leads to a central force problem in which the force is inversely proportional to the square of the distance between two atomic particles. The equations to be solved are

$$\ddot{r} - r\dot{\theta}^2 = \frac{k}{r^2} \quad \text{and} \quad 2\dot{r}\dot{\theta} + r\ddot{\theta} = 0$$

where r and θ are polar co-ordinates and k is a constant.

(i) Show that integrating the second equation leads to $r^2\dot{\theta} = h$, where h is a constant.

(ii) Hence show the first equation may be written $\ddot{r} - \frac{h^2}{r^3} = \frac{k}{r^2}$.

(iii) Show further that if $r = \frac{1}{u}$, then

$$\dot{r} = -h\frac{du}{d\theta} \quad \text{and} \quad \ddot{r} = -h^2u^2\frac{d^2u}{d\theta^2}$$

(iv) Hence form a differential equation for u in terms of θ.

(v) Solve this equation to find the general path of one particle relative to the other in the form

$$\frac{1}{r} = A\cos\theta + B\sin\theta - \frac{k}{h^2}$$

[MEI 1992]

11. A satellite of mass m and at a distance r from a planet experiences a central force per unit mass of k/r^2. Initially it is moving with speed V in a circular orbit radius R. Due to a firing of internal engines, it maintains the speed V but changes its direction so it is moving 30° to the radius, as shown. It thus goes into a new orbit.

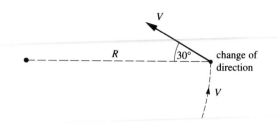

(i) Calculate V in terms of k and R.

(ii) Write down the value of $h = r^2\dot{\theta}$ for the new orbit.

(iii) Derive an expression for the total energy E for the new orbit. (The gravitational potential energy of a satellite when at distance r from the planet is $-km/r$.)

(iv) Calculate the speed of the satellite in terms of V when it is at perigee (nearest the planet) and apogee (furthest).

12. A satellite of mass m and distance r from the centre of a planet experiences a central force per unit mass of k/r^2. Initially it is travelling in a circular orbit with a constant speed V at a height H above the surface of the planet whose radius is R. The satellite's speed is suddenly reduced to αV $(0 < \alpha < 1)$ without a change of direction causing it to go into a new orbit.

(i) Write down the value of $h = r^2\dot{\theta}$ for the new orbit.

(ii) Show that the total energy of the satellite $E = -km/2a$, where $2a$ is the length of the major axis of the orbit. (The gravitational potential energy of the satellite is $-km/r$.)

(iii) Find the condition on α for the satellite to remain in orbit.

KEY POINTS

- Problems involving forces directed towards a fixed point are often best solved using polar co-ordinates. Unit vectors $\hat{\mathbf{r}}$ and $\hat{\boldsymbol{\theta}}$ are defined in the *radial* and *transverse* directions. These directions vary at different points, unlike those in the $\mathbf{i}$ and $\mathbf{j}$ directions.

- $\dfrac{d\hat{\mathbf{r}}}{dt} = \dot{\theta}\hat{\boldsymbol{\theta}} \qquad \dfrac{d\hat{\boldsymbol{\theta}}}{dt} = -\dot{\theta}\hat{\mathbf{r}}$

- The radial component of velocity is $\dot{r}$

- The transverse component of velocity is $r\dot{\theta}$

- The radial component of acceleration is $\ddot{r} - r\dot{\theta}^2$

- The transverse component of acceleration is $r\ddot{\theta} + 2\dot{r}\dot{\theta} = \dfrac{1}{r}\dfrac{d}{dt}(r^2\dot{\theta})$

- When the force is purely radial (i.e. a central force) $r^2\dot{\theta} = h$, a constant. This is equivalent to the conservation of angular momentum for a particle of constant mass.

- The constant h can also be written as rv_θ, i.e. r times the transverse component of velocity. This is particularly useful at the extremes of an orbit when the velocity is wholly transverse.

- For a particle in orbit under an attractive inverse square law $\dfrac{k}{r^2}$ per unit mass, the equation of motion in the radial direction is

  $-\dfrac{k}{r^2} = \ddot{r} - r\dot{\theta}^2$ which is best solved by putting $u = \dfrac{1}{r}$.

  The orbit is a conic with polar equation $r = \dfrac{l}{1 + e\cos\theta}$ where $l = \dfrac{h^2}{k}$

  This is a circle, ellipse, parabola or hyperbola depending on whether $e = 0$, $e < 1$, $e = 1$ or $e > 1$.

- Given the distance r_0 and speed v_0 at the closest point of the orbit, then

  $$e = \frac{h^2}{kr_0} - 1 = \frac{r_0 v_0^2}{k} - 1$$

- The gravitational potential energy of a planet of mass m distance r from a sun of mass M is $-\dfrac{GMm}{r}$

  The total energy (kinetic + potential) of the planet is

  $E = \tfrac{1}{2}mv^2 - \dfrac{GMm}{r}$

  For an elliptical orbit, $E = -\dfrac{km}{2a}$, where a is the semi-major axis of the ellipse.

- An attractive force proportional to distance, $-kr$, is best solved using xy co-ordinates.

5

Introduction to rotation and moments of inertia

Keep that wheel a-turning...

Popular song, 1920s

For Discussion

So far you have modelled moving objects as particles. In many circumstances this is reasonable, but how would you model the motion of the sails of a windmill or the other objects illustrated in the pictures above?

Do the two children on the roundabout have the same kinetic energy?

What is the kinetic energy of a rotating wheel?

The dynamics of a rigid body rotating about a fixed axis

You might not be able to answer all these questions fully now, but the issues involved should become clearer as you work through this chapter.

It is reasonable to treat a large object as a particle when every part of it is moving in the same direction with the same speed, but clearly this is not always the case. The particles in a rotating wheel have different velocities and accelerations and are subject to different forces.

The laws of particle dynamics which you have used so far need to be developed so that they can be applied to the rotation of large objects.

Definitions

You are already familiar with many aspects of rotation such as the angular speed and acceleration of a particle and you have also taken moments to determine the turning effect of a force, but it is as well to be clear about what is meant by some of the terms involved before continuing with the discussion.

Rigid bodies

Wheels can be modelled as rigid bodies. A *rigid body* is such that each point within it is always the same distance from any other point. You are not a rigid body but a hard chair is one (molecular vibrations being ignored).

The axis of rotation

When you lean back on your chair, it might rotate about a point, say A, at the end of one leg. You will have more control, however, if it rotates about the axis formed by the line joining the ends, A and B, of two legs.

The idea of an axis of rotation is important when considering the rotation of rigid bodies. When the only fixed point is A, the axis of rotation might be continually changing, any particle in the chair moves on the surface of a sphere with its centre at A. When the chair rotates about the fixed axis AB, however, each particle in it moves in a circle in a plane perpendicular to the axis (figure 5.1).

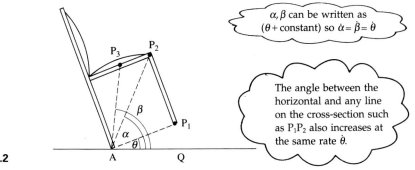

Figure 5.1

Angular speed

All the particles of a rigid body rotating about a fixed axis have the same *angular speed* $\dot{\theta}$ about the axis. This is defined as the rate of change of the angle, θ, between a fixed line in the body and a fixed direction in space, both of which are at right angles to the axis of rotation.

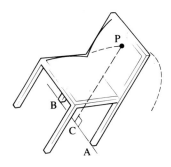

> α, β can be written as $(\theta + \text{constant})$ so $\dot{\alpha} = \dot{\beta} = \dot{\theta}$

> The angle between the horizontal and any line on the cross-section such as P_1P_2 also increases at the same rate $\dot{\theta}$.

Figure 5.2

In figure 5.2, the chair is rotating about an axis through A which is at right angles to the page. AP_1 is a fixed line in the body and AQ a fixed direction in space and both are at right angles to the axis, so the angular speed of the point P_1 is equal to $\dot{\theta}$. But notice that if the chair is rigid, the angle P_2AQ can be written as (θ + constant) so the angular speed of the point P_2 is also $\dot{\theta}$. The same argument applies to any other point in the plane of AP_1Q and similarly to points in all other planes of cross section of the chair perpendicular to the axis.

The moment of a force about an axis

You can turn the chair about the axis through A by pulling it at a point D in the centre of the top of the back. This produces a turning effect, or moment, about the axis. When you let go (if you haven't pulled it too far) the moment of the weight of the chair about the axis will return it to its normal position. Notice that these moments are about an axis rather than a point. When you have previously found moments about a point (*Mechanics 2*), they have really been about an axis through that point at right angles to the plane of the forces. The moment of the weight Mg about the point E in the two-dimensional diagram (figure 5.3a) is, in fact, the moment of Mg about the axis AB (figure 5.3b).

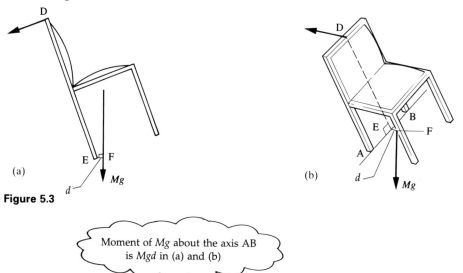

(a)

(b)

Figure 5.3

> Moment of Mg about the axis AB
> is Mgd in (a) and (b)

Couples

One way of rotating a rigid body is to use a *couple*.

- A couple has the same moment about any axis parallel to the axis of rotation but no linear resultant.
- A couple can be represented by equal and opposite parallel forces in a plane perpendicular to the axis of rotation.

Figure 5.4 illustrates two equal and opposite forces each of magnitude P and a distance d apart, which form a couple. The moment of the couple can be shown to be the same about any axis perpendicular to its plane (every couple has its moment).

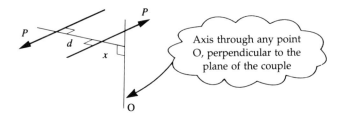

Axis through any point O, perpendicular to the plane of the couple

Figure 5.4

The anticlockwise moment about an axis perpendicular to the plane through some point O is

$$P(d + x) - Px = Pd$$

This has the same value whatever the position of O. This particular representation is not unique; any other pair of parallel forces in the plane could be used to represent the couple provided their moment about the axis of rotation is equal to Pd.

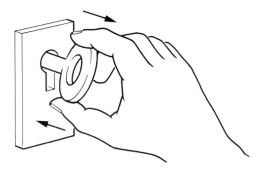

Figure 5.5

The key in figure 5.5 is being turned by a couple, there is no resultant force in any direction so the key only rotates. It isn't essential to apply only two forces in order to turn a key but, however you hold it, you will be applying forces which can be reduced to a couple. The frictional forces which prevent a pulley or a hinge from turning freely have a resultant moment about the axis of rotation but no linear resultant so they also reduce to a couple.

It is now possible to return to the discussion about the dynamics of a rigid body which rotates about a fixed axis. In this chapter you will be learning about its energy and its equation of motion. Other aspects such as angular momentum are treated in *Mechanics 6*.

The kinetic energy of a rigid body rotating about a fixed axis

Because kinetic energy is a scalar quantity, the kinetic energy of a rigid body, such as a wheel, can be found by calculating the energy of each of the separate particles and then adding them. For example, you could find the kinetic energy of the children on the roundabout by treating each one as a separate particle moving in a circle. They could be modelled as a simple rigid body like that in the next example.

EXAMPLE

A rigid body consists of two particles P_1 and P_2 of masses m_1 and m_2 attached to a light rod AB as shown. $AP_1 = r_1$ and $AP_2 = r_2$ and P_2 is at B.

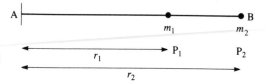

Find the kinetic energy of the body when it rotates with angular speed ω about an axis perpendicular to the rod (i) through A; (ii) through B.

Solution

(i) When the rod rotates with angular speed ω about an axis through A perpendicular to the rod, each particle moves in a circle, centre A.

P$_1$ has speed $v_1 = r_1\omega$ and P$_2$ has speed $v_2 = r_2\omega$. The total kinetic energy is, therefore,

$$\tfrac{1}{2}m_1v_1^2 + \tfrac{1}{2}m_2v_2^2 = \tfrac{1}{2}m_1r_1^2\omega^2 + \tfrac{1}{2}m_2r_2^2\omega^2$$
$$= \tfrac{1}{2}(m_1r_1^2 + m_2r_2^2)\omega^2$$

(ii) When the axis is through the end B, the particle P$_2$ at B does not move so has no kinetic energy. The particle P$_1$ now moves in a circle of radius $(r_2 - r_1)$, so the total kinetic is now $\tfrac{1}{2}m_1(r_2 - r_1)^2\omega^2$.

You can see that the kinetic energy depends not only on the mass of the body and the angular speed, but also on the distance of the particles of the body from the axis of rotation.

Now consider a more complex body rotating with an angular speed $\dot{\theta}$.

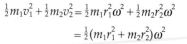

A typical particle of mass m_p, moving in its circle of radius r_p round the axis of rotation with angular speed $\dot{\theta}$, has a speed $r_p\dot{\theta}$ (figure 5.6). Its kinetic energy is $\tfrac{1}{2}m_pr_p^2\dot{\theta}^2$.

$\dot{\theta}$ is used rather than ω because it might vary with time

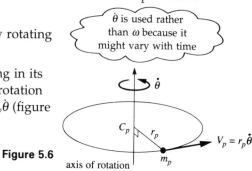

Figure 5.6

Summing over all particles gives the total kinetic energy as

$$\sum_{\text{all } p} \tfrac{1}{2}m_pr_p^2\dot{\theta}^2 = \tfrac{1}{2}\left(\sum_{\text{all } p} m_pr_p^2\right)\dot{\theta}^2 \quad \text{(since } \dot{\theta} \text{ is the same for all particles)}$$

Moment of inertia

The quantity $\displaystyle\sum_{\text{all } p} m_pr_p^2$ is called the *moment of inertia* of the body about the axis and is conventionally denoted by the letter I. Moment of inertia has dimensions ML^2 and so its SI unit is 1 kg m^2.

Thus the kinetic energy of a rigid body rotating about a fixed axis is $\tfrac{1}{2}I\dot{\theta}^2$. Notice that there is an analogy between this expression for kinetic energy and the kinetic energy, $\tfrac{1}{2}mv^2$, of a particle. The mass, m, can be replaced by I

and the speed, v, by the angular speed $\dot{\theta}$ (or ω) to give $\frac{1}{2}I\dot{\theta}^2$ (or $\frac{1}{2}I\omega^2$). You will see later that there are similar analogies between other quantities you have used for the motion of particles and those which apply to the rotation of a rigid body.

EXAMPLE

A wooden top has a moment of inertia of 2.4×10^{-5} kg m^2 about its axis. It starts spinning when a string wound round the spindle of the top is pulled with a constant force of 0.5 N. Assuming there is no loss of energy due to friction, find the angular speed attained by the top when the length of the string is 0.3 m.

Solution

The work done in pulling the string is $0.5 \times 0.3 = 0.15$ J, and as no energy is lost in the process, this is equal to the gain in kinetic energy of the top. So the angular speed attained is ω rad s^{-1}, where

$$\tfrac{1}{2} \times 2.4 \times 10^{-5}\omega^2 = 0.15$$

$$\Rightarrow \qquad \omega^2 = 12\,500$$

The angular speed is 112 rad s^{-1} correct to 3 significant figures.

The moment of inertia (I or $\sum m_p r_p^2$) of a body about an axis is a scalar quantity which depends on the manner in which the particles of the body are distributed about that particular axis. Its value varies according to the position and orientation of the axis.

Inertia is a word which is used to describe a resistance to change in motion; it is sometimes used in place of mass. The larger the mass, or inertia, of a particle the greater the amount of energy required to change its motion. In the same way, the energy required to change rotational motion is greater for bodies with large moments of inertia.

The kinetic energy of a rotating wheel

So what is the kinetic energy of a rotating wheel?

As usual when modelling mechanical systems, it is useful to begin with a simple case. The simplest model of a wheel is one in which the mass of the spokes or their equivalent is negligible and all the mass can be considered to be concentrated at the rim in a hoop or ring of radius r. Then every particle is the same distance, r, from the axle. The moment of inertia of the wheel about the axle is then

$$I = \sum m_p r_p^2$$

$$= \left(\sum m_p \right) r^2$$

$$= Mr^2$$

where M is the total mass.

When this wheel is rotating with angular speed $\dot{\theta}$ about an axis through its centre perpendicular to its plane, its kinetic energy is

$$\tfrac{1}{2}I\dot{\theta}^2 = \tfrac{1}{2}Mr^2\dot{\theta}^2$$

The kinetic energy of a more complex wheel can be found when you know its moment of inertia, I, about the axis of rotation. Rolling wheels are treated in *Mechanics 6*, but you will be interested to know that the kinetic energy of a rolling wheel of mass M is given by $\tfrac{1}{2}I\dot{\theta}^2 + \tfrac{1}{2}Mv^2$, where I is the moment of inertia about its axle (through the centre of mass) and v is the speed of the centre of mass.

Calculating moments of inertia

The moment of inertia of a rigid body about an axis is given by

$$I = \sum m_p r_p^2$$

The sum is calculated over all particles of the body and m_p denotes the mass of a typical particle which is a fixed perpendicular distance, r_p, from the axis. The axis can be anywhere, even outside the body, so long as r_p is constant for each particle, which is therefore restricted to motion in a circle of radius r_p relative to the axis.

You might have guessed that calculus methods are required to work out most moments of inertia. These are very similar to those you have used before where sums are involved, namely: subdivide the body into elementary parts for which you know the moment of inertia and then sum the parts.

EXAMPLE

Calculate the moment of inertia of a uniform circular disc of radius r, thickness t, and mass M about an axis through its centre perpendicular to its plane.

Decide on appropriate elements

Solution

The disc is divided into elementary rings. A typical ring has radius x, width δx and thickness t. Its volume is approximately $(2\pi x \delta x) \times t = 2\pi t x \delta x$

It is useful to use the density, ρ, of the disc so that the mass of each part can be obtained. It can be written in terms of M at the end of the calculation.

Area of ring
$= \pi(x + \delta x)^2 - \pi x^2$
$\approx 2\pi x \delta x$
for negligible δx^2

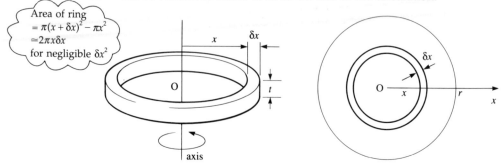

axis

Plan of disc

The mass of the ring is then approximately

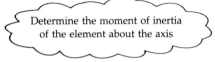
Find the mass of the element in terms of the density

$$\delta m = (2\pi t x \delta x) \times \rho = 2\pi t \rho x \delta x$$

For small δx every particle of such a ring is approximately the same distance, x, from the axis. The moment of inertia of the ring about the axis is therefore approximately

Determine the moment of inertia of the element about the axis

$$\delta m x^2 = (2\pi t \rho x \delta x) x^2$$
$$= 2\pi t \rho x^3 \delta x$$

and the moment of inertia of the whole disc about the axis is approximately

$$I = \sum (2\pi t \rho x^3 \delta x)$$

Sum for all elements

$$= 2\pi t \rho \sum (x^3 \delta x) \quad \text{(since } \rho \text{ and } t \text{ are the same for all rings)}$$

In the limit as $\delta x \to 0$, this gives

Write as an integral

$$I = 2\pi t \rho \int_0^r x^3 \, dx$$

Note the limits: x takes values between 0 and r.

Evaluate the integral

$$\Rightarrow \quad I = 2\pi t \rho \left[\frac{x^4}{4}\right]_0^r$$
$$= \tfrac{1}{2}\pi t \rho r^4$$

It is now necessary to replace ρ in terms of the mass of the disc, which has volume $\pi r^2 t$, so

Substitute for the density

$$M = \pi r^2 t \rho$$

Hence the moment of inertia of a disc about a perpendicular axis through its centre is

$$I = \tfrac{1}{2}(\pi t \rho r^2) r^2$$

or you could write
$$\frac{I}{M} = \frac{\pi t \rho r^4}{2\pi r^2 t \rho} = \tfrac{1}{2} r^2$$

$$\Rightarrow \quad I = \tfrac{1}{2}M r^2$$

Points worth noting at this stage are

- The moment of inertia of the disc ($\tfrac{1}{2}Mr^2$) is less than the moment of inertia (Mr^2) of a ring of the same mass and radius because most of the matter in the disc is nearer the axis.
- The thickness of the disc does not appear in the equation. The disc could be a solid cylinder of any length and the same formula would hold so long as the axis of rotation is the axis of the cylinder. Of course, the moment of inertia of a cylinder is greater than that of a thin disc of the same radius and density, but only because the mass is greater.
- Using the same argument, the moment of inertia of a hollow cylinder of radius r about its axis has the same form as that of the ring, that is Mr^2. Each particle of the cylinder is the same distance, r, from the axis.

EXAMPLE Find the moment of inertia of a thin rod of length $2a$ and mass M about an axis through its centre and perpendicular to the rod.

Solution

Imagine that the rod is subdivided into small elements of width δx. Assume that the area, A, of a cross-section of the rod is so small that every point on it can be regarded as being the same distance from the axis. Then the only variable is the distance x of the elementary portion of the rod from the axis.

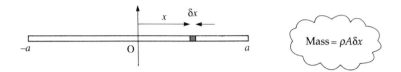

The mass of the element is $\rho A\delta x$, where ρ is the density, and its moment of inertia about the axis is approximately $(\rho A\delta x)x^2 = \rho Ax^2\delta x$.

The moment of inertia of the rod about the axis is therefore approximately $\sum \rho Ax^2\delta x$. In the limit as $\delta x \to 0$ this gives

$$I = \int_{-a}^{a} \rho Ax^2 \, dx$$

$$= \rho A \int_{-a}^{a} x^2 \, dx$$

$$\Rightarrow \quad I = \rho A \left[\frac{x^3}{3}\right]_{-a}^{a}$$

$$= \frac{\rho A}{3}[a^3 - (-a)^3]$$

$$= \tfrac{2}{3}\rho Aa^3$$

The mass of the rod is $M = 2aA\rho = 2\rho Aa$. Hence the moment of inertia of a rod about a perpendicular axis through its centre is

$$I = \tfrac{1}{3}(2\rho Aa)a^2 = \tfrac{1}{3}Ma^2$$

or write
$$\frac{I}{M} = \frac{2\rho Aa^3}{3 \times 2\rho Aa} = \frac{1}{3}a^2$$

Once the moment of inertia of a body such as a rod or a disc is known, it can be used to find moments of inertia of other bodies.

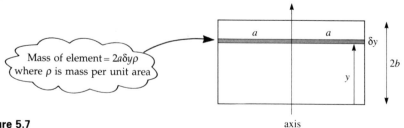

Mass of element $= 2a\delta y\rho$ where ρ is mass per unit area

Figure 5.7

axis

For example, the thin rectangular lamina shown in figure 5.7 can be thought of as the sum of a large number of elementary rods.

The mass of the elementary rod shown is $(2a\delta y)\rho$, where ρ is now the mass per unit area of the lamina. The moment of inertia of the elementary rod about the axis is then

$$\tfrac{1}{3}(2a\rho\delta y)a^2 = \tfrac{2}{3}a^3\rho\delta y$$

So the moment of inertia of the rectangular lamina about an axis of symmetry in its plane is

$$I = \int_0^{2b} \tfrac{2}{3}a^3\rho\,dy$$

$$= \tfrac{2}{3}a^3\rho \int_0^{2b} dy$$

$$= \tfrac{2}{3}a^3\rho 2b$$

$$= \tfrac{4}{3}ab\rho a^2$$

But the mass of the lamina is $M = 4ab\rho$. Hence

$$I = \tfrac{1}{3}Ma^2$$

Notice that this is independent of b and is in the same form as the moment of inertia of a thin rod about the axis. It is another case where the body is extended in the direction of the axis, leading to the same expression for the moment of inertia although, of course, the mass is greater.

Combining bodies

When a rigid body has several parts, its moment of inertia about an axis can be found by adding the moments of inertia of the separate parts about the same axis. This is a direct consequence of the definition of the moment of inertia as a sum taken over all particles of the body; it doesn't matter if the sum is taken separately for different groups of particles. The next example illustrates this principle.

EXAMPLE

A wheel of mass M is strengthened using a metal ring. It consists of a uniform disc of radius r and mass σ per unit area surrounded by a uniform solid ring (the rim) of radius r, negligible width, and mass $5r\sigma$ per unit length.
(i) Find the mass of the two parts of the wheel in terms of M.
(ii) Write down the moment of inertia of each part about the axle of the wheel.
(iii) Find the kinetic energy when the wheel is rotating with an angular speed ω.
(iv) Write this kinetic energy as a percentage of the kinetic energy of a hoop with the same mass and radius rotating at the same angular speed.

Solution

(i) The area of the disc is πr^2, so its mass is $M_1 = \sigma \pi r^2$.

The length of the ring is $2\pi r$, so its mass is

$$M_2 = 2\pi r(5r\sigma)$$

$$= 10\sigma\pi r^2$$

Hence
$$M = 11\sigma\pi r^2$$

$$\Rightarrow \quad M_1 = \frac{M}{11} \text{ and } M_2 = \frac{10M}{11}$$

(ii) The moment of inertia of the inside disc about the axle is

$$I_1 = \tfrac{1}{2}M_1 r^2$$

$$= \frac{Mr^2}{22}$$

The moment of inertia of the rim about the axle is

$$I_2 = M_2 r^2$$

$$= \frac{10Mr^2}{11}$$

(iii) The kinetic energy of the wheel is $\tfrac{1}{2}I\omega^2$, where $I = I_1 + I_2$.

$$I_1 + I_2 = \frac{Mr^2}{22} + \frac{10Mr^2}{11}$$

$$= \frac{21}{22}Mr^2$$

$$\Rightarrow \quad \text{kinetic energy} = \tfrac{1}{2}I\omega^2 = \frac{21}{44}Mr^2\omega^2$$

(iv) The moment of inertia of the hoop about its axis is Mr^2, so its kinetic energy is $\tfrac{1}{2}Mr^2\omega^2$.

The kinetic energy of the wheel is $\frac{21}{22} \times 100\%$ of the kinetic energy of the hoop, namely 95.5%.

Exercise 5A contains questions about moments of inertia and their use in calculating kinetic energy.

N O T E

The approach required when using integration methods for surfaces of revolution rather than volumes of revolution is described in Appendix 2, pages 178–180.

Exercise 5A

1. The diagrams show some models which have been made by joining equal light rods of length a with beads. The beads are each of mass m and can be treated as particles. Using the definition $I = \sum mr^2$, find the moments of inertia of each model about the given axes in terms of a and m, *and also in terms of a and the total mass M of the model.*

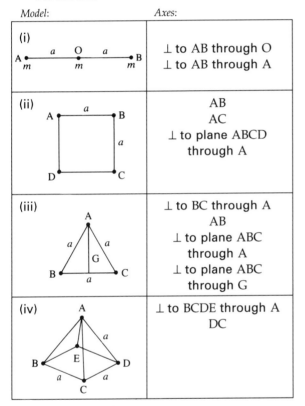

Model:	Axes:
(i) $\underset{m}{A} \bullet \overset{a}{\underline{\hspace{1cm}}} \underset{m}{\overset{O}{\bullet}} \overset{a}{\underline{\hspace{1cm}}} \underset{m}{\overset{B}{\bullet}}$	$\perp$ to AB through O $\perp$ to AB through A
(ii) (square $ABCD$, side a)	AB AC $\perp$ to plane ABCD through A
(iii) (triangle ABC, sides a)	$\perp$ to BC through A AB $\perp$ to plane ABC through A $\perp$ to plane ABC through G
(iv) (tetrahedron, edges a)	$\perp$ to BCDE through A DC

2. (i) Use the formula $\frac{1}{2}Mr^2$ (found on page 133) to calculate the moment of inertia about its axis of a flywheel which is a uniform disc of mass 10 kg and radius 0.25 m.

(ii) Find the kinetic energy required to change its angular speed from 50 to 100 rad s^{-1}.

3. A drum majorette twirls a baton of mass 0.4 kg and length 0.7 m in a circle around its centre. She says that the end moves at speeds up to 30 mph. Assuming the baton is a thin rod, find its kinetic energy in this case. (Note 1 mile$\approx$1.6 km)

4. A gyroscope has a moment of inertia of 0.01 kg m^2 about its axis of rotation and is set in motion by pulling a light string of length 0.5 m wrapped round the axis.
  (i) Find the work done when the string is pulled off the axis with a constant force of 50 N.
  (ii) Hence calculate the angular speed given to the gyroscope.

5. A uniform thin rod OA of mass M and length $2a$ is hinged at one end O so that it can rotate freely about an axis through O perpendicular to OA. An element of the rod of length δx is situated a distance x from O. ρ is the mass per unit length of the rod.

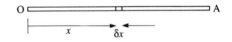

  (i) Write down the mass of the element.
  (ii) Write down the moment of inertia of the element about the axis.
  (iii) Form a suitable integral to calculate the moment of inertia of the rod about the axis and evaluate it.
  (iv) Write M in terms of a and ρ and hence show that the moment of inertia of the rod about the axis is $\frac{4}{3}Ma^2$.
  (v) Use your result to find the moment of inertia of a uniform rectangular lamina of width $2a$ about an axis along the edge which is perpendicular to this width.

6. Use the moment of inertia found in Question 5 for this question. A rod AB of length 0.8 m is hinged at A so that it can rotate freely in a vertical plane.

  (i) The rod is held with AB horizontal and let go. Assuming mechanical energy is conserved, find its angular speed when it makes an angle of $\frac{\pi}{3}$ with the downward vertical.
  (ii) The rod is now hanging vertically. Find the minimum velocity which must be given to the free end in order for it to make a complete revolution.

7. A door of mass 35 kg and width 0.8 m is slammed shut with an angular speed of 2.5 rad s^{-1}.
   (i) By modelling the door as a lamina and dividing it into vertical strips, or otherwise, find its moment of inertia about the axis through the hinges.
   (ii) Find the amount of energy dissipated when the door shuts.

8. A uniform solid sphere has radius r, density ρ and centre at the origin. It is divided into elementary discs perpendicular to the x axis so that a typical disc has thickness δx, radius y and centre at $(x,0)$, as shown in the diagram.

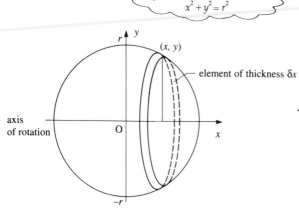

Equation of circular section:
$x^2 + y^2 = r^2$

element of thickness δx

axis of rotation

   (i) Write down an expression for the mass of a typical disc.
   (ii) Show that the moment of inertia of the sphere about the x axis is given by
$$I = \int_{-r}^{r} \tfrac{1}{2}\rho\pi y^4 \, dx$$
   (iii) Substitute an expression for y^2 in terms of x and so evaluate this integral.
   (iv) Write the mass, M, of the sphere in terms of ρ, π and r and hence show that the moment of inertia of the sphere about a diameter is given by $\tfrac{2}{5}Mr^2$.
   (v) Suggest a suitable integral for finding the moment of inertia of any solid of revolution about the x axis.

There is a complete solution to this question in Appendix 2, pages 178–180.]

9. (i) Assume the Earth is a uniform sphere of mass 6×10^{24} kg and radius 6400 km. Use the result of Question 8 part (iv) to estimate its moment of inertia and its kinetic energy of rotation about its axis. Given that the density of the Earth increases towards the centre, how would this compare with the true value?
   (iii) Assume the moon is a uniform sphere of mass 7.5×10^{22} kg and radius 1700 km and that it rotates once every 28 days. Estimate its kinetic energy of rotation about its axis.
   (iv) Now assume the moon is a particle which rotates round the centre of the Earth once every 28 days in a circle of radius 3.85×10^5 km. Find its kinetic energy due to this motion.
   (v) Find the ratio of the kinetic energy of the Earth to the total kinetic energy of the moon (not including their motion round the sun).

10. A potter is throwing clay on a wheel which turns at a constant rate. It starts as a solid cylinder of radius r and height h and gradually changes into a jar with the cross-section shown in the diagram.

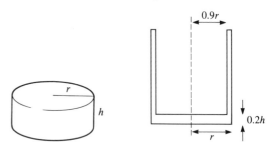

Assuming the mass, M kg, and the density of the clay are constant, find
   (i) the density of the clay;
   (ii) the height, in terms of h, of the jar when the thickness of the base is $0.2h$ and the inside radius is $0.9r$.
   (iii) Assuming the jar is a solid cylinder with another removed from the inside, find its moment of inertia about its axis of rotation in terms of M and r.
   (iv) Find the ratio of the kinetic energy of the jar to that of the original cylinder.

Moments of inertia of selected uniform bodies

	Body of mass M	Axis	Moment of inertia
	Hoop or hollow cylinder of radius r	Through centre perpendicular to circular cross-section	Mr^2
	Disc or solid cylinder, of radius r	Through centre perpendicular to circular cross-section	$\frac{1}{2}Mr^2$
	Thin rod of length $2l$	Through centre perpendicular to rod	$\frac{1}{3}Ml^2$
	Rectangular lamina	Edge perpendicular to sides of length $2l$	$\frac{4}{3}Ml^2$
	Solid sphere of radius r	Diameter	$\frac{2}{5}Mr^2$

The equation of motion for the rotation of a rigid body about a fixed axis

The expression for the kinetic energy of a wheel was obtained by summing the energies of individual particles. The same approach is useful for finding the equation of motion for the rotation of a rigid body about a fixed axis. This is the equivalent of Newton's Second Law for the linear motion of a particle, so first consider the equation of motion for a typical particle in the body. When the rigid body rotates about a fixed axis, each particle in the body moves in a plane in a circle with its centre on the axis, as shown in figure 5.8. Because the body is rigid and rotates about a fixed axis, all the particles in the body have the same angular velocity $\dot{\theta}$ at all times, so they also have the same angular acceleration $\ddot{\theta}$.

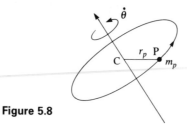

Figure 5.8

A typical particle P rotates in a circle of radius r_p about a point C on the axis. Its acceleration has radial and transverse components $-r_p\dot\theta^2$ and $r_p\ddot\theta$. These are shown in figure 5.9 together with the resultant force $\mathbf{F}_p$ which acts on the particle in the plane of the motion and is at an angle α to PC.

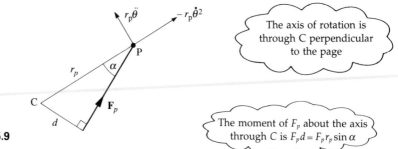

The axis of rotation is through C perpendicular to the page

The moment of F_p about the axis through C is $F_p d = F_p r_p \sin\alpha$

Figure 5.9

Newton's Second Law in the transverse direction (i.e. for circular motion, the tangential direction) gives

$$F_p \sin\alpha = m_p r_p \ddot\theta$$

Multiplying by r_p gives

$$F_p r_p \sin\alpha = m_p r_p^2 \ddot\theta$$

The equation of motion for the rotation of the whole rigid body can be found by summing both sides for all the particles.

$F_p r_p \sin\alpha$ is the moment of the force $\mathbf{F}_p$ about the axis of rotation. This is made up of components which include internal forces as well as the external forces on the body which happen to act on the particular particle.

By Newton's Third Law, the internal forces are equal and opposite. It is therefore to be expected that, when all the moments of all the forces are summed for all particles, the moments of the internal forces cancel and only the moments of the external forces remain.

Summing both sides of the equation for all particles therefore gives

$$\sum(\text{moments of external forces}) = \sum (m_p r_p^2 \ddot\theta)$$

$$= \left(\sum m_p r_p^2\right)\ddot\theta = I\ddot\theta$$

where $I = \sum m_p r_p^2$ as before.

There is no standard notation for a moment and M cannot be used because M is used for mass. In this book C is used to represent the moment of a force about a particular axis (sometimes called a torque) or the moment of a couple. So when the total moment of all forces about the axis is C, the equation of motion for rotation becomes $C = I\ddot{\theta}$.

This is the equivalent of Newton's Second Law for the rotation of a rigid body about a fixed axis. It is a very concise equation, and the beauty of it is that it can be compared with the equation $F = m\ddot{x}$ for the linear motion of a particle in a similar way to the analogy between kinetic energies met on page 130.

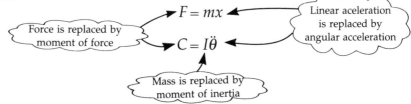

When using $F = m\ddot{x}$ for linear motion, it is important to take the positive direction of F in the direction of x increasing. In a similar way, moments can be clockwise or anticlockwise and when you use the equation $C = I\ddot{\theta}$, it is important to remember that the moments of the forces should be positive in the same sense as that of increasing θ.

It is also important to note that, although the moment of inertia of a body about an axis is a scalar quantity, its value depends on the position and direction of the axis and this should always be stated.

Wheels in machines

If you visit a place where old machines are conserved, you will see that they often have a large wheel as part of the driving mechanism. The next example demonstrates why these large wheels are useful.

EXAMPLE

Two wheels have the same mass M. One can be modelled by a large hollow cylinder of radius R and the other by a smaller solid cylinder of radius r. When they are rotating, they are each subject to a frictional couple of constant magnitude C. While the wheels are being driven, there is a break in power which lasts for a time t. Assuming all units are compatible, find expressions for
(i)   the angular retardation of each wheel while the power is off;
(ii) the reduction in the angular velocity of each wheel during this time.
The wheels have the same initial angular velocity, and the radius of the larger is twice that of the smaller.
(iii) Show that the percentage reduction in the angular velocity during the time the power is off, is eight times greater for the smaller wheel.
(iv) When the wheels are rotating with the same angular speed, show that the kinetic energy of the larger is eight times that of the smaller.

Solution

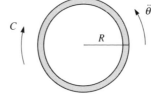

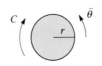

(i) When a wheel has an angular acceleration $\ddot{\theta}$ as a result of the action of a couple $-C$, the equation of motion gives

$$-C = I\ddot{\theta}$$

$$\Rightarrow \quad \ddot{\theta} = \frac{-C}{I}$$

The moment of inertia of the hollow cylinder about its axis is MR^2.

Its acceleration is $\dfrac{-C}{MR^2}$.

The moment of inertia of the solid cylinder about its axis is $\frac{1}{2}Mr^2$.

Its acceleration is $\dfrac{-2C}{Mr^2}$.

(ii) For constant angular acceleration α, the new angular velocity, ω, of a wheel after t seconds is

$$\omega = \omega_0 + \alpha t$$

So the reduction in the angular speed is

$$\omega_0 - \omega = \frac{Ct}{I} \quad \left(\alpha = \frac{-C}{I} \right)$$

The reductions in the angular speeds of the wheels are

$$\frac{Ct}{MR^2} \quad \text{and} \quad \frac{2Ct}{Mr^2}$$

(iii) When the initial angular speeds are the same, the percentage reductions in the angular speeds are proportional to the actual reductions.
These are in the ratio (larger : smaller) of

$$\frac{Ct}{MR^2} : \frac{2Ct}{Mr^2}$$

$$= r^2 : 2R^2$$

$$= r^2 : 2 \times 4r^2 \qquad (R = 2r)$$

$$= 1 : 8$$

The angular speed is reduced for both wheels, but the percentage reduction is 8 times greater for the smaller wheel.

(iv) When it rotates with angular speed ω, the kinetic energy of a wheel is $\frac{1}{2}I\omega^2$.
The ratio of the kinetic energy of the larger wheel to that of the smaller is

$$\tfrac{1}{2}(MR^2)\omega^2 : \tfrac{1}{2}(\tfrac{1}{2}Mr^2)\omega^2$$

$$= R^2 : \tfrac{1}{2}r^2$$

$$= 4r^2 : \tfrac{1}{2}r^2$$

$$= 8 : 1$$

The solution to part (iii) of the foregoing example shows that the use of the larger wheel in the driving mechanism leads to a smaller change in angular speed and so enables the machine to keep working at a steadier rate when there are fluctuations in the power. Such fluctuations are inevitable for many machines. An example is the engine of a car which incorporates a relatively large wheel, called a 'flywheel', to help smooth out the effects of the intermittent firing in the cylinders. The answer to part (iv) of the example shows, however, that the amount of energy required to set the larger wheel spinning is much greater than that required to make the smaller wheel spin with the same angular speed. This could be a disadvantage and it is also likely to take longer because the angular acceleration is less for the same torque. When the time required to set a wheel in motion is at a premium, as in the case of a racing car, the flywheel is lighter.

When there is surplus energy in a system, flywheels can be used to store energy. There is a bus design which incorporates a flywheel that is activated when braking takes place. When the bus stops some of its energy is stored in the flywheel and this can be used to boost the power of the engine when the bus starts again. Some toys are designed with flywheels to enable them to go on moving much further by themselves after a child stops pushing.

The work done by a couple or the moment of a force

When a rigid body rotates about its axis through a small angle $\delta\theta$, the point of application of a typical force F moves through an arc of length $r\delta\theta$, as shown in figure 5.10.

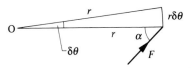

Figure 5.10

The work done by the force is then approximately

$$F\sin\alpha \times r\delta\theta = (F \times r\sin\alpha)\delta\theta$$

This can also be written as

$$(\text{moment about axis}) \times \delta\theta$$

The total work done by this force when the body rotates through an angle θ is thus

$$\sum \text{moment} \times \delta\theta$$

In the limit as $\delta\theta \to 0$, this becomes $\int (\text{moment})\, d\theta$.

Summing for all the forces, the work done by the total moment, C, is

$$\int C\, d\theta$$

This again demonstrates the equivalence between the equations for rotation and those for linear motion. It is comparable to $\int F\, dx$ with F replaced by C and x by θ.

When the equation $C = I\ddot{\theta}$ is integrated with respect to θ it gives

$$\int C\,d\theta = \int I\ddot{\theta}\,d\theta$$

$$= I\int \ddot{\theta}\frac{d\theta}{dt}\,dt$$

$$= I\int \ddot{\theta}\dot{\theta}\,dt$$

But
$$\ddot{\theta}\dot{\theta} = \dot{\theta}\ddot{\theta} = \dot{\theta}\frac{d\dot{\theta}}{dt} = \frac{d}{dt}\left(\tfrac{1}{2}\dot{\theta}^2\right)$$

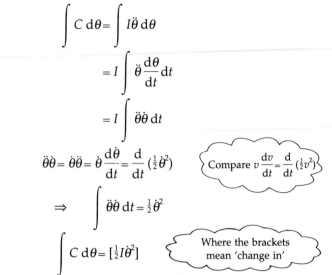

Compare $v\dfrac{dv}{dt} = \dfrac{d}{dt}\left(\tfrac{1}{2}v^2\right)$

$$\Rightarrow \int \ddot{\theta}\dot{\theta}\,dt = \tfrac{1}{2}\dot{\theta}^2$$

and
$$\int C\,d\theta = \left[\tfrac{1}{2}I\dot{\theta}^2\right]$$

Where the brackets mean 'change in'

This is the work–energy equation for a body rotating about a fixed axis.

NOTE

This equation has been obtained by integrating the equation of motion. Conversely, the energy equation can be differentiated with respect to t to give the equation of motion.

Conservation of energy

The next example illustrates how the use of $\int C\,d\theta$ to find the work done by gravity leads to the expression for the loss in potential energy of the rotating body. The work–energy equation then becomes the equation for the conservation of mechanical energy of the body.

EXAMPLE

A rigid body of mass M is free to rotate about a horizontal axis through a point A at a distance l from its centre of mass G. Its moment of inertia about the axis is I. The body is displaced through an angle α and let go.

(i) Find, by using $\int C\,d\theta$, the work done by gravity when AG falls into the vertical position and show that this is equal to the loss in potential energy of a particle with the same mass which falls through the same height as G.

(ii) Find an expression for the angular speed when AG is vertical.

Solution

The weight is the force which makes the body rotate about the axis. When it is in the position shown, the moment of the weight about the axis in the direction of increasing θ is $-Mgl\sin\theta$.

The work done by gravity when θ decreases from α to 0 is therefore

$$\int_\alpha^0 -Mgl\sin\theta\,d\theta = [Mgl\cos\theta]_\alpha^0$$

$$= Mgl\,(1-\cos\alpha)$$

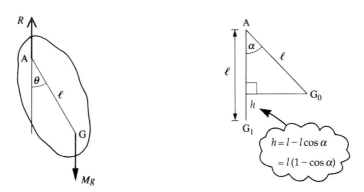

$$h = l - l\cos\alpha$$
$$= l\,(1-\cos\alpha)$$

But the height fallen by G is $h = l\,(1-\cos\alpha)$. Hence the work done by gravity is equal to Mgh, the loss in potential energy of a particle of mass M which has fallen the same height as G.

(ii) By the work–energy equation, or the principle of conservation of energy, the kinetic energy gained when AG is vertical is given by

$$\tfrac{1}{2}I\dot\theta^2 = Mgh$$

The angular speed at this point is then

$$\sqrt{\left(\frac{2Mgh}{I}\right)} = \sqrt{\left[\frac{2Mgl}{I}(1-\cos\alpha)\right]}$$

The assumption that the potential energy of a large body is the same as that of a particle of equal mass situated at G follows from the definition of G. If Oy is the co-ordinate axis in the vertical direction, then

$$M\bar y = \sum (m_p y_p) \quad \Rightarrow \quad Mg\bar y = \sum (m_p g y_p)$$

A winch is a useful device for applying a force to a moving object. For example, winches are used for towing gliders into the air and as lifting devices on boats. The photograph on the next page shows a cylinder with a rope wrapped round it, which is used to raise bags of flour in an old working mill.

The well bucket in the next example is raised and lowered using a similar device.

EXAMPLE

A bucket of mass m for drawing water from a well is attached by a light rope to a cylinder of mass M and radius r. The rope is wound round the cylinder using a light handle and the bucket is then allowed to fall freely from rest. What is its speed when it has fallen a height h
(i)   when there is no resistance to its motion?
(ii) when there is a constant resistive couple of magnitude C?

Solution

(i)   Consider the energy of the cylinder and the bucket and let the zero level of potential energy be the initial position of the bucket. The tension T in the rope does no work because it is an internal force.

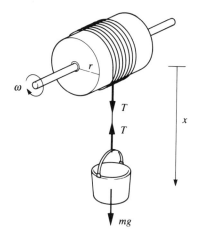

Suppose the speed of the bucket is v when the angular speed of the cylinder is ω. While the cylinder rotates through an angle θ, the bucket falls a height x where

$$x = r\theta$$

Differentiating $\quad \Rightarrow \quad v = r\omega$

After falling a height h from rest, the total energy of the bucket is $\frac{1}{2}mv^2 - mgh$.

The cylinder also has a kinetic energy due to its rotation of

$$\tfrac{1}{2}I\omega^2 = \tfrac{1}{2}(\tfrac{1}{2}Mr^2)\omega^2$$
$$= \tfrac{1}{4}Mr^2\omega^2$$

So the total energy of the bucket and cylinder when the bucket has fallen a height h is

$$\tfrac{1}{4}Mv^2 + \tfrac{1}{2}mv^2 - mgh$$

The initial energy is zero, so by the principle of conservation of energy

$$\tfrac{1}{4}Mv^2 + \tfrac{1}{2}mv^2 - mgh = 0$$

$$\Rightarrow \quad \tfrac{1}{4}(M + 2m)v^2 = mgh$$

$$\Rightarrow \quad v^2 = \frac{4mgh}{(2m + M)}$$

$$\Rightarrow \quad v = \sqrt{\left(\frac{4mgh}{2m + M}\right)}$$

(ii) When there is a resistive couple C, the work done against the couple is

$$\int C \, d\theta = C\theta$$

$$= \frac{Ch}{r} \quad (x = h = r\theta)$$

Then the gain in kinetic energy is the difference between the work done by gravity and the work done against the resistance

$$\Rightarrow \quad \tfrac{1}{2}mv^2 + \tfrac{1}{4}Mv^2 = mgh - \frac{Ch}{r}$$

$$\Rightarrow \quad \tfrac{1}{4}(2m + M)v^2 = \frac{(mgrh - Ch)}{r}$$

$$\Rightarrow \quad v^2 = \frac{4(mgr - C)h}{(2m + M)r}$$

$$v = \sqrt{\left[\frac{4(mgr - C)h}{(2m + M)r}\right]}$$

Investigation

The rope on the winch illustrated on page 146 probably has a density of the same order as the wooden cylinder and is clearly not of negligible thickness. Given that the diameter of the winch is about 0.3 m and the density of wood is about 600 kg m^{-3}, use measurements on the picture to estimate the velocity acquired by a bag of flour of mass 8 kg if it falls through a height of 10 m while attached to the end of the rope. You might decide to ignore resistances to motion, but do not ignore the rope.

The next example illustrates the use of the work–energy principle when the moment of a force about the axis is not constant.

EXAMPLE

A shed door of mass 30 kg, width 0.76 m and height 1.8 m is standing open at 90° when a gust of wind hits it. The wind is initially perpendicular to the door and does not change its direction as the door shuts. The resultant force on the door due to the wind acts through its centre and is of magnitude 10 N per m² of door 'facing' the wind. A constant frictional couple of 2 N m opposes the motion of the door.

(i) Assuming the door is a lamina, find its moment of inertia about its axis of rotation.

(ii) Find the moment of the force of the wind about the hinges when the door is open at an angle θ.

(iii) Use the work–energy principle to find the angular speed of the door when it shuts.

Solution

(i) The moment of inertia is given by $\frac{4}{3}Ml^2$. In this case

$$2l = 0.76$$

$$\Rightarrow \quad l = 0.38$$

$$\Rightarrow \quad \text{moment of inertia} = \frac{4}{3} \times 30 \times (0.38)^2 \text{ kg m}^2$$

$$= 5.776 \text{ kg m}^2$$

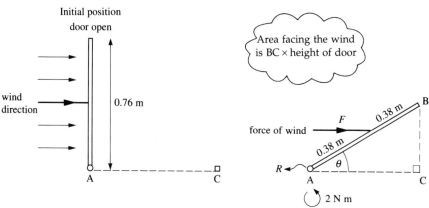

(ii) When the door is open at an angle θ, the area facing the wind is

$$0.76 \sin \theta \times 1.8 = 1.368 \sin \theta \text{ m}^2$$

The force due to the wind is then $13.68 \sin \theta$ N. It acts through the centre, so its moment about the axis of rotation is

$$13.68 \sin \theta \times 0.38 \sin \theta = 5.1984 \sin^2 \theta \text{ Nm}$$

(iii) There is a frictional couple of 2 N m resisting the motion, so the resultant moment in the direction of increasing θ is

$$-(5.1984 \sin^2 \theta - 2) \text{ N m}$$

The work in joules done in turning the door from $\theta = \dfrac{\pi}{2}$ radians to $\theta = 0$ is

$$\int_{\pi/2}^{0} -(5.1984 \sin^2 \theta - 2) \, d\theta = \int_{0}^{\pi/2} +(5.1984 \sin^2 \theta - 2) \, d\theta$$

$$= \int_0^{\pi/2} 5.1984 \times \tfrac{1}{2}(1 - \cos 2\theta) - 2 \, d\theta$$

$$= [2.5992\theta - 1.2996 \sin 2\theta - 2\theta]_0^{\pi/2}$$

$$= 0.5992 \times \tfrac{\pi}{2} = 0.9412$$

By the work–energy principle, this is equal to the gain in kinetic energy of the door, $\tfrac{1}{2}I\dot\theta^2$.

From part (i) the moment of inertia about the axis of rotation is 5.776 kg m².

$$\Rightarrow \quad 0.9412 = \tfrac{1}{2} \times 5.776 \times \dot\theta^2$$

$$\Rightarrow \quad \dot\theta^2 = 0.3259$$

The angular speed is 0.571 rad s^{-1} (to 3 significant figures)

Exercise 5B

1. The flywheel of a car can be modelled as a disc of diameter 0.2 m and thickness 0.02 m from which another concentric disc of diameter 0.07 m has been removed. The density of the wheel is 7800 kg m^{-3}.
  (i) Find its moment of inertia about its axis.
  The starter motor of the engine of the car makes the flywheel rotate at 50 rev s^{-1} in 3 s starting from rest. Find
  (ii) its angular acceleration (assumed constant);
  (iii) the average couple required to accelerate the flywheel;
  (iv) the kinetic energy of the flywheel after 3 s.

2. A garden gate can be modelled as a rectangular lamina of width 0.9 m and mass 20 kg. It is kept shut by a spring mechanism which applies a couple equal to $4(1 + \theta)$ N m when the gate is opened through an angle of θ radians. The gate is opened 1.5 radians and then allowed to shut naturally.
  (i) Find the moment of inertia of the gate about its hinges.
  (ii) Find the work done in opening the gate 1.5 radians.
  (iii) Assuming there is no loss of energy due to friction, find the angular speed of the gate just before it shuts.

3. A microwave oven has a turntable which can be modelled as a disc of mass 0.9 kg and radius 0.16 m rotating at $\tfrac{\pi}{6}$ rad s^{-1}.
  (i) The turntable has an angular retardation of 0.4 rad s^{-2}. Find the magnitude of the frictional couple acting.
  A cylindrical cake of mass 0.6 kg and radius 0.9 m is placed centrally on the turntable and cooked in a light plastic container for 6 minutes at a power of 650 W.
  (ii) Assuming the same constant frictional couple, find:
    A: the kinetic energy, A, given to the cake and turntable;
    B: the work done B, in keeping the turntable rotating at $\tfrac{\pi}{6}$ rad s^{-1};
    C: the energy, C, required to cook the cake.
  (iii) What is the ratio of the energies A:B:C?

4. The front wheel of a bicycle is rotating freely at 20 rad s^{-1} when the brakes are applied. The wheel, which can be modelled as a hoop of mass 2 kg and radius 0.3 m, is brought to rest in 0.08 s by two brake blocks applied to its rim. Find
  (i) the angular deceleration of the wheel (assumed constant);
  (ii) the angle turned through before coming to rest;

(iii) the total frictional force applied by the brake blocks;

(iv) the total contact force between the blocks and the wheel, given that the coefficient of friction is 0.9.

When the wheel is wet the coefficient of friction is reduced to 0.3. Assuming the same contact force, find the angle turned through while the wheel comes to rest in this case.

5. Chris wishes to find the resistance to the motion of a bicycle and decides to start by investigating the resistance to the front wheel. When an object of 50 g mass is placed on the valve it turns from rest in position A through 160° before coming to rest at B. OA is horizontal.

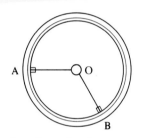

(i) Assuming the valve has a mass of 10 g and is 28.5 cm from the centre of the wheel, find the loss in potential energy between A and B.

(ii) Show that the resistive couple is about 0.0205 N m.

When Chris flicks the wheel into motion (without the object), it comes to rest in 41 s after making 28 revolutions.

(iii) Find the angular retardation (assumed constant) of the wheel, and its moment of inertia about its axle given the same resistive couple.

(iv) The wheel has a mass of 2 kg and its radius is 30 cm. Find the moment of inertia of a hoop with these measurements.

Comment on these results.

6. A uniform rectangular shop sign ABCD has mass 1.8 kg and sides AB = 0.4 m and BC = 0.3 m. The sign is hung along the edge AB,

which is horizontal, and is free to rotate about this edge.

(i) Calculate the moment of inertia of the sign about AB.

(ii) Assuming that there are no forces acting other than gravity, and that $g = 10$ m s^2, show that, when the sign makes a small angle θ with the vertical, $\ddot{\theta} \approx -50\theta$.

(iii) Hence find the period of small oscillations of the shop sign.

7. A toy engine of total mass 75 g contains a uniform flywheel of mass 15 g and radius 1.2 cm. The wheels of the engine are of radius 1 cm and they are attached to the flywheel with gears which ensure that the flywheel rotates 20 times as fast as they do. The engine is pushed 4 cm along the floor with a force of 20 N and let go.

(i) Find the angular speed of the flywheel when the speed of the engine is v m s^{-1}.

(ii) Assuming that the kinetic energy of the toy consists only of the rotational energy of the flywheel and the linear energy of the whole engine, find the speed of the engine when it is released.

8. A pulley wheel is a disc of mass M and radius a, and can turn freely about a horizontal axis through its centre. Particles of mass M and $\frac{1}{2}M$ hang vertically over the pulley at the ends of a rough, light string. When the system is set in motion, the string does not slip over the pulley.

(i) By considering the equations of motion of the pulley and the two particles, show that the acceleration of the particles is $\frac{1}{4}g$.

(ii) Find the difference in tension between the two portions of the string.

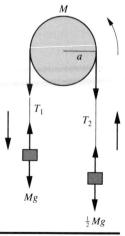

Further calculations of moments of inertia

Combining or dividing regular bodies

Once calculus methods have been used to find a few basic moments of inertia, it is possible to work out many others using the methods described below and the two theorems which follow.

You have already met the idea of elongating bodies in the direction of the axis. The elongated body is formed by combining identical elementary parts which all have the same moment of inertia about the axis. Conversely, when a body can be divided into two or more parts which obviously have the same moment of inertia about a given axis, it is possible to calculate the moments of inertia of the parts quite easily. For example, a sphere can be divided into two hemispheres using any plane through its centre. To each particle P_1 in one half there corresponds a similarly placed particle P_2 in the other half with the same moment of inertia about the axis. These are shown in figure 5.11.

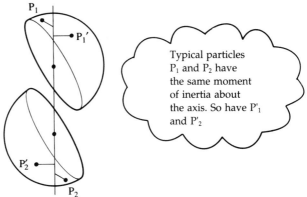

Typical particles P_1 and P_2 have the same moment of inertia about the axis. So have P'_1 and P'_2

Figure 5.11

The moment of inertia of each hemisphere about the axis is therefore

$$\tfrac{1}{2}(\tfrac{2}{5}Mr^2) = \tfrac{2}{5}\left(\frac{M}{2}\right)r^2$$

$$\Rightarrow \qquad = \tfrac{2}{5}M_1 r^2$$

where M_1 is the mass of each hemisphere. Again, the form of the expression does not change although the mass does.

The perpendicular axes theorem

You know the moment of inertia of a rectangular lamina about an axis of symmetry in its plane. The moments of inertia of the lamina in figure 5.12 are

$$I_y = \tfrac{1}{3}Ma^2 \quad \text{about the } y \text{ axis (AB has length } 2a)$$

and similarly

$$I_x = \tfrac{1}{3}Mb^2 \quad \text{about the } x \text{ axis (BC has length } 2b)$$

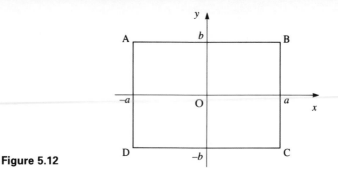

Figure 5.12

But what is its moment of inertia about the z axis through O perpendicular to its plane? This can be found by applying a very useful result known as the perpendicular axes theorem for a lamina, which can be stated as follows:

- The moment of inertia of a lamina about an axis which is perpendicular to the plane of the lamina and passes through a point O in its plane, is equal to the sum of the moments of inertia about two perpendicular axes in the plane of the lamina which also pass through O.

Using the notation above, $I_z = I_x + I_y$.

You will see from the proof which follows that it is essential for the body to be a lamina.

A particle P of mass m situated at the point (x, y) of any lamina in the xy plane is a distance r from O (and hence the z axis), where $r^2 = x^2 + y^2$ (figure 5.13).

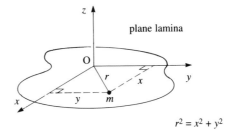

Figure 5.13

The moment of inertia of P about the z axis through O is therefore $m(x^2 + y^2)$. Hence the moment of inertia of the whole lamina about the z axis is

$$I_z = \sum m(x^2 + y^2)$$

$$= \sum mx^2 + \sum my^2$$

But $\sum mx^2$ is the moment of inertia, I_y, about the y axis and $\sum my^2$ is the moment of inertia, I_x, about the x axis. It follows that

$$I_z = I_x + I_y$$

This is true for any lamina in the xy plane.

Note that the origin and the x and y axes must always be in the plane of the lamina.

For the rectangle this gives $I_z = \frac{1}{3}Ma^2 + \frac{1}{3}Mb^2$
$$= \frac{1}{3}M(a^2 + b^2) \quad \text{or} \quad \frac{1}{3}Md^2$$

where d is the distance from the centre to a vertex of the rectangle.

EXAMPLE

Use the perpendicular axes theorem to find the moment of inertia of a thin circular disc of radius r about a diameter.

Solution

By symmetry the moment of inertia about every diameter is the same, say I_d. This means that, when the disc is in the xy plane with its centre at the origin,

$$I_x = I_y = I_d$$

But I_z is the moment of inertia about the axis through O perpendicular to the disc and this is $\frac{1}{2}Mr^2$. Hence

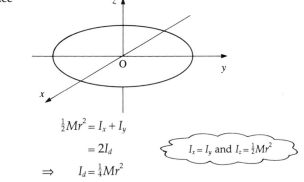

$$\frac{1}{2}Mr^2 = I_x + I_y$$

$$= 2I_d \qquad \boxed{I_x = I_y \text{ and } I_z = \frac{1}{2}Mr^2}$$

$$\Rightarrow \quad I_d = \frac{1}{4}Mr^2$$

The parallel axes theorem

In many circumstances, for example when a disc rotates about a diameter, or when two rigid bodies are combined to form another, the axis of rotation may not be in the most convenient position for the calculation of a moment of inertia. The parallel axes theorem can be used in these circumstances because it gives a relationship between the moments of inertia of the same rigid body about different parallel axes.
The theorem states that:

- The moment of inertia of a rigid body of mass M about a fixed axis through a point A is equal to its moment of inertia about a parallel axis through its centre of mass G plus Md^2, where d is the perpendicular distance between the axes:

$$I_A = I_G + Md^2$$

Figure 5.14 shows two parallel axes which are a fixed distance, d, apart. The axis GR passes through G and the other axis, AQ, passes through a point A such that AG is perpendicular to both axes and hence of length d. The moments of inertia of the body about these two axes are denoted by I_G and I_A. Consider a typical particle P of mass m_p and suppose that the plane through P perpendicular to the two axes meets GR at O and AQ at S as shown. Then $OS = GA = d$.

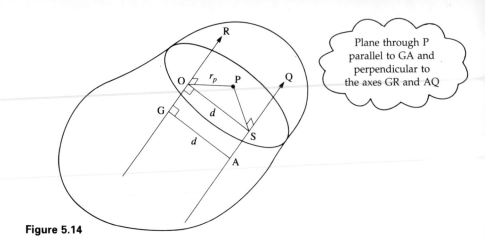

Figure 5.14

Figure 5.15 shows the plane through P. Co-ordinate axes have been taken through O parallel and perpendicular to OS. The co-ordinates of P relative to these axes are (x_p, y_p) and $r_p^2 = x_p^2 + y_p^2$ as usual.

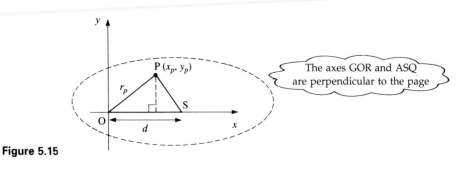

Figure 5.15

$$PS^2 = (d - x_p)^2 + y_p^2$$
$$= x_p^2 + y_p^2 + d^2 - 2dx_p$$
$$= r_p^2 + d^2 - 2dx_p$$

The moment of inertia of the particle P about the axis ASQ is

$$m_p PS^2 = m_p(r_p^2 + d^2 - 2dx_p)$$

The moment of inertia of the whole body about the axis AQ is

$$I_A = \sum m_p(r_p^2 + d^2 - 2dx_p) \quad \text{(summed over all particles P)}$$

$$= \sum m_p r_p^2 + \sum m_p d^2 - \sum 2dm_p x_p$$

$$= I_G + Md^2 - 2d \sum m_p x_p$$

remember, r_p is the distance of P from the axis GR

Now x_p is equal to the x co-ordinate of P relative to a new three-dimensional co-ordinate system with origin at G and GA as its x axis (figure 5.16).

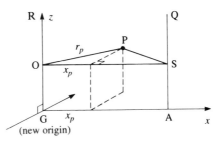

Figure 5.16

But, by the definition of the centre of mass, $\sum m_p x_p = M\bar{x}$ and $\bar{x}$ is zero because G is at the origin. So $2d \sum m_p x_p = 0$. Hence

$$I_A = I_G + Md^2$$

The parallel axes theorem can be used to extend your repertoire of moments of inertia, but remember that one of the axes must be through the centre of mass. A consequence of this theorem is that the moment of inertia, and hence also the rotational kinetic energy, are least when the axis is through G. However, the minimum kinetic energy possible for a given body and a given value of ω depends on the orientation of the axis.

EXAMPLE

Use the parallel axes theorem to find the moments of inertia of
(i)   a thin uniform rod of mass M and length h about a perpendicular axis through its end;
(ii) a thin uniform solid disc of mass M and radius r about
      (a) an axis perpendicular to its plane through a point on its circumference;
      (b) a tangent.
(iii) Which of the above is equally applicable to a solid cylinder?

Solution

(i)   For the rod $I_G = \frac{1}{3}M\left(\frac{h}{2}\right)^2$ and the axes are a distance $\frac{h}{2}$ apart. Hence

$$I_A = I_G + M\left(\frac{h}{2}\right)^2$$

$$= \frac{1}{3}M\left(\frac{h}{2}\right)^2 + M\left(\frac{h}{2}\right)^2$$

$$= \frac{1}{3}Mh^2$$

(ii) (a) In this case $I_G = \frac{1}{2}Mr^2$ and the axes are a distance r apart.

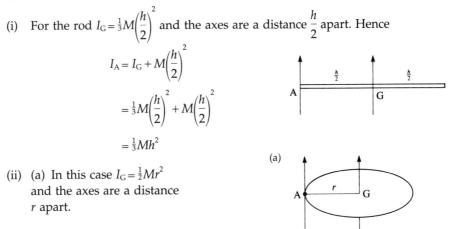

(a)

Hence
$$I_A = I_G + Mr^2$$
$$= \tfrac{1}{2}Mr^2 + Mr^2$$
$$= \tfrac{3}{2}Mr^2$$

(b) Now
$$I_G = \tfrac{1}{4}Mr^2$$
$$\Rightarrow \quad I_A = \tfrac{1}{4}Mr^2 + Mr^2$$
$$= \tfrac{5}{4}Mr^2$$

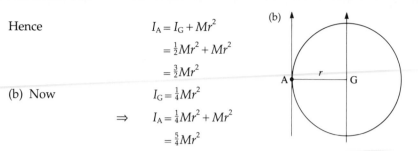

(iii) The result is applicable to a cylinder so long as this is formed by elongating the body in a direction parallel to the axes of rotation. It therefore applies in part (iia) but not in the other cases.

The next example illustrates how the parallel axes theorem can be used in conjunction with calculus methods to find the moment of inertia of a solid, in this case a cylinder, about an axis which is not an axis of symmetry.

EXAMPLE

Find the moment of inertia of a uniform solid cylinder of mass M, radius r and height h about an axis which is perpendicular to the axis of the cylinder and which passes through the centre of one end.

Solution

Choose axes as shown in the diagram so that the required moment of inertia is about the x axis. Let ρ be the density of the cylinder, so that $M = \pi r^2 h \rho$. Subdivide the cylinder into elementary discs as shown.

A typical disc has thickness δy and centre at $C(0, y)$.
Its mass δm is approximately $\rho \pi r^2 \delta y$.

Mass of disc $\delta m = \rho \pi r^2 \delta y$

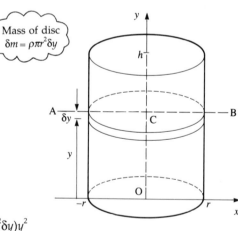

The moment of inertia of the element about an axis through C parallel to the x axis (i.e. a diameter) is $\tfrac{1}{4}\delta mr^2$.

Moment of inertia of disc about $AB = \tfrac{1}{4}\delta mr^2$

C is the centre of mass of the disc, so by the parallel axes theorem, its moment of inertia about the x axis is

$$\tfrac{1}{4}\delta mr^2 + \delta my^2 = \tfrac{1}{4}(\rho \pi r^2 \delta y)r^2 + (\rho \pi r^2 \delta y)y^2$$
$$= \tfrac{1}{4}\rho \pi r^2(r^2 + 4y^2)\delta y$$

For the whole cylinder, the moment of inertia about the x axis, I_x is approximately

$$\sum \tfrac{1}{4}\rho \pi r^2(r^2 + 4y^2)\delta y = \tfrac{1}{4}\rho \pi r^2 \sum (r^2 + 4y^2)\delta y$$

In the limit as $\delta y \to 0$, this gives

$$I_x = \tfrac{1}{4}\rho \pi r^2 \int_0^h (r^2 + 4y^2)\,dy$$

$$= \tfrac{1}{4}\rho\pi r^2 \left[r^2 y + 4\frac{y^3}{3} \right]_0^h$$

$$= \tfrac{1}{4}\rho\pi r^4 h + \tfrac{1}{3}\rho\pi r^2 h^3$$

$$= \tfrac{1}{4}Mr^2 + \tfrac{1}{3}Mh^2 \quad (M = \rho\pi r^2 h)$$

It is interesting to note that this has the same form as the sum of the moments of inertia of a disc of radius r about its diameter and a rod of length h about a perpendicular axis through its end: $h = 0$ gives the disc and $r = 0$ gives the rod.

Summary

- $I = \sum mr^2$
- Perpendicular axes theorem: $I_z = I_x + I_y$ for a lamina only
- Parallel axes theorem: $I_A = I_G + Md^2$

$$= I_G + M(AG)^2$$

when the centre of mass and AG is perpendicular to the axes

Exercise 5C

1. The model shown in the diagram is made up of light rods of length a and beads of mass m.

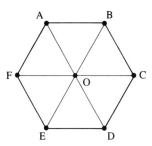

(i) Calculate in terms of m and a its moment of inertia about the axis:
   (a) through O perpendicular to the plane ABCDEF;
   (b) AOD;
   (c) through O and the mid-point of BC.
(ii) Verify that the perpendicular axes theorem is true for this model.
(iii) Use the results of part (i) and the parallel axes theorem to find the moment of inertia about
   (a) an axis through A perpendicular to the plane ABCDEF;
   (b) BC.

Check your answers by considering the moments of inertia for individual beads.

2. An odd number $(2n + 1)$ of beads each of mass m and negligible size are joined in a straight line by equal light rods of length a. The total mass of the model is M.
(i) Show that the moment of inertia I_M about an axis perpendicular to the line through the middle bead is $\dfrac{n}{3}(n+1)Ma^2$.

$$\left[\text{Hint: } \sum_{r=1}^{n} r^2 = \frac{n}{6}(n+1)(2n+1) \right]$$

(ii) Use the parallel axes theorem to show that the moment of inertia I_E about the end bead is given by $I_E = I_M + n^2 Ma^2$. Hence find I_E.
(iii) Verify that the answer to part (ii) is the same as the result obtained from first principles.
(iv) Write down the length of the line of beads. The number of rods now increases and the length of each rod decreases in such a way that the total length is always $2l$. Show that in the limit as $n \to \infty$ and $a \to 0$, the moments of inertia you found in parts (i) and (ii) approach the equivalent moments of inertia for a rod of mass M and length $2l$.

Exercise 5C continued

3. The following table can be completed by using the perpendicular and parallel axes theorems and the idea of elongating the body parallel to the axis. Copy and complete it. You might find it useful to keep a copy for future reference.

Uniform body of mass M	Axis	Moment of inertia
Hoop radius r	Through centre $\perp$ to plane	Mr^2
Hollow cylinder radius r	Through centre $\perp$ to circular cross-section	
Thin ring radius r	Diameter	
Thin ring radius r	Through edge $\perp$ to plane	
Thin ring radius r	Tangent	
Disc of radius r	Through centre $\perp$ to plane	$\frac{1}{2}Mr^2$
Solid cylinder radius r	Through centre $\perp$ to circular cross-section	
Thin disc radius r	Diameter	
Thin disc radius r	Through edge $\perp$ to disc	
Thin disc radius r	Tangent	
Thin rod	Parallel to rod at distance d	Md^2
Thin rod of length $2l$	Through centre $\perp$ to rod	$\frac{1}{3}Ml^2$
Thin rod of length $2l$	Through end $\perp$ to rod	
Rectangular lamina	In plane, through centre $\perp$ to sides length $2l$	$\frac{1}{3}Ml^2$
Rectangular lamina	Edge $\perp$ to sides length $2l$	
Rectangular lamina sides $2a$ and $2b$	Through centre to plane	
Rectangular block sides $2a$, $2b$ and $2c$	Through centre parallel to sides of length $2c$	
Solid sphere radius r	Diameter	$\frac{2}{5}Mr^2$
Solid sphere radius r	Tangent	
Hollow sphere radius r	Diameter	$\frac{2}{3}Mr^2$
Hollow sphere radius r	Tangent	

4.

Sectors	(a)	(b)

Write down the moment of inertia of a thin disc of mass M and radius r about an axis through its centre perpendicular to its plane.

The disc is divided into equal sectors as shown in the diagrams on page 158.

(i) By using ideas of symmetry, find the moments of inertia of the shaded sectors about axes through O perpendicular to their planes
   (a) in terms of M;
   (b) in terms of the mass m of the particular sector.
(ii) Write down the moment of inertia of the disc about its diameter AB. Symmetry can still be used to find the moments of inertia of some of the shaded sectors about OA. Write down the moments of inertia of those that are possible and explain why the others cannot be found in this way.

5. The surface of a uniform hollow sphere of mass M and radius r can be represented in three dimensions by the equation:
$x^2 + y^2 + z^2 = r^2$.

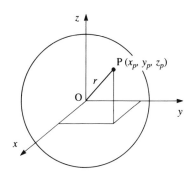

(i) Write down the distance from the z axis of a typical particle of mass m_p situated at the point (x_p, y_p, z_p) and hence show that the moment of inertia of the whole sphere about the z axis is given by $I_z = \sum m_p(x_p^2 + y_p^2)$.
(ii) Write down similar equations for the moments of inertia, I_x and I_y, of the sphere about the x axis and the y axis.
(iii) The moment of inertia of the sphere about any diameter is I.

Explain why $3I = I_x + I_y + I_z$ and hence show that $I = \frac{2}{3}Mr^2$, as given in question 3.

6. A uniform thin rod AB of mass M and length $2a$ is free to rotate about an axis through its centre O which is inclined at an angle α to the rod AB as shown.

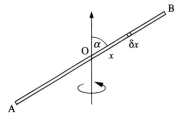

An element of the rod of length δx is situated a distance x from O. ρ is the mass per unit length of the rod.
(i) Write down the mass of the element.
(ii) Write down the moment of inertia of the element about the axis.
(iii) Form a suitable integral to calculate the moment of inertia of the rod about the axis and evaluate it.
(iv) Write M in terms of a and ρ and hence show that the moment of inertia of the rod about the axis is the same as that of a rod of the same mass and of length $2a \sin \alpha$ which is perpendicular to the axis. Check your answer in the cases $\alpha = \frac{\pi}{2}$ and $\alpha = 0$.
(v) Use your result to find the moment of inertia of a parallelogram about an axis in its plane which passes through its centre and bisects a side of length $2a$, as shown in the diagram.

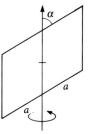

(vi) How will the moment of inertia of any lamina be affected by a shearing in the direction of the axis of rotation as shown in the diagram?

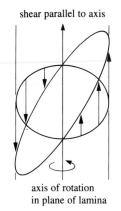

shear parallel to axis

axis of rotation
in plane of lamina

7. A lamina of mass M in the form of an isosceles triangle OAB is shown in the diagram. $OA = OB$, $AB = 2a$ and the height $OC = h$. The lamina is divided into elementary strips of width δx parallel to AB as shown.

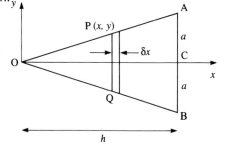

(i) Write down the mass of a typical strip PQ of width δx in terms of σ, the mass per unit area of the lamina, and the co-ordinates (x, y) of P.

(ii) By treating the strip as a thin rod, write down its moment of inertia about the x axis.

(iii) Write y in terms of x and form an integral to find the moment of inertia of the triangle about the x axis.

(iv) Find the moment of inertia about the x axis in terms of M and a.

(v) Give a reason why the moment of inertia of the same elementary strip about the y axis is $2x^2\sigma y\delta x$.

(vi) Find the moment of inertia of the triangle about the y axis.

(vii) Use the perpendicular axes theorem to find the moment of inertia of the

triangle about an axis through O perpendicular to its plane.

(viii) Use the parallel axes theorem to find the moment of inertia of the triangle about an axis perpendicular to its plane through its centre of mass.

8. (i) The arc of the circle AB shown in the

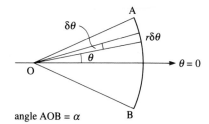

angle AOB = α

diagram above is of mass m and is symmetrical about the initial line $\theta = 0$. It is divided into elements of length $r\delta\theta$. Show by integration that its moment of inertia about the axis $\theta = 0$ is

$$\tfrac{1}{2}mr^2\left(1 - \frac{\sin \alpha}{\alpha}\right)$$

(ii) The sector of the circle shown in the diagram below is of mass M and radius and it is symmetrical about the initial line $\theta = 0$. It is divided into elementary arcs of radius r and width δr. Use the result of part (i) to show that its moment of inertia about the axis $\theta = 0$ is

$$\tfrac{1}{4}Ma^2\left(1 - \frac{\sin \alpha}{\alpha}\right).$$

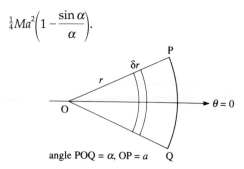

angle POQ = α, OP = a

9. A uniform solid cone of mass M, base radius r and height h is placed with its vertex at O and its axis of symmetry along the x axis, as shown. A typical elementary disc of thickness δx is at the point P(x, y).

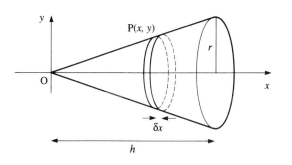

and base radius r about an axis through its vertex perpendicular to its axis of symmetry.

13. A point O on a plane lamina is used as the origin in a system of co-ordinates and the lamina is in the xy plane. A typical particle, P, of the lamina has mass m and co-ordinates $(x, y, 0)$. I_z and I_x are the moments of inertia of the lamina about the z axis and the x axis respectively.

Using ρ for the density of the disc:
(i) Write down the approximate mass of the elementary disc in terms of y, ρ and δx.
(ii) Show that the moment of inertia of the disc about the x axis is approximately $\frac{1}{2}\pi\rho y^4 \delta x$.
(iii) Form a suitable integral to find the moment of inertia of the cone about the x axis.
(iv) Write y in terms of x and evaluate the integral.
(v) Write the mass of the cone in terms of r, h and ρ and hence find the moment of inertia of the cone about the x axis.

10. The region between the parabola $y^2 = 4ax$, the x axis and the line $x = b$ is rotated completely about the x axis.
(i) Find the volume of revolution so formed and hence write its mass M in terms of its density ρ.
(ii) Use the method outlined in Question 9 to calculate its moment of inertia about the x axis in terms of M and a.

11. The ellipse $\dfrac{x^2}{a^2} + \dfrac{y^2}{b^2} = 1$ is rotated completely about the x axis, forming a solid of revolution of mass M.
(i) Find the moment of inertia of the solid about the x axis in terms of M.
(ii) Write down a similar expression for the moment of inertia of the solid formed when the ellipse is rotated about the y axis.
Comment on your results.

12. Find by integration the moment of inertia of a uniform solid cone of mass M, height h

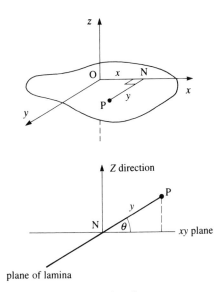

(i) Explain why $I_z = \sum m(x^2 + y^2)$ summed over all particles and write I_x in terms of another sum.
(ii) The lamina is now turned through an angle θ about the x axis. By considering the new moment of inertia of the typical particle, P, about the z axis, show that the moment of inertia about the z axis of the lamina in its new position is given by

$$\sum m(x^2 + y^2 \cos^2 \theta)$$

and so is equal to $I_z - I_x \sin^2\theta$.

(iii) A helicopter blade of mass m is modelled as a rectangle of length a and width b. One long edge, AB, is attached to the axle of the rotating mechanism and is a distance c from the axle. The blade can be turned about AB during flight. Find expressions for the moment

of inertia of the blade about the axle when AB is horizontal and the blade is (a) horizontal and (b) inclined to the horizontal at an angle θ.

14. A uniform hollow sphere of mass M, has internal radius a, external radius b, and density ρ.
    (i) Find, in terms of ρ, a and b, expressions for the masses of solid spheres of radius a and b and their moments of inertia about a diameter.
    (ii) Show that the moment of inertia of the hollow sphere about an axis through its centre is $\dfrac{8\pi\rho}{15}(b^5 - a^5)$.
    (iii) The table shows a model of the earth with different densities at different depths.

Distance from centre in km	Density in kg m^{-3}
< 1250	13 000
1250–	10 000
3500–	5500
5700–	4000
6200–6400	2900

Assuming the Earth is a sphere, use this model to find its moment of inertia about its axis.
    (iv) The mean density of the Earth is 5500 kg m^{-3}. Show that when it is assumed to be a uniform sphere, the moment of inertia of the Earth about its axis is overestimated by about 12% compared with the above model.

15. (i) Use the result of Question 14(ii) to show that the moment of inertia about a diameter of a hollow sphere of mass M, internal radius a and external radius b is

$$\frac{2M(b^5 - a^5)}{5(b^3 - a^3)}$$

    (ii) Verify that
$$b^5 - a^5 = (b - a)(b^4 + ab^3 + a^2b^2 + a^3b + a^4)$$
and by factorising $(b^3 - a^3)$ in a similar way, find an alternative expression for the moment of inertia of the hollow sphere about its diameter.
    (iii) Now let $a \to b$ and hence show that the moment of inertia about a diameter of a thin hollow sphere of radius b is $\frac{2}{3}Mb^2$.

The compound pendulum and the radius of gyration

A compound pendulum is any rigid body which oscillates freely about a fixed horizontal axis. No external forces or couples act on the pendulum apart from its weight and a supporting force at the axis. The next example shows that such a pendulum performs approximate simple harmonic motion when the oscillations are small.

Moments of inertia all have dimensions $[ML^2]$, so when the moment of inertia of an unspecified body of mass M is required it is often denoted by Mk^2. The quantity, k, has the dimensions of length and is called the *radius of gyration* of the body about the axis concerned. For example, the radius of gyration of a uniform solid sphere of radius r about an axis through its centre is obtained from

$$Mk^2 = \tfrac{2}{5}Mr^2$$

$$\Rightarrow \quad k = \sqrt{\left(\frac{2}{5}\right)}r = \frac{\sqrt{10}}{5}r$$

EXAMPLE

A compound pendulum, of mass M and centre of mass G, is free to rotate about a fixed horizontal axis through a point A. AG is perpendicular to the axis of rotation and of length h. The moments of inertia of the pendulum about the fixed axis and a parallel axis through G are I_A and I_G respectively. Find expressions for

(i)   the period of small oscillations of the pendulum;
(ii) the length of a simple pendulum with the same period (called the *simple equivalent pendulum*);
(iii) the period in terms of the radius of gyration, k, of the pendulum about the axis through G.
When such a pendulum has a period of 2 seconds, it is called a *seconds pendulum*.
(iv) Show that, for a seconds pendulum, k cannot be greater than about $\tfrac{1}{2}$.
(v)   Find the value of h for which the period is a minimum.

Solution

(i)

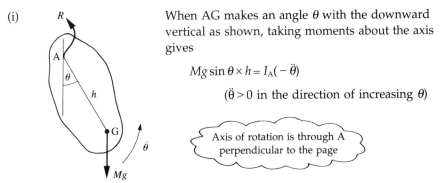

When AG makes an angle θ with the downward vertical as shown, taking moments about the axis gives

$$Mg \sin\theta \times h = I_A(-\ddot{\theta})$$

$$(\ddot{\theta} > 0 \text{ in the direction of increasing } \theta)$$

Axis of rotation is through A perpendicular to the page

where I_A is the moment of inertia about the fixed axis.

$$\Rightarrow \quad \ddot{\theta} = -\frac{Mgh}{I_A}\sin\theta$$

Hence, for small θ

$$\ddot{\theta} = -\frac{Mgh}{I_A}\theta$$

This represents simple harmonic motion with period

$$T = 2\pi\sqrt{\left(\frac{I_A}{Mgh}\right)} \qquad \text{①}$$

By the parallel axes theorem $I_A = I_G + Mh^2$. Therefore, the period for small oscillations is

$$T = 2\pi\sqrt{\left(\frac{I_G + Mh^2}{Mgh}\right)}$$

(ii) The period for small oscillations of a simple pendulum of length l is

$2\pi\sqrt{\left(\frac{l}{g}\right)}$ so the length of an equivalent simple pendulum is

$$l = \frac{I_G + Mh^2}{Mh} \quad \text{or} \quad l = \frac{I_A}{Mh} \quad \text{(from ①)}$$

(iii) The radius of gyration about the axis through G is k, so

$$I_G = Mk^2$$
$$\Rightarrow \quad I_A = Mk^2 + Mh^2$$

The period is then

$$T = 2\pi\sqrt{\left[\frac{M(k^2 + h^2)}{Mgh}\right]} \quad \text{(from ①)}$$

$$\Rightarrow \quad T = 2\pi\sqrt{\left(\frac{k^2 + h^2}{gh}\right)} \qquad \text{②}$$

(iv) For a given pendulum the length h, and hence the position of A, can usually be calculated so that it has the correct period. A seconds pendulum has a period of 2 s.

In this case, $T = 2$ and squaring both sides of ② gives

$$4 = 4\pi^2\frac{(k^2 + h^2)}{gh}$$

$$\Rightarrow \quad gh = \pi^2(k^2 + h^2)$$

$$\Rightarrow \quad h^2 - \left(\frac{g}{\pi^2}\right)h + k^2 = 0$$

This has real roots for h provided $\left(\frac{g}{\pi^2}\right)^2 \geqslant 4k^2$.

$$\Rightarrow \quad k \leqslant \frac{1}{2}\left(\frac{g}{\pi^2}\right)$$

But $\dfrac{g}{\pi^2} = 0.994 \approx 1$, so k cannot be greater than about $\frac{1}{2}$.

(v) Consider the function

$$f(h) = \frac{(k^2 + h^2)}{h} = \frac{k^2}{h} + h$$

The period is least when $f(h)$ is a minimum. Differentiating gives

$$f'(h) = -k^2 h^{-2} + 1$$

so the minimum occurs when $k^2 = h^2$, that is $h = k$.

The period is least when AG is equal to the radius of gyration about the axis through G. (You can check that $f''(k) > 0$ to ensure a *minimum*.)

Bodies with variable density

Some bodies are made of several materials of differing densities. The moments of inertia of these can be found by dealing with each part separately (for example, see Exercise 5C, question 10). Others have densities which vary continuously and, provided the functions are relatively simple, calculus methods can be used to determine their moments of inertia.

EXAMPLE

A star is modelled as a sphere of radius a whose density is given in terms of the distance r from the centre by the function

$$\rho = \rho_0 \left(1 - \frac{r^2}{a^2} \right)$$

(i)   Given that ρ_0 is constant, find the total mass M of the sphere.
(ii) Find its moment of inertia about a diameter.

Solution

Imagine that the sphere is divided into elementary shells and a typical shell has radius r and thickness δr.
(i)   The mass of the shell is

$$4\pi r^2 \rho \delta r = \frac{4\pi \rho_0}{a^2} (a^2 r^2 - r^4) \delta r \qquad \text{①}$$

$$\Rightarrow \quad M = \int_0^a \frac{4\pi \rho_0}{a^2} (a^2 r^2 - r^4)\, dr$$

$$= \frac{4\pi \rho_0}{a^2} \left[a^2 \frac{r^3}{3} - \frac{r^5}{5} \right]_0^a$$

$$= \frac{8\pi \rho_0}{15} a^3 \qquad \text{②}$$

(ii) The moment of inertia of a hollow shell of mass m about a diameter is $\frac{2}{3}mr^2$, so the moment of inertia of an elementary shell is

$$\frac{2}{3} \times \frac{4\pi\rho_0}{a^2}(a^2r^2 - r^4)\delta r \times r^2 \quad \text{(from ①)}$$

$\Rightarrow$ moment of inertia of sphere $= \displaystyle\int_0^a \frac{8\pi\rho_0}{3a^2}(a^2r^4 - r^6)\,dr$

$$= \frac{8\pi\rho_0}{3a^2}\left[a^2\frac{r^5}{5} - \frac{r^7}{7}\right]_0^a$$

$$= \frac{16\pi\rho_0}{105}a^5$$

$$= \frac{2}{5}Ma^2 \quad \text{(from ②)}$$

A baseball bat is thicker at one end than the other, but its thickness is always small compared with its length so it is possible to model it by a rod of varying density as illustrated in the next example.

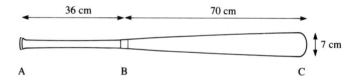

36 cm   70 cm   7 cm

A   B   C

EXAMPLE

A baseball bat has a uniform handle AB of length 0.36 m and mass $M_1 = 0.136$ kg. The remainder of the bat is modelled as a thin rod BC of length 0.7 m and mass per unit length $1.43\,(x^2 + x + 0.25)$ kg m^{-1}, where x is the distance from B. Assuming the handle can also be modelled as a thin rod, find
(i) the mass M kg of the bat;
(ii) the distance of its centre of mass, G, from B;
(iii) its moment of inertia about a perpendicular axis through B;
(iv) its moment of inertia about a perpendicular axis through the end A.

Solution

The diagram shows the essential features of the bat.

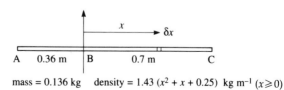

A   0.36 m   B   0.7 m   C

mass = 0.136 kg   density = $1.43\,(x^2 + x + 0.25)$ kg m^{-1} $(x \geqslant 0)$

(i) The mass of an element of BC of length δx is $1.43\,(x^2 + x + 0.25)\delta x$ kg.

$$\Rightarrow \quad \text{mass of BC} = \int_0^{0.7} 1.43 \ (x^2 + x + 0.25) \ dx \ \text{kg}$$

$$= \left[1.43 \left(\frac{x^3}{3} + \frac{x^2}{2} + 0.25x \right) \right]_0^{0.7} \text{kg}$$

$$= 0.764\,097 \ \text{kg}$$

$$\Rightarrow \quad \text{total mass } M = 0.764 + 0.136 = 0.900 \ \text{kg (to 3 significant figures)}$$

(ii) To find the distance, $\bar{x}$, of G from B, take moments about an axis through B. Then

$$\left(\sum \delta m \right) \bar{x} = \sum (\delta mx) \quad \text{(summed for all particles of the bat)}$$

For AB
$$\sum (\delta mx) = -M_1 \times \tfrac{1}{2}\text{AB}$$

$$= -0.136 \times 0.18$$

$$= -0.024\,48$$

and for BC
$$\sum (\delta mx) = \sum 1.43 \ (x^2 + x + 0.25)x\delta x$$

In the limit as $\delta x \to 0$, this becomes

$$\int_0^{0.7} 1.43 \ (x^2 + x + 0.25)x \ dx$$

$$= \int_0^{0.7} 1.43 \ (x^3 + x^2 + 0.25x) \ dx$$

$$= 1.43 \left[\frac{x^4}{4} + \frac{x^3}{3} + 0.25 \frac{x^2}{2} \right]_0^{0.7}$$

$$= 0.336\,92$$

Combining AB and BC gives

$$0.9\bar{x} = -0.024\,48 + 0.336\,92$$

$$\Rightarrow \quad 0.9\bar{x} = 0.312\,44$$

$$\bar{x} = 0.347\,15$$

The centre of mass is about 35 cm from B

(iii) The moment of inertia, I_1, of the uniform rod AB about a perpendicular axis through its centre is $\frac{1}{3}M_1(0.18)^2$. So, by the parallel axes theorem, its moment of inertia about the end B is

$$I_1 = \tfrac{1}{3}M_1(0.18)^2 + M_1(0.18)^2$$

$$= \tfrac{4}{3}(0.136)(0.18)^2$$

$$= 0.005\,875 \ \text{kg m}^2$$

The moment of inertia, I_2, of the part BC about the axis through B is

$$I_2 = \lim_{\delta x \to 0} \left(\sum \delta m x^2 \right)$$

$$= \int_0^{0.7} 1.43 \, (x^2 + x + 0.25)x^2 \, dx$$

$$= \int_0^{0.7} 1.43 \, (x^4 + x^3 + 0.25x^2) \, dx$$

$$= 1.43 \left[\frac{x^5}{5} + \frac{x^4}{4} + 0.25 \frac{x^3}{3} \right]_0^{0.7}$$

$$= 0.174\,78$$

hence: $\quad I_1 + I_2 = 0.180\,66$

The moment of inertia of the bat about a perpendicular axis through B is about 0.18 kg m².

(iv) By the parallel axes theorem

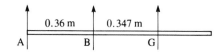

$$I_B = I_G + M(BG)^2$$

and $\quad I_A = I_G + M(AG)^2$

$\Rightarrow \quad I_A = I_B - M(BG)^2 + M(AG)^2$

$\Rightarrow \quad I_A = I_B + M(AG^2 - BG^2)$

$$= 0.180\,66 + 0.9 \, (0.379\,54)$$

$$= 0.180\,66 + 0.341\,593$$

$$= 0.522\,25$$

The moment of inertia about the end, A, is about 0.52 kg m².

NOTE

The parallel axes theorem
Part (iv) of the last example illustrates the necessity for one of the axes to be through the centre of mass, G, when using the parallel axes theorem.
In general, the moments of inertia about parallel axes through two points A and B where AG and BG are perpendicular to the axes are related by the equation

$$I_A = I_B + M(AG^2 - BG^2)$$

(Note also that $AG^2 - BG^2 \neq AB^2$). The moment of inertia of the bat about the player's elbow or shoulder (assuming the arm and wrist are kept rigid) can be calculated using this result.

1. The following objects have variable density. For each one, find in terms of the given constants (a, k, h, p, q, r)
   (i) its mass;
   (ii) the position of its centre of mass relative to O;
   (iii) its moment of inertia about the given axis.

Object	Density	Axis
Rod OA of length a	$2x$	Through O perpendicular to rod
Disc centre O, radius a	kr	Through O perpendicular to disc
Rod OA of length a	$p + qx$	Through O perpendicular to rod
Thin cylindrical pipe, height h, base centre O, radius r	$k(h + y)$	y axis

2. A thin uniform rod of mass M and length $2a$ performs approximate simple harmonic oscillations about a perpendicular axis through one end.
   (i) By taking moments about the axis, write down the equation of motion of the rod.
   (ii) Find the period of small oscillations of the rod.
   (iii) Find the length of a simple pendulum with the same period.

3. A clock pendulum consists of a uniform thin rod of length $2l$ and mass m, to the end of which is attached a thin disc of radius l and mass $2m$.
   (i) Find the moment of inertia of the pendulum about an axis perpendicular to the disc through the free end of the rod.
   (ii) By taking moments about the axis, find the equation of motion and hence the period of small oscillations of the pendulum.
   (iii) Find the value of l if this period is 2 seconds (as for a clock).

4. A solid circular cylinder of radius a and height h is formed by pouring a resin into a mould and allowing it to set. The resin settles unevenly so that the density of the resulting cylinder is 3ρ at the base and decreases uniformly to ρ at the top.
   (i) Write the density as a function of the distance from the base.
   (ii) Find the mass, M, of the cylinder.
   (iii) Find the position of the centre of mass of the cylinder.
   (iv) Find the moment of inertia of the cylinder about its axis of symmetry.

[MEI 1993]

5. (i) Prove, by integration, that the moment of inertia of a thin uniform rod of mass M and length $2a$ about an axis through the centre of the rod and perpendicular to the rod is $\frac{1}{3}Ma^2$.
   (ii) Three uniform rods AB, BC and AC have lengths $8a$, $6a$ and $10a$, respectively. Each rod has mass m per unit length. The rods are fastened together to make a triangular frame ABC. Show that the moment of inertia of the frame about a fixed smooth horizontal axis, passing through A and perpendicular to the frame, is $960ma^3$.
   (iii) The frame is released from rest with AB horizontal and C vertically above B. Show that when AB is vertical, the angular speed of the frame is $\sqrt{\left(\dfrac{7g}{20a}\right)}$.

(iv) State which point of the frame has the greatest speed at the instant when AB is vertical, and find this speed in terms of a and g.

[Cambridge 1992]

6. A uniform rod AB, of mass $2m$ and length $2l$, is rigidly attached to a uniform rod BC, of mass m and length l, in such a way that angle ABC is a right angle. Show that the moment of inertia of the frame ABC about an axis through B perpendicular to the plane of the frame is $3ml^2$.

The frame can rotate freely about a fixed, smooth, horizontal axis through B and perpendicular to the plane of the frame. The frame is released from rest in the position in which AB is horizontal and C is below B. At time t, AB makes an angle θ with the horizontal. Show that

$$l\dot{\theta}^2 = \tfrac{1}{3}g(4\sin\theta + \cos\theta - 1)$$

Find, in terms of g and l.
(i) the greatest value of $\dot{\theta}^2$;
(ii) the greatest speed of A.

[Cambridge 1993]

7. A uniform solid sphere of mass $5m$ and radius a is fixed to the end of a thin uniform rod of mass $3m$ and length $4a$ so that the centre of the sphere lies on the extended axis of the rod. The other end of the rod is freely pivoted at a point in such a way that the system can swing in a vertical plane under the influence of a vertical gravitational field g.
(i) Find the moment of inertia of the system about its pivot.
(ii) If the system is balanced vertically above the pivot and given a slight displacement, find its angular speed at its lowest point.

[MEI 1983]

8. (i) A uniform circular disc of radius a and thickness b has a mass m. Derive an expression for the moment of inertia of this disc about an axis through its centre and perpendicular to its plane.

(ii) A flywheel is to be made with a radius of 25 cm, a thickness of 5 cm and a moment of inertia about its axis of 1.8 kg m^2. The flywheel is to be constructed with a central disc made from an aluminium alloy, density 2800 kg m^{-3}, and an outer rim of steel, density 7900 kg m^{-3}. Find the radius at which the junction between the central alloy disc and the steel rim must occur.

[MEI 1983]

9. (i) Show that the moment of inertia of a uniform rod, of length $2a$ and mass m, about an axis through a point a distance h from its centre of gravity and perpendicular to the plane containing the rod and the point, is $\tfrac{1}{3}m(a^2 + 3h^2)$.

(ii) A uniform equilateral triangular lamina of side $2l$ has mass M. By considering the lamina as a series of thin rods, determine the moment of inertia of the lamina about an axis through a vertex and perpendicular to its plane.

(iii) (a) The triangle is enlarged by a factor k, with its mass per unit area remaining the same as before. Show that the moment of inertia of the new triangle about an axis through its vertex and perpendicular to its plane is $\tfrac{5}{3}Mk^4l^2$.

(b) Find the period of small oscillations about the same axis when this lamina is freely suspended, and show that this is proportional to $\sqrt{k}$.

[Oxford 1991]

10.

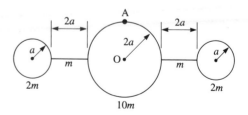

The diagram shows two small uniform hollow spheres of mass $2m$ and radius a and a larger hollow uniform sphere of mass $10m$ and radius $2a$. Each of the smaller spheres is joined to the larger one by a small solid uniform rod of mass m and length $2a$. The line containing the rods passes through the centres of the three spheres.

(i) Show that the moment of inertia of the system about an axis through the centre O of the larger sphere and perpendicular to the line containing the rods is $148ma^2$.

(ii) A wire is attached to the highest point A of the larger sphere and the system is suspended in equilibrium from this wire so that the vertical through O is along the wire. The wire is such that when the line containing the rods is turned through a small angle θ about the wire then a restoring couple of magnitude $mk\theta$, where k is a positive constant, is exerted on the system. Find the period of small oscillations of the system about the vertical axis containing the wire.

[Oxford 1993]

11. A rough uniform rod AB, of mass m and length $2r$, can rotate freely in a vertical plane about a fixed smooth horizontal axis through the point C of the rod, where $AC = \frac{1}{2}r$. A small bead of mass $3m$ can slide on the rod. Initially the rod is held in a horizontal position with the bead at the end A of the rod. The rod is released from rest and after a time t the rod makes an angle θ with the horizontal. For the part of the motion in which the bead remains at rest relative to the rod,

(i) show that $\dot{\theta}^2 = \dfrac{3g \sin \theta}{2r}$;

(ii) show that the magnitude of the frictional component of the force of the rod on the bead is $\frac{21}{4} mg \sin \theta$;

(iii) differentiate the expression for $\dot{\theta}^2$ with respect to t and hence, or otherwise, show that $\ddot{\theta} = \dfrac{3g}{4r} \cos \theta$;

(iv) find the magnitude of the normal component of the force of the rod on the bead.

(v) The coefficient of friction between the bead and the rod is $\frac{4}{5}$. Find, to the nearest degree, the value of θ at the instant when the bead is about to slide off the rod.

[Cambridge 1993 adapted]

12. A thin uniform rod, of mass m and length $6a$, can rotate freely in a vertical plane about a fixed horizontal axis through one end A of the rod. A uniform circular disc, of mass $12m$ and radius a, is clamped to the rod so that its centre C lies on the rod and its plane coincides with the plane in which the rod can rotate. Given that $AC = x$, find, in terms of m, a and x, expressions for

(i) the distance of the centre of mass of the system from A;

(ii) the moment of inertia of the system about the axis of rotation.

(iii) Show that T, the period of small oscillations of the system, is given by

$$T^2 g(a + 4x) = 8\pi^2 (3a^2 + 2x^2)$$

(iv) Hence show that the minimum value of T occurs when $x = a$.

[AEB 1994]

Investigation

How long does a clock pendulum need to be? (Or why are grandfather clocks too tall for the shelf?)

A pendulum which oscillates with a period of 2 seconds is called a seconds pendulum (the clock ticks every half period). Investigate the dimensions of a seconds pendulum when it is

(i) a simple pendulum with a bob;

Investigation continued

(ii) a disc of radius r attached to a light rod so that the centre of the disc is a distance d from the axis of rotation. What is r if $d = r$? Find d in terms of r in other cases.

(iii) a rod pivoted about a point distance d from its centre. (d must be real and the pivot must be on the rod.) Note that $\pi^2 \approx g$.

Investigate the periods of other types of clock pendulum.

KEY POINTS

- The angular speed of a rigid body about an axis is $\dot\theta$, where θ is the angle between a fixed line in the body and a fixed direction in space (both perpendicular to the axis of rotation).
- The moment of a couple is the same about any axis perpendicular to its plane.

Equivalent quantities

Rotation of a rigid body about a fixed axis		Linear motion	
* Moment of inertia about axis:	$I = \sum mr^2$	Mass:	m
* Moment of force about axis:	C	Force:	F
* Angular displacement:	θ	Displacement:	x or s
* Equation of motion:	$C = I\ddot\theta$	$F = m\ddot{x}$	
* Kinetic energy:	$\frac{1}{2}I\dot\theta^2$ or $\frac{1}{2}I\omega^2$	$\frac{1}{2}m\dot{x}^2$ or $\frac{1}{2}mv^2$	
* Work:	$\int C\,d\theta$	$\int F\,dx$ or $\int F\,ds$	

Moments of inertia

- The moment of inertia of a body about an axis depends on the position and direction of the axis, which should always be stated.
- Perpendicular axes theorem: $I_z = I_x + I_y$ for a lamina only
- Parallel axes theorem:

$$I_A = I_G + Md^2$$
$$= I_G + M(AG)^2 \text{ for AG perpendicular to the axes}$$

- To find a moment of inertia by integration:
  * Divide the body into elements for which the moment of inertia is known.
  * Find the mass of a typical element in terms of the density.
  * Determine the moment of inertia of the element about the axis.
  * Sum for all elements and write the sum as an integral.
  * Evaluate the integral.
  * Write the density in terms of the total mass and replace it.
- When the density varies, it should be written in terms of a suitable variable and included in the integration.

Appendix 1:
Polar equations of conics

This appendix summarises the equations of *conics*—the generic name for the ellipse, parabola and hyperbola—in polar co-ordinates. It gives sufficient coverage for your needs in understanding motion under central forces.

The ellipse, hyperbola and parabola are important curves in mechanics: for example, Chapter 4 shows that planets move in ellipses around the sun. They would be the curves produced if you were to saw through a cone in different directions. Hence the curves are known as conic sections or simply conics.

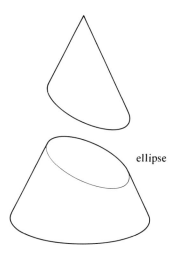

ellipse

These curves are normally defined in terms of a fixed point, the *focus*, and a fixed line, the *directrix*. A conic is the curve traced out by a point P which moves so that

$$\frac{\text{distance of P from the focus}}{\text{distance from the directrix}} = \text{constant}$$

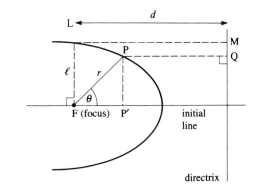

Figure A1.1

In figure A1.1, P is any point on the conic. Then the ratio FP/PQ is a constant known as the *eccentricity*, which is usually denoted e. The value of the eccentricity determines the shape of the curve, as you will see.

Polar equations

To find the polar equation of a conic using the definition above, take the focus as the origin F of the polar co-ordinates, and take the initial line to be perpendicular to the directrix. In figure A1.1, FL is the 'y axis', (it is perpendicular to the initial line) and L is on the conic.

Denote the eccentricity by e. Then, by definition, since P is on the conic

$$FP = e(PQ)$$

Taking (r, θ) as the polar co-ordinates of P,

$$FP = r$$

and

$$PQ = d - FP' = d - r\cos\theta$$

$$\Rightarrow \quad r = e(d - r\cos\theta)$$

$$\Rightarrow \quad r(1 + e\cos\theta) = ed$$

L is also on the conic, so $FL = e(LM)$. Denote FL by l. Then

$$l = ed$$

Hence the polar equation of a conic is

$$r(1 + e\cos\theta) = ed = l$$

$$\Rightarrow \quad r = \frac{l}{(1 + e\cos\theta)}$$

The shape of the curve depends entirely on the value of the eccentricity e: l (known as the semi-lactus rectum) is just a scaling factor.

$e<1$: ellipse or circle

The denominator $1 + e\cos\theta$ is always positive (because $|\cos\theta| \leqslant 1$). Then r has its least value, $FA = \dfrac{l}{(1+e)}$, at A when $\theta = 0$ and its greatest, $FA' = \dfrac{l}{(1-e)}$, at A' when $\theta = \pi$. The result is a closed symmetrical curve as shown in figure A1.2. The long axis A'A is called the *major axis* and its mid-point O is the geometrical centre of the ellipse. OA, the *semi-major axis*, has a length a where

$$a = \tfrac{1}{2}\left[\frac{l}{(1+e)} + \frac{l}{(1-e)}\right]$$

$$= \frac{l}{(1-e^2)}$$

or $\quad l = a(1 - e^2)$

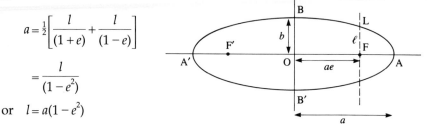

Figure A1.2

The distance of the centre from the focus is $OF = OA - FA = ae$ (by substituting the values of OA and FA from above).

The curve is, in fact, also symmetrical about the orthogonal axis, BB', through O. This is the *minor axis*, whose length is normally denoted by $2b$. This is made clearer if you derive the Cartesian equation of the ellipse referred to O as origin (see example). It is

$$\frac{x^2}{a^2} + \frac{y^2}{b^2} = 1$$

The symmetry of the ellipse means that there is another focus, F', and another directrix the same distance the other side of the centre from the originals. The identical curve is traced out if the ellipse is defined relative to these. The polar equation relative to this second focus, F', as origin is

$$r = \frac{1}{(1 - e \cos \theta)}$$

The closer e is to 1, the longer and thinner is the ellipse, while as e approaches zero, the nearer it gets to a circle. When $e = 0$ you can see the polar equation reduces to $r = l$, a circle radius l. Both foci F and F' have come together at the centre.

The planetary orbits are ellipses, but mostly their eccentricities are small (that of the Earth is only about 0.016) so the orbits are near circular. Halley's comet, which is seen every 76 years, orbits the sun in a long, thin ellipse with an eccentricity of about 0.97.

e >1: hyperbola

With e greater than 1, you can see that there are always values of θ for which the denominator is zero. Take as an example

$$r = \frac{1}{(1 + 2 \cos \theta)}$$

The denominator is zero when $\cos \theta = -\frac{1}{2}$, i.e. $\theta = \pm \frac{2\pi}{3}$. As θ starts from zero and gets bigger, r increases and becomes indefinitely large as θ approaches $\frac{2\pi}{3}$. It is similar if you go round the other way with θ becoming negative down to $-\frac{2\pi}{3}$. The resulting curve, a hyperbola, is shown in figure A1.3.

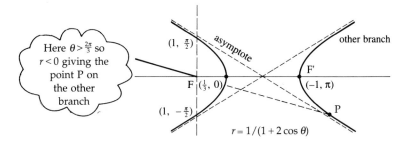

Figure A1.3

The curve approaches infinity along lines parallel to $\theta = \pm\frac{2\pi}{3}$. Such lines are known as *asymptotes*.

When θ is between these two limits on the left-hand side of figure A1.3 r is negative and a totally separate branch of the hyperbola is given: a mirror image of the first, approaching the asymptotes at their other ends. Thus a hyperbola consists of two branches. If in a mechanics problem you find that a particle follows a hyperbolic orbit, you have to decide which branch it follows: it is not likely to disappear and reappear on the other branch!

The two branches are symmetrical and hence there is another focus F', as shown. As with the ellipse, a negative value of e in the equation gives the same curve referred to the other focus as origin.

$e=1$: parabola

Although, in a sense, the parabola is a limiting case of both the ellipse and the hyperbola, it is different from both. For example, like the hyperbola it is an open curve (r tends to infinity as θ approaches π), but unlike the hyperbola it has just a single branch. Any quadratic curve (such as $y = x^2$) is a parabola—the path of a projectile under uniform gravity (with no resistance) is an example you have met already.

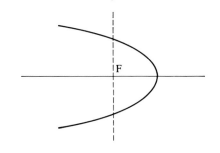

Figure A1.4

EXAMPLE

You are given that the equation $r = \dfrac{l}{(1 + e\cos\theta)}$ represents an ellipse (i.e. $e < 1$).

(i) Show that its equation in Cartesian co-ordinates is

$$\frac{(x - ae)^2}{a^2} + \frac{y^2}{b^2} = 1$$

where $a = \dfrac{l}{(1 - e^2)}$ and $b = \dfrac{l}{\sqrt{(1 - e^2)}}$

(ii) Write down the translated equation of the ellipse if the origin is moved to the geometrical centre and confirm that a and b represent the lengths of the semi-major and semi-minor axes.

Solution

(i)
$$r = \frac{l}{(1 + e\cos\theta)}$$

$$\Rightarrow \quad r + er\cos\theta = l$$

$$\Rightarrow \quad r = l - er\cos\theta$$

Substituting $r = \sqrt{(x^2 + y^2)}$ and $r\cos\theta = x$ and squaring gives

$$x^2 + y^2 = (l - ex)^2$$

$$= l^2 - 2lex + e^2x^2$$

$$\Rightarrow \quad x^2(1 - e^2) + 2lex + y^2 = l^2$$

Since $e \neq 1$, divide by $(1 - e^2)$ to give

$$x^2 + \frac{2el}{1 - e^2}x + \frac{y^2}{1 - e^2} = \frac{l^2}{1 - e^2}$$

Eliminate l using $l = a(1 - e^2)$:

$$x^2 + 2aex + \frac{y^2}{1 - e^2} = a^2(1 - e^2)$$

Complete the square in x to give

$$(x + ae)^2 + \frac{y^2}{1 - e^2} = a^2 - a^2e^2 + a^2e^2$$

$$= a^2$$

Divide by a^2, and substitute $b^2 = a^2(1 - e^2)$ to give

$$\frac{(x + ae)^2}{a^2} + \frac{y^2}{b^2} = 1$$

(ii) This is the Cartesian equation of an ellipse referred to the focus as origin. Translation along the x axis a distance ae will refer the ellipse to the geometrical centre as origin. This gives

$$\frac{x^2}{a^2} + \frac{y^2}{b^2} = 1$$

This is the standard Cartesian equation of an ellipse. When $x = 0$, $y = \pm b$, when $y = 0$, $x \pm a$.

Appendix 2:
Dividing three-dimensional bodies into elements

When a three-dimensional body is divided into elements, care should be taken when deciding the 'width' of a typical element. For a solid body formed by rotation about the x axis, δx is appropriate, but for a hollow body the length of the arc δs should be used. The reason for this is indicated in the following argument.

Figure A2.1 shows a typical element, which you might always have considered to be approximately cylindrical. But if it is near a point where the surface crosses the x axis, a more appropriate approximation is part of a cone.

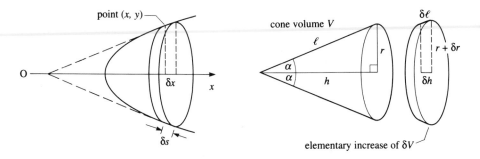

Figure A2.1 ⠀⠀⠀⠀⠀⠀⠀⠀⠀⠀⠀⠀⠀⠀⠀⠀⠀⠀⠀**Figure A2.2**

The cone in figure A2.2 has radius r, height h, slant height l and semi-vertical angle α. Its volume V is given by

$$V = \tfrac{1}{3}\pi r^2 h$$
$$= \tfrac{1}{3}\pi h^3 \tan^2 \alpha$$

$$\Rightarrow \quad \frac{\mathrm{d}V}{\mathrm{d}h} = \pi h^2 \tan^2 \alpha = \pi r^2 \quad \text{(assuming } \alpha \text{ is constant)}$$

So for a small increase δh in the height, the corresponding increase in volume, $\delta V \approx \pi r^2 \delta h$, which translates into $\pi y^2 \delta x$ for the element of a volume of revolution which is approximately conical. Notice that this is the same expression which you would use were you to assume it is cylindrical. The curved surface area A is given by

$$A = \pi r l$$
$$= \pi l^2 \sin \alpha$$

$$\Rightarrow \quad \frac{\mathrm{d}A}{\mathrm{d}l} = 2\pi l \sin \alpha = 2\pi r$$

For a small increase in the height, the surface area increases by $\delta A \approx 2\pi r \delta l$, which translates into $2\pi y \delta s$ for a surface of revolution.

The following two cases illustrate the use of these two results.

The moment of inertia of a solid sphere about a diameter

The sphere can be formed by rotating the region inside the circle $x^2 + y^2 = r^2$ about the x axis. The moment of inertia about the x axis is then found by subdividing the sphere into elementary discs, as shown in figure A2.3.

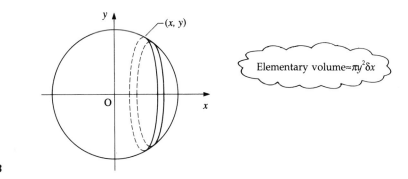

Elementary volume $\approx \pi y^2 \delta x$

Figure A2.3

A typical disc at the point (x, y) has radius y and thickness δx. Its mass, δm, is approximately $\rho \pi y^2 \delta x$, so its moment of inertia about the x axis is

$$\tfrac{1}{2}\delta m r^2 = \tfrac{1}{2}(\rho \pi y^2 \delta x) y^2$$
$$= \tfrac{1}{2}\rho \pi y^4 \delta x$$

Thus the moment of inertia of the sphere about the x axis is

$$I = \int_{-r}^{r} \tfrac{1}{2}\rho \pi y^4 \, dx$$

Notice that, apart from the limits, this could be used for any volume of revolution

$$= \tfrac{1}{2}\rho \pi \int_{-r}^{r} (r^2 - x^2)^2 \, dx$$

$$= \tfrac{1}{2}\rho \pi \int_{-r}^{r} (r^4 - 2r^2 x^2 + x^4) \, dx$$

$$= \tfrac{1}{2}\rho \pi \left[r^4 x - 2r^2 \frac{x^3}{3} + \frac{x^5}{5} \right]_{-r}^{r}$$

$$= \tfrac{1}{2}\rho \pi [(r^5 - \tfrac{2}{3}r^5 + \tfrac{1}{5}r^5) - (-r^5 + \tfrac{2}{3}r^5 - \tfrac{1}{5}r^5)]$$

$$\Rightarrow \quad I = \tfrac{1}{2}\rho \pi (\tfrac{16}{15}r^5)$$

$$= \tfrac{8}{15}\rho \pi r^5$$

But the mass M of the sphere is $\tfrac{4}{3}\pi r^3 \rho$, therefore

$$\frac{I}{M} = \frac{8\rho \pi r^5}{15} \times \frac{3}{4\pi r^3 \rho}$$

$$= \frac{2r^2}{5}$$

$$\Rightarrow \quad I = \tfrac{2}{5}M r^2$$

By symmetry, this is the moment of inertia of the sphere about any axis through its centre.

The moment of inertia of a hollow sphere about a diameter

Divide the sphere into elementary rings, as shown in figure A2.4. A typical ring has a 'width' $\delta s = r\delta\theta$ and radius $r\cos\theta$, so its area is approximately $2\pi(r\cos\theta) \times (r\delta\theta)$.

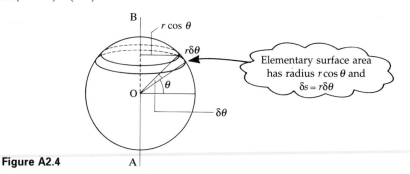

Figure A2.4

Using σ as the mass per unit area, the moment of inertia of the ring about the axis AB is

$$\text{mass} \times (\text{radius})^2 = (2\pi r^2\cos\theta\, r\delta\theta\sigma) \times (r\cos\theta)^2$$

The moment of inertia of the sphere about AB is

$$\int_{-\frac{\pi}{2}}^{\frac{\pi}{2}} 2\pi r^4\sigma\cos^3\theta\, d\theta = 2\pi r^4\sigma \int_{-\frac{\pi}{2}}^{\frac{\pi}{2}} \cos^3\theta\, d\theta$$

$$= 2\pi r^4\sigma \int_{-\frac{\pi}{2}}^{\frac{\pi}{2}} (1-\sin^2\theta)\cos\theta\, d\theta$$

$$= 2\pi r^4\sigma \int_{-\frac{\pi}{2}}^{\frac{\pi}{2}} (\cos\theta - \sin^2\theta\cos\theta)\, d\theta$$

$$= 2\pi r^4\sigma[\sin\theta - \tfrac{1}{3}\sin^3\theta]_{-\frac{\pi}{2}}^{\frac{\pi}{2}} \quad \text{(substitute } u = \sin\theta)$$

$$= 2\pi r^4\sigma \times \tfrac{4}{3}$$

The mass of the sphere is $M = 4\pi r^2\sigma$, so the moment of inertia is $\tfrac{2}{3}Mr^2$.

6

The rotation of a rigid body

The diagrams and the whole business I got the Nobel Prize for came from that piddling around with the wobbling plate.

Richard Feynman

For Discussion

Newton's laws apply to the motion of particles. How do you think they apply to the motion of large bodies?

What happens to a rotating space station, light years from any stars, if its rockets cut out? Does it continue to rotate? Does it move in a straight line?

Why does it feel good it you hit a tennis ball in the right place on the racket, but hurt your wrist when you hit it on the frame?

Chapter 5 in *Mechanics 5* covered the equation of motion and the energy of a rigid body rotating about a fixed axis. In this chapter you will complete the study of fixed-axis rotation and apply the techniques you have learnt to the motion of rolling bodies. Later, in Chapter 9, you will learn how some of the vector methods you have met can be applied to the more general motion of rigid bodies. You should find that the vector theory enables you to appreciate the significance of the simplifications being made in the present chapter.

The motion of the centre of mass of a rigid body

Activity

Find the centre of mass of an object such as a table tennis bat and mark it so that it is clearly visible (for example by using luminous paint). Throw the bat carefully across the room and describe the motion of the bat and of its centre of mass.

Any change in position of a rigid body usually combines a translation with a rotation and both these aspects must be considered in order to describe the motion of such a body in full. It is particularly important to study the motion of the centre of mass. The next section shows that for any rigid body or other system of particles with constant mass, the centre of mass of the system moves as though it is a single particle equal in mass to the whole system, obeying Newton's Second Law with all the forces concentrated on that particle.

Throughout this chapter, the motion of a point will be described in terms of position or displacement vectors. For example, when $\mathbf{r}_P$ is the position vector of P relative to a fixed origin O, the velocity of P is $\dfrac{d\mathbf{r}_p}{dt} = \dot{\mathbf{r}}_p$ and its

acceleration is $\dfrac{d^2\mathbf{r}_p}{dt^2} = \ddot{\mathbf{r}}_p$

A typical particle in a rigid body of total mass M has position vector $\mathbf{r}_p$ referred to a fixed origin O. The mass of the particle is m and it is subject to a resultant force $\mathbf{F}_p$ (figure 6.1). In general, this is a combination of external and internal forces. Newton's Second Law for the typical particle is given by

$$\mathbf{F}_p = m_p\ddot{\mathbf{r}}_p$$

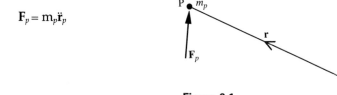

Figure 6.1

Summing this over all the particles of the system gives

$$\sum_{\text{all }p} \mathbf{F}_p = \sum m_p\ddot{\mathbf{r}}_p$$

By Newton's Third Law, every internal force acting on one particle in the system is counteracted by an equal and opposite force acting on another particle. The only forces remaining on the left-hand side of the equation are the external forces. Let the sum of the external forces to be $\sum \mathbf{F}_{ext}$. Then

$$\sum \mathbf{F}_{ext} = \sum (m_p \ddot{\mathbf{r}}_p) \qquad \text{①}$$

The position of the centre of mass of the system relative to O is $\mathbf{r}_G$, where

$$M\mathbf{r}_G = \sum_{all\, p} (m_p \mathbf{r}_p) \quad \left(M = \sum m_p \right)$$

Differentiating twice with respect to t gives

$$M\ddot{\mathbf{r}}_G = \frac{d^2}{dt^2} \left(\sum m_p \mathbf{r}_p \right)$$

$$= \sum (m_p \ddot{\mathbf{r}}_p) \quad \text{(all } m_p \text{ are constant)}$$

Equation ① then becomes the equation of motion for G:

$$\sum \mathbf{F}_{ext} = M\ddot{\mathbf{r}}_G \quad \text{(for constant } M\text{)}$$

Thus the *centre of mass*, G, of a rigid body or any system of particles of constant mass moves as though it is a single particle with mass equal to the total mass which obeys Newton's Second Law with all the external forces concentrated at G.

If you threw a table tennis bat across the room, as described in the activity on page 182, you might have observed that its centre of mass moved along an approximate parabolic path as it would if it were a particle. The picture shows an example of such motion taken in a darkened room using a stroboscope.

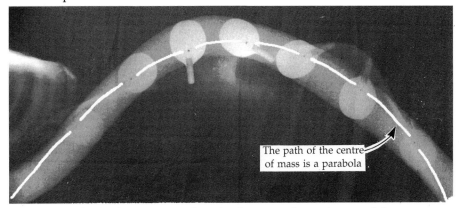

The path of the centre of mass is a parabola

Even when the body is not rigid, as in the case of a swimmer doing a somersault from the top diving board, the centre of mass moves in a path determined only by the sum of the external forces. Though in the case of a diver, of course, a force such as the air resistance might change according to his or her body position. This result for the motion of the centre of mass and the one which follows later about the effect of impulses together justify many of the simplifying assumptions you might have made in earlier work when you treated large bodies such as tennis balls, or even cars and trains, as particles.

Reaction, angular momentum and impulse

The reaction at the support

When considering the rotation of a rigid body about a fixed axis in *Mechanics 5* (pages 139–141), moments were taken about the axis, so the forces between the axis and the body did not have to be considered. If you have ever used an electric drill, however, you will know that there are often quite considerable forces required to support a rotating object.

You can now use the equation for the motion of the centre of mass, in conjunction with the equations you have already learnt for the rotation of the body, in order to find the reaction at the support.

Consider the external forces acting on a body rotating about a fixed horizontal axis through a point A. G is the centre of mass of the body and AG is perpendicular to the axis so that G moves in a vertical circle with centre A.

Figure 6.2 shows the forces at the axis reduced to two components, Y in the direction GA and X at right angles to GA in the direction of increasing θ. The axis of rotation is at right angles to the page.

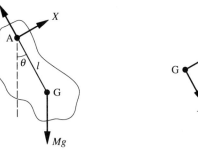

Figure 6.2   External forces   Acceleration of G

There is no resultant external force in the direction of the axis because G moves in the plane at right angles to it.

It is now possible to use the result for the motion of the centre of mass: G moves in a circle of radius l about the fixed point A, so the radial and tangential components of its acceleration are $-l\dot{\theta}^2$ and $l\ddot{\theta}$. The equations for the linear motion of G are

$$Mg\cos\theta - Y = -Ml\dot{\theta}^2 \qquad\qquad ①$$

$$\text{and} \qquad X - Mg\sin\theta = Ml\ddot{\theta} \qquad\qquad ②$$

You already know how to find expressions for $\dot{\theta}^2$ and $\ddot{\theta}$ using the energy equation and the equation for the rotation of the body.

When the zero level for potential energy is taken through A, the total energy of the body in the position shown is

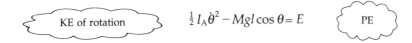

KE of rotation $\frac{1}{2} I_A \dot{\theta}^2 - Mgl \cos \theta = E$ PE

where I_A is a moment of inertia of the body about the axis through A, and E is a constant. Substituting in ① gives

$$Y = Mg \cos \theta + Ml\dot{\theta}^2$$

$$= Mg \cos \theta + \frac{2Ml}{I_A}(E + Mgl \cos \theta)$$

or $\quad Y = Mg \cos \theta \left(1 + \frac{2Ml^2}{I_A}\right) + \frac{2MlE}{I_A}$

The equation of motion for the rotation of the body is

$$C = I_A \ddot{\theta}$$

where C is the sum of the moments of the external forces about the axis, or, more concisely, the *resultant moment* about the axis. Hence

$$-Mgl \sin \theta = I_A \ddot{\theta}$$

Substituting this in ② gives

$$X = Mg \sin \theta + Ml\ddot{\theta}$$

$$= Mg \sin \theta \left(1 - \frac{Ml^2}{I_A}\right)$$

NOTE

This section illustrates the methods for finding the reactions at the support. You are advised to work out each example using these methods and not to use the results as formulae.

EXAMPLE

Jim is rotating freely in a rigid posture on a horizontal bar. His mass m is 75 kg, his moment of inertia I about the bar is 90 kg m^2 and his centre of mass is 1.1 m from the bar. Initially, Jim is at rest vertically above the bar and then lets himself fall.

(i)   Find the components of the reaction at the bar in the radial and transverse directions when he has turned through an angle θ.

(ii) Determine the values of θ when each of these components is zero.

(iii) Show that the bar has to be able to support a force greater than five times Jim's weight.

Solution

The diagram shows the forces acting on Jim when he has rotated through an angle θ

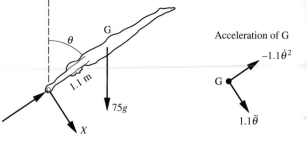

Acceleration of G

(i) His centre of mass G moves in a circle of radius 1.1 m so the component of its acceleration in the radial direction along AG is $-1.1\dot{\theta}^2$ as shown. Its equation of motion in this direction is

$$Y - 75g\cos\theta = -75 \times 1.1\dot{\theta}^2$$
$$\Rightarrow \quad Y = 75g\cos\theta - 75 \times 1.1\dot{\theta}^2 \qquad \text{①}$$

Jim has fallen a vertical height of $(1.1 - 1.1\cos\theta)$ m, so by the principle of conservation of energy:

$$75g \times 1.1(1 - \cos\theta) = \tfrac{1}{2}I\dot{\theta}^2 = 45\dot{\theta}^2$$

Substituting for $\dot{\theta}^2$ in ① gives

$$Y = 75g\cos\theta - 75 \times 1.1 \times \frac{75g}{45} \times 1.1(1 - \cos\theta)$$
$$\Rightarrow \quad Y = 75g(3.017\cos\theta - 2.017) \qquad \text{②}$$

The acceleration of Jim's centre of mass in the transverse direction perpendicular to AG is $1.1\ddot{\theta}$. Its equation of motion in this direction is

$$75g\sin\theta + X = 75 \times 1.1\ddot{\theta}$$
$$\Rightarrow \quad X = 75 \times 1.1\ddot{\theta} - 75g\sin\theta \qquad \text{③}$$

The equation of motion for rotation about A is

$$75g \times 1.1\sin\theta = I\ddot{\theta} = 90\ddot{\theta}$$

Substituting in ③ gives

$$X = 75 \times 1.1 \times \frac{75g}{90} \times 1.1\sin\theta - 75g\sin\theta$$
$$\Rightarrow \quad X = 75g\sin\theta(1.0083 - 1) \qquad \text{④}$$

The components of the reactions at the axis are

$$X = 75g \times 0.0083\sin\theta$$

and $\quad Y = 75g(3.017\cos\theta - 2.017)$

(ii) The transverse component X is zero when $\theta = 0$ and when $\theta = \pi$, at the top and bottom of the motion. The radial component Y is zero when

$$3.017\cos\theta = 2.017$$
$$\Rightarrow \quad \theta = 0.8385$$

that is, when Jim's body makes an angle of about 48° with the upward vertical. When the angle is less than this, the reaction on Jim is positive and away from the bar. For greater angles, Y is negative and the force on Jim is towards the bar.

(iii) To find the greatest force required to support Jim, it is necessary to consider how X and Y change during the motion. When he is below the horizontal, $\cos\theta < 0$, so the component Y is negative but its magnitude is greater than $2.017 \times 75g$ and increases as θ increases.

The maximum value of X is $0.0083 \times 75g$. Hence $|Y|$ is much greater than X, so the force on the bar is approximately $-Y$, which has its maximum value at the bottom of the motion. In this position, $\cos\theta = -1$ and

$$|Y| = 75g \times 5.034$$

The force on the bar is more than five times Jim's weight.

The angular momentum of a rigid body rotating about a fixed axis

When Newton's first two laws are written in terms of the linear momentum of a particle they are equivalent to the following two statements:

- The linear momentum of a particle is constant if no resultant force acts on it.
- The resultant force acting on the particle is equal to the rate of change of its linear momentum.

But how do these laws apply to rigid bodies? Think, for example, of the motion of a rotating space station which has escaped from all external forces. Do you expect the momentum of every particle to be constant?

Clearly this cannot be the case if the space station is rotating, but the continued rotation of individual particles in the space station does not mean that Newton's laws are contradicted. Although external forces are non-existent, all the particles in the station have forces acting upon them due to their interaction with each other. These internal forces are essential for parts of the space station to retain their rigid shape.

So the linear momentum of each particle is not constant, but momentum is a vector and so it is possible to find its moment about the axis of rotation. This moment of momentum is called the *angular momentum of the particle about the axis*. It is, in fact, the total angular momentum of all particles about an axis through the centre of mass, which is the quantity that remains unchanged in the absence of external forces. Similarly, when a rigid body rotates about a fixed axis and there is no resultant moment about the axis, the total angular momentum about the axis is constant. This is shown below.

Figure 6.3 shows a typical particle P of mass m_p rotating with angular speed $\dot\theta_p$ in a circle of radius r_p about a fixed axis. Its momentum is $m_p v_p = m_p r_p \dot\theta_p$ along the tangent so the moment of its momentum about the axis is:

$$(m_p v_p)r_p = m_p r_p^2 \dot\theta_p$$

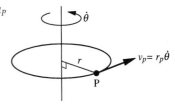

Figure 6.3

Summing for all particles gives the total angular momentum, L, about the axis, namely

$$L = \sum_{\text{all } p} m_p r_p^2 \dot{\theta}_p$$

For a rigid body, all the particles have the same angular speed, $\dot{\theta}$, about the axis, so the expression for angular momentum becomes

$$L = I\dot{\theta}$$

where $I = \sum m_p r_p^2$ is the moment of inertia of the body about the axis. This expression for angular momentum is analogous to mv for linear motion, as you might expect.

When a rigid body rotates about a fixed axis, the conservation of angular momentum in the absence of external forces follows from the equation of motion:

$$C = I\ddot{\theta}$$

where C is the total moment of forces about the axis. When there are no external forces $C = 0$,

$$\Rightarrow \qquad I\ddot{\theta} = 0$$

$$\Rightarrow \qquad I\dot{\theta} = \text{constant}$$

This is the equivalent of Newton's First Law of motion. The Second Law has already been written in the form $C = I\ddot{\theta}$, but it too can be written in terms of the angular momentum of the body:

$$C = \frac{d}{dt}(I\dot{\theta}) = \frac{dL}{dt}$$

This is an important result which can be stated as follows.

For a rigid body rotating about a fixed axis:

- when there is no external moment of forces about the axis, its angular momentum about the axis is constant
- the resultant moment of forces about the axis of rotation is equal to the rate of change of angular momentum of the body about the axis.

The impulse of a couple about an axis

Integrating the equation for C gives

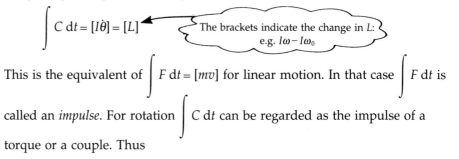

$$\int C \, dt = [I\dot{\theta}] = [L]$$

The brackets indicate the change in L:
e.g. $I\omega - I\omega_0$

This is the equivalent of $\int F \, dt = [mv]$ for linear motion. In that case $\int F \, dt$ is called an *impulse*. For rotation $\int C \, dt$ can be regarded as the impulse of a torque or a couple. Thus

- the sum of the moments of impulses about the axis of rotation is equal to the change in angular momentum of the body about the same axis.

EXAMPLE

In a party game a flat board in the shape of a space ship is made to spin round a fixed vertical axis when hit by a small 'meteorite' of mass m. The board is always stationary and facing the player when a meteorite is thrown. Meteorites, which can stick to the board, are not removed between throws. The moment of inertia of the board about its axis is I.

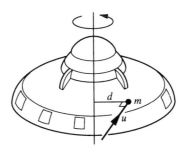

(i)   Assuming that a meteorite is thrown at the empty board and sticks after hitting it at right angles with speed u at a distance d from the axis, find the initial angular velocity of the board.

(ii) The board slows down under the action of a constant frictional couple C and the winner in the game is the person who makes it spin for the longest time. Is this a fair game?

(iii) Someone suggests that it would be easier to determine the winner by counting the number of revolutions made by the board before coming to rest. Would the game be fair in this case?

Solution

(i)   During the short period of time taken by the impact, the impulse of the frictional couple is negligible, so the change in total angular momentum about the axis is negligible. In other words, total angular momentum is conserved. You can compare this with conservation of linear momentum when two objects collide.

The angular momentum of the meteorite about the axis just before it hits the board is $mu \times d$. The new moment of inertia of the meteorite and board about the axis after the collision is $I + md^2$, so the angular momentum after the collision is $(I + md^2)\omega$, where ω is the angular speed. Hence

$$(I + md^2)\omega = mud \qquad \Rightarrow \qquad \omega = \frac{mud}{(I + md^2)} \qquad \text{①}$$

(ii) The impulse of the constant frictional couple in time t is Ct and this is equal to the loss of angular momentum of the board and meteorite during that time. When t is the time for the board to stop rotating:

$$Ct = mud$$

$$\Rightarrow \qquad t = \frac{mud}{C}$$

This depends only on the speed of the meteorite and the position where it hits the board, and is independent of the moment of inertia of the board when it is hit. So it does not matter how many other meteorites are there already. Provided there is sufficient room on the board, the game is fair.

(iii) Suppose that I_n is the moment of inertia of the board about its axis when there are n meteorites attached to it. After another hit this will become $I_n + md^2$. The initial angular speed after this additional meteorite hits the space ship is

$$\omega = \frac{mud}{(I_n + md^2)} \quad \text{(see ①)}$$

When the board has turned through an angle θ, the work done against the couple is $C\theta$ and this is equal to the loss in kinetic energy. Hence

$$C\theta = \tfrac{1}{2}(I_n + md^2)\omega^2$$

$$\Rightarrow \quad \theta = \frac{(mud)^2}{2C(I_n + md^2)}$$

This is dependent on the value of I_n so it will be more difficult to make the board turn through a given angle as the game progresses. In this case, the game is not fair.

The effects of impulses

When you hit a ball with a racket or other bat, there is an impulse on the racket or bat and possibly also an impulsive reaction on your hand. It feels most comfortable when the impulse on your hand is as small as possible.

Techniques similar to the ones you have used to analyse forces, can be used to find the effects of impulses on different parts of a rotating body.

For a rigid body rotating about a fixed axis, the equation of motion for the centre of mass can be written as

$$\sum \mathbf{F}_{ext} = \frac{d}{dt}(M\dot{\mathbf{r}}_G)$$

where $\sum \mathbf{F}_{ext}$ denotes the sum of all external forces. Integrating with respect to t gives

$$\sum \int \mathbf{F}_{ext}\, dt = [M\dot{\mathbf{r}}_G]$$

So, for the motion of the centre of mass

The brackets indicate the change in linear momentum $M\dot{\mathbf{r}}_G$

$$\sum \mathbf{J}_{ext} = [M\dot{\mathbf{r}}_G]$$

where $\sum \mathbf{J}_{ext}$ is the sum of impulses of all external forces.

This shows that when impulses act on a body, the centre of mass, G, behaves as a particle of mass M with all the impulses concentrated at G. Also

$$\int C\, dt = [I\dot{\theta}]$$

where C is the sum of the moments of all external forces about the axis.

When large forces act for a short time, this integral is replaced by the sum of the moments of the impulses about the axis of rotation.

$$\Rightarrow \quad \sum (\text{moment of } \mathbf{J}) = [I\dot{\theta}]$$

These results can be summarised as follows.

- The sum of the external impulses is equal to the change in momentum of the whole mass concentrated at the centre of mass.
- The sum of the moments of external impulses about a fixed axis is equal to the change in the angular momentum, $I\dot{\theta}$, about the axis.

EXAMPLE

A door of mass M kg and width $2l$ m is moving with angular velocity $\dot{\theta}$ rad s^{-1} when it bangs shut.
(i) Find the impulsive reaction at the hinge.
(ii) Are the hinges or the door jamb likely to suffer most from the banging door?
(iii) Evaluate the impulse on the hinge for a door of mass 30 kg, and width 0.76 m moving with angular speed 1.3 rad s^{-1}.

Solution

All the motion is at right angles to the door, so the impulsive reactions on the door are perpendicular to it, as shown in the diagram (though the sense of J_2 could be opposite to the direction shown).

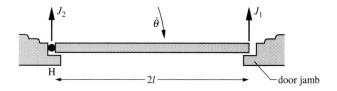

(i) Taking moments about the axis through the hinges H gives

$$J_1 \times 2l = I\dot{\theta}$$

$$\Rightarrow \quad J_1 = \frac{I\dot{\theta}}{2l}$$

where I is the moment of inertia about the axis.

The velocity of the centre of mass is $l\dot{\theta}$ at right angles to the door, so its momentum is $Ml\dot{\theta}$ in the same direction.

The total impulse is

$$J_1 + J_2$$

$$\Rightarrow \quad J_1 + J_2 = Ml\dot{\theta}$$

$$\Rightarrow \quad J_2 = Ml\dot{\theta} - \frac{I\dot{\theta}}{2l}$$

$$\Rightarrow \quad J_2 = \left(Ml - \frac{I}{2l} \right)\dot{\theta}$$

Assuming the door can be modelled as a rectangular lamina, then

$$I = \tfrac{4}{3}Ml^2$$

$$\Rightarrow \quad J_2 = \tfrac{1}{3}Ml\dot\theta$$

$$\text{and} \quad J_1 = \tfrac{2}{3}Ml\dot\theta$$

(ii) Although $J_1 > J_2$, the impulse on the door jamb is spread throughout its length. The hinges bear the effect of J_2 and for standard hinges these are much shorter in total length than the door jamb. Consequently, the hinges are more likely to suffer damage from a banging door.

(iii) When $M = 30$, $2l = 0.76$ and $\dot\theta = 1.3$

$$J_2 = \tfrac{1}{3} \times 30 \times 0.38 \times 1.3$$

The impulse on the hinges is 4.94 Ns (to 3 significant figures).

An application: the centre of percussion

The position of the centre of percussion (or sweet spot) of a piece of sports equipment is an important factor in its design. This is the point at which a body must be hit in order to minimise impulsive reactions.

Figure 6.4 shows a body consisting of two spheres joined by a thin rod that is suspended by a light string and then struck at different points. In (a) it is struck at the centre of percussion relative to the fixed end of the string and rotates smoothly about this point. When it is struck at the centre of mass (b), the body moves parallel to itself as far as the string will allow. The effects of striking above and below the centre of mass are shown in (c) and (d) respectively.

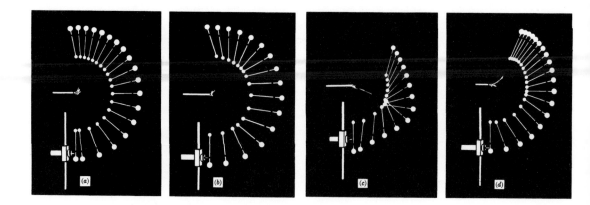

Figure 6.4

For a piece of sports equipment, the best place to hit the ball depends on the position of the axis about which it is required to rotate and this might vary during the game: for example, when using different strokes at tennis.

The next example shows how the relationship between the positions of the centre of percussion for a particular axis and the centre of mass of a racket can be found.

EXAMPLE

The racket in the diagram has mass M and is free to rotate about a fixed smooth axis at A which is parallel to the plane of the racket and perpendicular to its handle. The moment of inertia of the racket about a parallel axis through its centre of mass, G, is Mk^2.

When the racket is at rest, it is hit at the point P with an impulse J_1 perpendicular to its plane and to the axis of rotation. $AG = h$ and $AP = d$.
(i) Find the value of d which ensures that the impulsive reaction, J_2, at A is zero.
(ii) Show that this value of d is equal to the length of a simple pendulum which would perform small oscillations about the axis through A with the same period as the racquet.

Solution

(i) Suppose the initial angular speed of the racket after being hit is ω. The initial speed of G is then $h\omega$.

The impulse–momentum equation for the centre of mass is

$$J_1 - J_2 = Mh\omega \qquad \text{①}$$

By the parallel axes theorem, the moment of inertia of the racket about the axis through A is $Mk^2 + Mh^2$. Taking moments about the axis through A gives

$$J_1 d = M(k^2 + h^2)\omega \qquad \text{②}$$

When $J_2 = 0$, ① gives

$$J_1 = Mh\omega$$
$$\Rightarrow \qquad Mh\omega d = M(k^2 + h^2)\omega$$
$$\Rightarrow \qquad d = \frac{(k^2 + h^2)}{h}$$

This is the distance of the centre of percussion from the axis through A.

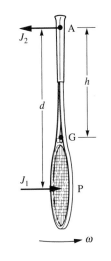

The plane of the racket and the axis through A are perpendicular to the page

(ii) When the racket hangs from A and is then displaced though a small angle, the equation of motion for its rotation is

$$I_A\ddot{\theta} = M(k^2 + h^2)\ddot{\theta} = -Mgh\sin\theta$$

For small θ this becomes

$$\ddot{\theta} = -\frac{gh}{(k^2 + h^2)}\theta$$

This is the equation for simple harmonic motion of period

$$2\pi\sqrt{\left(\frac{k^2 + h^2}{gh}\right)}$$

which is the same as the period of a simple pendulum of length

$$l = \frac{(k^2 + h^2)}{h} = d$$

Exercise 6A

1. A rigid body of mass M kg and centre of mass G is free to rotate about a horizontal axis through a point A. Its moment of inertia about a parallel axis through G is $0.15M$ kg m². AG is perpendicular to the axis and of length 0.5 m. Write your answers in terms of M and g as appropriate.
   (i) Find the moment of inertia of the body about the axis through A.
   The body is held at rest with AG at an angle α to the downward vertical and let go. In the two cases (a) $\alpha = \frac{\pi}{2}$ and (b) $\alpha = \frac{\pi}{3}$:
   (ii) use the principle of conservation of energy to find the square of the angular speed when AG is vertical;
   (iii) by considering the equation of motion for the centre of mass, find the vertical component of the reaction at A when AG is vertical.

2. A rigid body of mass M and centre of mass G is free to rotate about a horizontal axis through a point A. Its moment of inertia about the axis of rotation is Ma^2, AG is perpendicular to the axis and of length $\frac{a}{2}$. When the body is hanging at rest with AG verticle it is hit at G by a horizontal impulse $Ma\omega$ perpendicular to the axis of rotation.

   (i) By taking moments about the axis, find an expression for the initial angular speed ω_0.
   (ii) By considering the change in momentum of the centre of mass, find the impulsive reaction at the axis.
   (iii) Use the principle of conservation of energy to find an expression for $a\dot{\theta}^2$ when AG has turned through an angle θ.
   (iv) Differentiate your expression for $a\dot{\theta}^2$ with respect to t to find an expression for $a\ddot{\theta}$ in the same position.
   (v) Verify that your answer for $a\ddot{\theta}$ is the same as you would obtain using the equation of motion for rotation about the axis.
   (vi) Write down the components of the acceleration of G parallel and perpendicular to GA when AG makes an angle θ with the downward vertical. Hence find the components of the reaction at A in these directions.
   (vii) Verify that the resultant reaction at A is vertical when AG is vertical.
   (viii) Find ω if the resultant reaction is also vertical when AG is horizontal.

3. A rod AB, of mass m and length $4l$, is free to rotate in a vertical plane about a

horizontal axis through a point C at a distance l from A. When it is hanging vertically, the rod is hit at B by a horizontal impulse J which is perpendicular to the axis of rotation.

(i)   Use the parallel axis theorem to show that the moment of inertia of the rod about the axis is $\frac{7}{3}ml^2$.

(ii) Find an expression for the initial angular speed of the rod.

(iii) Use the principle of conservation of energy to find an expression for the least value of J which will cause the rod to make complete revolutions about the axis of rotation.

4. A door of mass m opens against a wall as shown in the diagram. In order to stop the door hitting the wall, a doorstop is to be attached to the wall at the same height as the centre of mass of the door and a distance x from the hinge. The door opens and is moving with angular speed ω when it is brought to rest by the doorstop. The door can be modelled as a uniform thin rectangular lamina of width a.

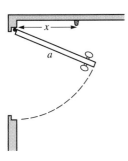

(i)   By considering the change in angular momentum when the door hits the doorstop, find the impulse of the stop on the door.

(ii) By considering the change in linear momentum of the centre of mass of the door, find the impulse at the hinge.

(iii) Show that the impulse at the hinge is zero when $x = \frac{2}{3}a$

5. A smooth rod AB, of mass M and length $2a$, has a small ring of mass $\frac{1}{6}M$ threaded on it. The rod is free to turn in a horizontal plane

about a vertical axis through A. The ring is held at the midpoint of the rod while it is set in motion with angular speed ω and then released.

(i)   By considering the forces acting on the ring, explain why it will not stay in its initial position relative to the rod.

(ii) Find the angular momentum of the rod and the ring at the instant the ring is released.

(iii) Explain why angular momentum is conserved once the ring is released and show that the angular speed of the rod and the ring when the ring reaches B is $\frac{3}{4}\omega$.

6. Two gear wheels are such that, when they are engaged, their angular speeds are inversely proportional to their radii. One has a radius a and moment of inertia pa^2 about its axis of rotation. The other gear wheel has radius b and moment of inertia qb^2. The first is rotating with angular speed ω when it engages with the second which is initially at rest.

(i)   By considering the change in angular momentum of each wheel separately, find the impulse between the teeth of the gear wheels when they engage.

(ii) Find the angular speed of each wheel.

(iii) Why is the angular momentum not conserved?

(iv) Find the energy lost when the gears engage.

7. This question describes a simplified model of a device used to de-spin a satellite. A uniform circular disc of mass $12m$ and radius a lies on a smooth horizontal table and is free to rotate about a fixed vertical axis through its centre. A light wire is attached to a point on the rim of the disc and is wound round this rim. A particle of mass m is attached to the free end of the wire and is initially attached to the rim.

When the disc is rotating with angular speed ω in the opposite sense to that in which the wire is wound the particle is released so that the wire unwinds and remains taut. The length of the wire is chosen so that it is completely unwound at the instant that the disc stops rotating. The particle is then moving at right angles to the wire.

Use the principles of conservation of angular momentum and energy to find the length of the wire.

8. A square board, of side $2a$ m and mass M kg, is to be used to estimate the speed of bullets. It is freely hinged about one horizontal edge and hangs at rest in a vertical plane. A bullet of mass m kg travelling horizontally with speed V m s^{-1} hits the board at its centre and becomes embedded in it. The board then rotates through an angle α before coming to rest.
   (i)   Show that the initial angular speed of the board is $\dfrac{3mV}{(4M+3m)a}$.
   (ii) Show also that
         $$V^2 = \frac{2ga}{3m^2}(M+m)(4M+3m)(1-\cos\alpha).$$
   (iii) Evaluate V when $M=5$, $m=0.005$, $a=0.1$, $\alpha=0.2$ radians.

9. A cricket bat of mass M kg is pivoted about a horizontal axis through a point A on the handle as shown. The axis is perpendicular to the page, G is the centre of mass and $AG=a$ m.

   The period of small oscillations about the axis is found to be T s.
   (i)   Show that the moment of inertia of the bat about the axis is $\dfrac{MgaT^2}{4\pi^2}$.

   The bat is hanging vertically when it is hit with a horizontal impulse at the point B a distance b m below G.
   (ii) Prove that the impulsive reaction at A is zero when $a+b=\dfrac{gT^2}{4\pi^2}$.

10. A trap-door in the form of a uniform rectangular lamina, ABCD, has mass M and dimensions $AB=2a$ and $BC=2b$. The

trap-door is freely hinged about a horizontal axis through AB. The diagram shows the forces acting on the trap-door after it has rotated from a horizontal rest position through an angle θ. X and Y are the components of the reaction on the trap-door at the hinge.

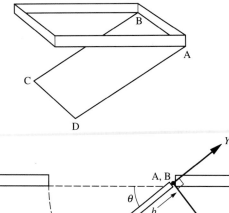

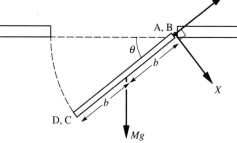

(i)   By using the principle of conservation of energy, show that
      $$\dot\theta^2 = \frac{3g}{2b}\sin\theta$$
(ii) By using the equation of motion for rotation about the axis AB, find an expression for $\ddot\theta$.
(iii) By considering the equations of motion for the centre of mass of the trap-door, find expressions for X and Y.
(iv) Find the values of X and Y when $\theta=\frac{\pi}{2}$ and compare these with the corresponding values when the trap-door is hanging vertically at rest.

                                   [MEI 1992 adapted]

11. A thin ruler of mass m and length 0.3 m is held in a horizontal position perpendicular to the rough edge of a table with two-thirds of its length overhanging the edge. It is then released.
    (i)   Show that, in the period before slipping occurs, when the ruler has turned

through an angle θ, its angular speed is given by $\dot{\theta}^2 = 10g \sin \theta$.

(ii) By considering the equation for the rotation of the ruler about the edge of the table, find an equation for $\ddot{\theta}$ in terms of θ.

(iii) Show that the second equation can be obtained by differentiating the first with respect to t.

(iv) By considering the equation for the linear motion of the centre of mass of the ruler, find the frictional force and normal reaction at the edge of the table. Hence show that the ruler begins to slip when $\tan \theta = \frac{1}{2}\mu$, where μ is the coefficient of friction between the ruler and the table edge.

12. A uniform circular disc, centre O, of mass m and radius a, can turn in a vertical plane about a horizontal axis through the point A on its circumference and perpendicular to the plane of the disc. It is released from rest with AO making an angle $\frac{\pi}{4}$ with the downward vertical.

(i) Show that when AO makes an angle θ $(<\frac{\pi}{4})$ with the downward vertical

$$\dot{\theta}^2 = \frac{g}{3a}(4\cos\theta - 2\sqrt{2})$$

(ii) Find the components of the reaction at A, parallel and perpendicular to OA in this position.

[Oxford 1994]

13. A trap-door ABCD of mass M is freely hinged along the edge AB, and AD is of length $2a$. It is initially horizontal and a small gnome of mass m is standing on its

centre. The trap-door then opens and swings downwards.

(i) Assuming the gnome stays at the centre of the trap-door, show that when it makes an angle θ with the horizontal

$$\dot{\theta}^2 = \frac{6(M+m)g\sin\theta}{(4M+3m)a}$$

(ii) The coefficient of friction between the gnome and the door is μ. Show that the gnome will not slip until

$$\tan\theta = \frac{\mu M}{(10M + 9m)}$$

14. A uniform rod AB, of length $2a$ and mass m, has a particle of mass m attached to the end B. Initially the rod is at rest in a vertical position with the end A in contact with a rough horizontal table. The rod is slightly disturbed from this position and after a time t it is inclined at an angle θ with the vertical.

(i) Assuming that the end A remains in contact with the table and has not slipped, show that

$$a\dot{\theta}^2 = \frac{9}{8}g(1 - \cos\theta)$$

(ii) Find an expression for $a\ddot{\theta}$ in terms of θ and g.

(iii) Find the components of the reaction between the table and the rod in the directions parallel and perpendicular to the rod. Hence show that the frictional force between the rod and the table is

$$\tfrac{27}{16}mg\sin\theta(3\cos\theta - 2)$$

(iv) Show also that the normal reaction exerted on the rod by the table is

$$\tfrac{1}{16}mg(9\cos\theta - 5)(9\cos\theta - 1)$$

[Cambridge 1992]

Investigation

Investigate the position of the centre of percussion for a piece of sports equipment. You could do this by modelling it mathematically or by doing an experiment.

The rotation of a rigid body about a moving axis

The quotation at the head of this chapter comes at the end of Richard Feynman's description of a difficult period in his work. One day someone threw a plate in the air and he noticed how it wobbled. With his insatiable curiosity, he started to work out the motion of the plate and found a satisfyingly simple result from a 'very complicated equation'. As any mathematician would, he then searched for a simpler 'more fundamental way, by looking at the forces or dynamics' to show why it moved as it did. He found a way and this event led to a turning point in his research when he decided to pursue whatever interested him.

HISTORICAL NOTE

Richard Feynman (1918 to 1988) was awarded the Nobel Prize for physics in 1965. His other accomplishments included life drawing, cracking safes and playing the bongos.

Richard Feynman was not the only person to find rotation interesting. The motion of fairground rides and gyroscopes, for example, are a source of great fascination for many people. This is a good reason for relating Feynman's story, but for present purposes it also illustrates how important it is to use a sensible approach in order to avoid complicated equations and erroneous assumptions. If you apply only the equations you have used for motion about a fixed axis to a plate thrown in the air, you will never discover anything about the wobble.

Angular momentum about a moving axis

It is shown on pages 295–296 that when a rigid body rotates about a moving axis through a point A, the results already found for a fixed axis still hold when

    (i) A is moving with constant velocity;

or (ii) A is G, the centre of mass of the body.

In other words, provided A satisfies one of the above conditions:

* when there is no resultant moment about the axis through A, total angular momentum is conserved;
* the resultant moment about the axis through A is equal to the rate of change of angular momentum about the same axis.

You might have seen the effect of the first result when an ice skater is rotating about a fixed point on the ice. She starts rotating with her arms spread out and then brings them in towards her body, and as she does so she speeds up. When the distance from the axis of the particles in her arms becomes smaller, the moment of inertia decreases and the magnitude of $\dot{\theta}$ increases so that constant angular momentum is maintained.

During the remainder of this chapter you will see how these principles can be used to determine further results for the rotation of a rigid body about an axis which moves so that it is always in the same direction: for example, when a ball rolls in a straight line. Although they are still true when referred to a point moving with constant speed, you will see later that, when using the *kinetic energy* of a rolling body, it is essential to consider motion relative to the centre of mass.

Rotation about an axis with constant direction

Conditions for $C = I\ddot{\theta}$

You have already used the result $C = I\ddot{\theta}$ for rotation about a fixed axis. Now you can use it more generally, but it is reliable only under certain conditions which you will appreciate on reading the appropriate vector analysis. These are

- The direction of the axis must never change.
- Unless the axis moves with constant velocity, you should always take moments about an axis through the centre of mass of the body.
- When the axis is not fixed, there might also be problems unless the body is suitably symmetric. The theory behind this is indicated on pages 291, 292 and in Exercise 9D, Question 6. You might have heard of the need to 'balance' car wheels to ensure that they do not wobble.

Rolling bodies

The results obtained in this section can be applied to the motion of rolling bodies so long as the above conditions are satisfied. When a body rolls under the influence of gravity, for example, the axis accelerates, so in this case it is essential to take moments about an axis through the centre of mass. It is also useful to remember that, even when a body has an axis of symmetry, it does not necessarily roll in such a way that the direction of the axis remains constant. A cone, for example, moves round in a circle when rolling on a horizontal plane. However, you can assume that all the rolling bodies you will meet in this book do roll so that the axis remains in a constant direction.

When an object rolls on a surface without slipping, the point of contact of any circle of cross-section which touches the surface is instantaneously at rest. When the centre of such a circle, of radius r, is moving with speed $\dot{x}$ and the angular speed of the body is $\dot{\theta}$, the velocity of the point of contact, P, relative to the centre of the circle is $r\dot{\theta}$. Its actual velocity is $\dot{x} - r\dot{\theta}$, as shown in figure 6.5.

So the condition for rolling without slipping is $\dot{x} - r\dot{\theta} = 0$.

$$\Rightarrow \qquad \dot{x} = r\dot{\theta}$$

The velocity of P is $\dot{x} - r\dot{\theta}$

Figure 6.5

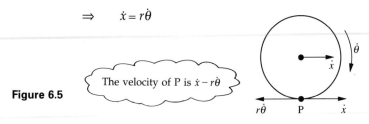

EXAMPLE

An empty milk bottle of mass 0.25 kg and maximum radius 4 cm rolls without slipping down a draining board inclined at 2° to the horizontal. Starting from rest it takes 2 s to roll 0.5 m.
(i) Find the moment of inertia of the bottle about its axis of symmetry.
(ii) Find the least value of the coefficient of friction which is required to prevent the bottle from slipping.

Solution

(i) The diagram shows the forces acting on the bottle in a cross-section through the centre of mass G. F is the frictional force.

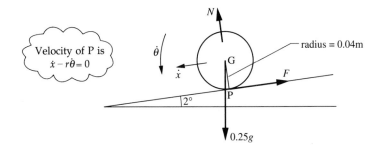

For rolling without slipping, $\dot{x} = r\dot{\theta} = 0.04\,\dot{\theta}$, so the acceleration of G is $\ddot{x} = 0.04\,\ddot{\theta}$ and its equation of motion is

$$0.25g \sin 2° - F = 0.25\ddot{x}$$

$$\Rightarrow \qquad 0.25g \sin 2° - F = 0.25 \times 0.04\,\ddot{\theta}$$

$$\Rightarrow \qquad F = 0.0855 - 0.01\,\ddot{\theta} \qquad\qquad ①$$

The bottle rolls 0.5 m in 2 s from rest with acceleration $0.04\,\ddot{\theta}$:

$$\Rightarrow \qquad 0.5 = \tfrac{1}{2} \times 0.04\,\ddot{\theta} \times 2^2 \quad (s = \tfrac{1}{2}at^2)$$

$$\Rightarrow \qquad \ddot{\theta} = 6.25$$

$$\Rightarrow \qquad F = 0.0855 - 0.0625 = 0.0230 \quad \text{(from ①)}$$

Taking moments about G the equation of motion for rotation is

$$F \times 0.04 = I\ddot{\theta}$$

$$\Rightarrow \qquad 0.0230 \times 0.04 = 6.25\,I$$

$$\Rightarrow \qquad I = 1.47 \times 10^{-4}$$

The moment of inertia of the bottle about its axis is 1.47×10^{-4} kg m². This may seem rather small, but not when you think that a mass of 1 kg at a distance of 1 m from the axis has a moment of inertia of 1 kg m².

(ii) G has no acceleration perpendicular to the plane so the normal reaction

$$N = 0.25g \cos 2° \quad = 2.4485$$

There is no slipping provided $F \leqslant \mu N$ and $F = 0.0230$. Hence the least value of the coefficient of friction μ is $0.0230/2.4485 = 0.0094$. This shows that very little friction is required for the bottle to roll without slipping.

Work and energy

You learnt in *Mechanics 5* that the work done by a force $\mathbf{F}_p$ acting at a point P with position vector $\mathbf{r}_p$ relative to a fixed origin O is

$$\int \mathbf{F}_p . \mathrm{d}\mathbf{r}_p = \int \mathbf{F}_p . \dot{\mathbf{r}}_p \, \mathrm{d}t$$

and when $\mathbf{F}_p$ is the resultant force acting on a particle at P, the work done is equal to the increase in kinetic energy of the particle:

$$\int \mathbf{F}_p . \dot{\mathbf{r}}_p \, \mathrm{d}t = [\tfrac{1}{2} m_p |\dot{\mathbf{r}}_p|^2]$$

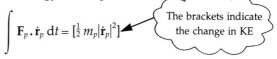

The brackets indicate the change in KE

The work–energy equation for a rigid body can be found by summing this over all particles. The work done by the internal forces cancels so

$$\sum \int \mathbf{F}_{\mathrm{ext}} . \dot{\mathbf{r}} \, \mathrm{d}t = \sum_{\mathrm{all}\,p} [\tfrac{1}{2} m_p |\dot{\mathbf{r}}_p|^2]$$

$$= \left[\sum_{\mathrm{all}\,p} (\tfrac{1}{2} m_p |\dot{\mathbf{r}}_p|^2) \right]$$

where $\mathbf{F}_{\mathrm{ext}}$ is a typical external force acting at a point $\mathbf{r}$.

As you might expect, the work done by the external forces is equal to the change in the total kinetic energy of the system. This result can be used for a rigid body when you know more convenient ways of expressing the total kinetic energy and the work done by the forces.

The kinetic energy of a rigid body

The total kinetic energy of a rigid body is $\sum \tfrac{1}{2} m_p |\dot{\mathbf{r}}_p|^2$, where m_p and $\dot{\mathbf{r}}_p$ are the mass and velocity of a typical particle, P, in the body. Figure 6.6 shows such a particle in a body which is rotating with angular speed $\dot{\theta}$ about an axis through a point A. The particle moves in a circle of radius r relative to the axis and $\boldsymbol{\rho}_p$ is its position relative to A.

Referring to a fixed origin, O

$$\mathbf{r}_p = \boldsymbol{\rho}_p + \mathbf{r}_A$$

Hence $\quad \dot{\mathbf{r}}_p = \dot{\boldsymbol{\rho}}_p + \dot{\mathbf{r}}_A$

$$\Rightarrow \quad \dot{\mathbf{r}}_p . \dot{\mathbf{r}}_p = (\dot{\boldsymbol{\rho}}_p + \dot{\mathbf{r}}_A) . (\dot{\boldsymbol{\rho}}_p + \dot{\mathbf{r}}_A)$$

$$\Rightarrow \quad |\dot{\mathbf{r}}_p|^2 = |\dot{\boldsymbol{\rho}}_p|^2 + 2\dot{\boldsymbol{\rho}}_p . \dot{\mathbf{r}}_A + |\dot{\mathbf{r}}_A|^2$$

$$= r^2\dot{\theta}^2 + 2\dot{\boldsymbol{\rho}}_p . \dot{\mathbf{r}}_A + |\dot{\mathbf{r}}_A|^2$$

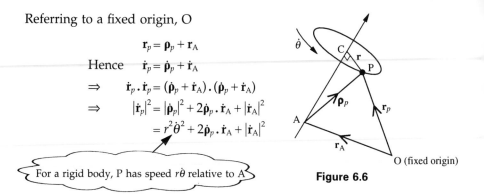

For a rigid body, P has speed $r\dot{\theta}$ relative to A

Figure 6.6

Adding the kinetic energies for all the particles gives

$$\sum (\tfrac{1}{2} m_p |\dot{\mathbf{r}}_p|^2) = \tfrac{1}{2}\left(\sum m_p r^2\right)\dot{\theta}^2 + \left(\sum m_p \dot{\boldsymbol{\rho}}_p\right) . \dot{\mathbf{r}}_A + \tfrac{1}{2}\sum m_p |\dot{\mathbf{r}}_A|^2$$

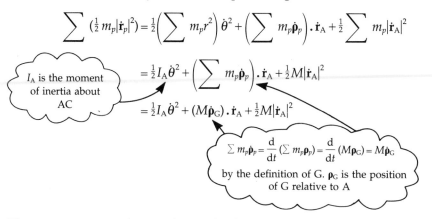

I_A is the moment of inertia about AC

$$= \tfrac{1}{2} I_A \dot{\theta}^2 + \left(\sum m_p \dot{\boldsymbol{\rho}}_p\right) . \dot{\mathbf{r}}_A + \tfrac{1}{2} M |\dot{\mathbf{r}}_A|^2$$

$$= \tfrac{1}{2} I_A \dot{\theta}^2 + (M\dot{\boldsymbol{\rho}}_G) . \dot{\mathbf{r}}_A + \tfrac{1}{2} M |\dot{\mathbf{r}}_A|^2$$

$$\sum m_p \dot{\boldsymbol{\rho}}_p = \frac{\mathrm{d}}{\mathrm{d}t}(\sum m_p \boldsymbol{\rho}_p) = \frac{\mathrm{d}}{\mathrm{d}t}(M\boldsymbol{\rho}_G) = M\dot{\boldsymbol{\rho}}_G$$

by the definition of G. $\boldsymbol{\rho}_G$ is the position of G relative to A

There are two special cases that make this more manageable:

- A is fixed so that $\dot{\mathbf{r}}_A = \mathbf{0}$.
- A is the centre of mass G, so that $\boldsymbol{\rho}_G$ and hence $\dot{\boldsymbol{\rho}}_G$ are always zero.

In the first case, the total kinetic energy is $\tfrac{1}{2} I_A \dot{\theta}^2$, where I_A is the moment of inertia about the axis through A (see *Mechanics 5*, pages 129–132).

In the second case, the total kinetic energy is

$$\tfrac{1}{2} I_G \dot{\theta}^2 + \tfrac{1}{2} M |\dot{\mathbf{r}}_G|^2$$

$$\text{or} \quad \tfrac{1}{2} I_G \dot{\theta}^2 + \tfrac{1}{2} M v_G^2$$

where v_G replaces $|\dot{\mathbf{r}}_G|$ as the speed of the centre of mass.

Thus the total kinetic energy is equal to the energy of rotation about an axis through the centre of mass plus the kinetic energy of a particle of equal mass moving with the speed of the centre of mass.

The work done by an external force

The work done by an external force F is $\displaystyle\int \mathbf{F} . \mathrm{d}\mathbf{r}_p = \int \mathbf{F} . \dot{\mathbf{r}}_p \, \mathrm{d}t$, where $\dot{\mathbf{r}}_p$ is the velocity of the point, P, at which the force acts.

When $\boldsymbol{\rho}_p$ is the position of this point relative to the point A on the axis of rotation, as shown in figure 6.7

$$\mathbf{r}_p = \boldsymbol{\rho}_p + \mathbf{r}_A \quad \text{and} \quad \dot{\mathbf{r}}_p = \dot{\boldsymbol{\rho}}_p + \dot{\mathbf{r}}_A$$

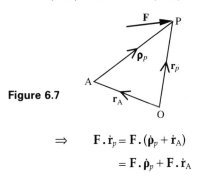

Figure 6.7

$$\Rightarrow \quad \mathbf{F} \cdot \dot{\mathbf{r}}_p = \mathbf{F} \cdot (\dot{\boldsymbol{\rho}}_p + \dot{\mathbf{r}}_A)$$
$$= \mathbf{F} \cdot \dot{\boldsymbol{\rho}}_p + \mathbf{F} \cdot \dot{\mathbf{r}}_A$$

When the body is rotating with angular speed $\dot{\theta}$, about an axis through A which is in a constant direction, $\dot{\boldsymbol{\rho}}_p = r\dot{\theta}$ in the direction of the tangent to the circle of radius r which is described by the point P (figure 6.8). Hence

$$\mathbf{F} \cdot \dot{\boldsymbol{\rho}}_p = r\dot{\theta} \times \text{component of } \mathbf{F} \text{ parallel to the tangent}$$

$$= \dot{\theta} \times \text{moment of } \mathbf{F} \text{ about the axis through A}$$

$$= C\dot{\theta}$$

The work done by the force is therefore

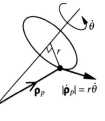

$$\int (\mathbf{F} \cdot \dot{\boldsymbol{\rho}}_p + \mathbf{F} \cdot \dot{\mathbf{r}}_A)\, dt = \int C\dot{\theta}\, dt + \int \mathbf{F} \cdot \dot{\mathbf{r}}_A\, dt$$

$$= \int C\, d\theta + \int \mathbf{F} \cdot \dot{\mathbf{r}}_A\, dt$$

Figure 6.8

When A, and therefore the axis, is fixed this simplifies to $\int C\, d\theta$. When A is

G the work done is made up of two parts. The first part is due to the moment, C, of the force about the axis through G; the second part is

$$\int \mathbf{F} \cdot \dot{\mathbf{r}}_G\, dt = \int \mathbf{F} \cdot d\mathbf{r}_G$$

This is the work done by the force *as though it acts at G*. These results can also be proved using vector products.

The work–energy equation

When the results in the last two sections are combined they give the work–energy equation for a rigid body.

Provided moments are taken about an axis through the centre of mass:

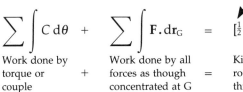

$$\sum \int C\, d\theta \;+\; \sum \int \mathbf{F} \cdot d\mathbf{r}_G \;=\; [\tfrac{1}{2} I_G \dot{\theta}^2 \;+\; \tfrac{1}{2} M v_G^2]$$

| Work done by torque or couple | + | Work done by all forces as though concentrated at G | = | Kinetic energy of rotation about axis through G | + | Kinetic energy of whole mass concentrated at G |

Conservation of energy

The weight of a body has no moment about G, so the work done by gravity is

$$0 + \int M\mathbf{g}.d\mathbf{r}_G = \text{loss in potential energy}$$

When there are no other forces acting, mechanical energy is conserved.

Frictional forces and conservation of energy for a rolling body

When a ball or cylinder is placed on an inclined plane, it tends to roll rather than slip. The frictional force acts up the plane to prevent slipping and as a result has a moment about the centre of mass which makes the body roll. The next example shows how the equations of motion can be used to demonstrate two useful results for rolling without slipping: conservation of energy and the fact that the friction does no work.

EXAMPLE

A rigid body rolls, without slipping, down the line of greatest slope of a plane inclined at an angle α to the horizontal in such a way that the axis of rotation through its centre of mass remains horizontal.
(i)   Find the total work done on the body when its centre of mass has rolled a distance x from rest and hence show that
     (a) the work done by the frictional force is zero;
     (b) mechanical energy is conserved.
(ii) Write down the energy equation for the body and hence express the velocity, $\dot{x}$, of the centre of mass in terms of x.

Solution

The diagram shows the forces acting on the cross-section through G, which is assumed to be a circle of radius a. F_r is the resultant of the frictional forces acting at the points of contact between the body and the plane, and N is the resultant normal reaction. These must be in the plane of G, because otherwise there would be a turning effect on the axis of rotation. N acts through G because it is at right angles to the tangent of the circular cross section. F_r acts up the plane to prevent sliding and is responsible for the rotation of the body, being the only force having a moment about the axis.

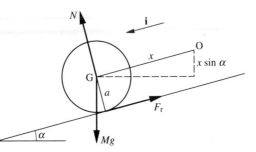

(i)   Assume that the initial position of G is O and $\overrightarrow{OG} = x\mathbf{i}$. The velocity of G is then $\dot{x}\mathbf{i}$.

The total work done by the forces is

$$\int C \, d\theta + \int \mathbf{F} \cdot (\dot{x}\mathbf{i}) \, dt$$

In this case $C = F_r a$, and the resultant force parallel to the plane is $Mg \sin \alpha - F_r$

$$\Rightarrow \quad \sum \mathbf{F} \cdot (\dot{x}\mathbf{i}) = (Mg \sin \alpha - F_r)\dot{x}$$

The total work done is therefore

$$\int F_r a \, d\theta + \int (Mg \sin \alpha)\dot{x} \, dt - \int F_r \dot{x} \, dt = F_r a\theta - F_r x + Mgx \sin \alpha \quad \text{(for constant } F_r)$$

(a) When the body has rolled without slipping through an angle θ

$$x = a\theta$$

$$\Rightarrow \quad F_r a\theta - F_r x = 0$$

Hence the work done by friction is zero. This is not surprising. When the body is rolling without slipping, the frictional forces always act through points which are instantaneously at rest so their points of application do not move. Notice also that the normal reaction does no work because G does not move perpendicular to the plane.

(b) The total work done is therefore $Mg(x \sin \alpha)$, which is the decrease in the potential energy of the body. The total work done is also equal to the increase in kinetic energy of the body, so

increase in KE = decrease in PE

$$\Rightarrow \quad \text{total mechanical energy is conserved}$$

(ii) The energy equation is

$$\tfrac{1}{2} I\dot{\theta}^2 + \tfrac{1}{2} M\dot{x}^2 = Mgx \sin \alpha$$

also

$$\dot{x} = a\dot{\theta}$$

$$\Rightarrow \quad \tfrac{1}{2} I\left(\frac{\dot{x}^2}{a^2}\right) + \tfrac{1}{2} M\dot{x}^2 = Mgx \sin \alpha$$

$$\Rightarrow \quad \dot{x}^2 = \frac{2Mgxa^2 \sin \alpha}{(I + Ma^2)}$$

$$\Rightarrow \quad \dot{x} = \sqrt{\left[\frac{2Mgxa^2 \sin \alpha}{(I + Ma^2)}\right]}$$

N O T E

This example illustrates a method. You are advised not to quote the last result as a formula but to use the energy equation for each problem separately whenever it applies.

EXAMPLE

The yo-yo (meaning come-come) originates from a Filipino fighting weapon. One recorded in the sixteenth century weighed about 1.8 kg and had a 6 m thong. A hypothetical yo-yo of this mass might be carved from a block of wood and consist of two discs of radius 10 cm and thickness 4 cm joined by a cylinder of radius 2 cm and width 4 cm, as shown in the diagram.

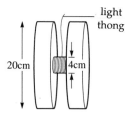

(i)   Find the mass and moment of inertia of each part of the yo-yo about the axis of symmetry and hence the total moment of inertia about the axis.

(ii) The thong is wound round the cylinder and the yo-yo is dropped from a tree. Assuming there are no frictional forces, find its velocity just before the thong tightens.

(iii) Compare this with the speed of the yo-yo when it is dropped 6 m without rotation.

Solution

(i)   The volume of each disc is $400\pi \text{ cm}^3$ and the volume of the cylinder is $16\pi \text{ cm}^3$ giving a total of $816\pi \text{ cm}^3$.

The mass of each disc is therefore $\dfrac{400}{816} \times 1.8 \text{ kg} = 0.882 \text{ kg}$ and the mass of the cylinder is approximately $1.8 - 2 \times 0.882 = 0.036 \text{ kg}$.

The moment of inertia of a solid cylinder about its axis of symmetry is $\frac{1}{2}Mr^2$. Therefore

$$\text{the moment of inertia of each disc} = 0.441 \times 0.1^2 \text{ kg m}^2$$
$$= 0.004\,41 \text{ kg m}^2$$
$$\text{and the moment of inertia of the cylinder} = 0.018 \times (0.02)^2 \text{ kg m}^2$$
$$= 0.000\,0072 \text{ kg m}^2$$

The total moment of inertia about the axis of symmetry is $0.008\,83 \text{ kg m}^2$.

(ii) The direction of the axis of the yo-yo is fixed, so when the angular velocity is $\dot{\theta}$, the kinetic energy of rotation is

$$\tfrac{1}{2}I\dot{\theta}^2 = \tfrac{1}{2} \times 0.008\,83\,\dot{\theta}^2$$
$$= 0.004\,41I\dot{\theta}^2$$

The kinetic energy due to the motion of the centre of mass is

$$\tfrac{1}{2}M\dot{x}^2 = 0.9\,\dot{x}^2$$

While the yo-yo is falling, the speed of the point P where the thong starts to wind round the yo-yo is zero because the vertical part of the thong is stationary. At the end of the fall, P is 0.02 m from the centre of the yo-yo,

as shown in the diagram. Hence

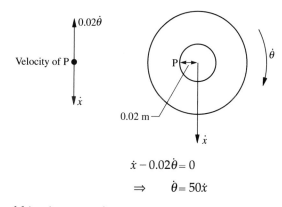

$$\dot{x} - 0.02\dot{\theta} = 0$$
$$\Rightarrow \quad \dot{\theta} = 50\dot{x}$$

so the total kinetic energy is

$$11.025\dot{x}^2 + 0.9\dot{x}^2 = 11.925\dot{x}^2 \text{ J}$$

When the yo-yo has fallen 6 m, its loss of potential energy is $1.8g \times 6 = 10.8g$ J. Assuming that the thong is light, the principle of conservation of energy gives

$$11.925\dot{x}^2 = 10.8g$$

The speed of the yo-yo is therefore 2.98 m s^{-1}.

(iii) When the yo-yo is dropped without rotation its energy equation is

$$\tfrac{1}{2}Mv^2 = Mgh$$
$$\Rightarrow \quad v^2 = 12g$$

In this case, the speed of the yo-yo is therefore 10.8 m s^{-1}, which is over three times as fast as the speed with rotation.

The effects of impulses on rolling bodies

It is shown on page 190 that the motion of the centre of mass when the body is subjected to impulses is given by

$$\sum \mathbf{J}_{\text{ext}} = [M\dot{\mathbf{r}}_{\text{G}}]$$

It also follows from the results for angular momentum quoted on page 198 that when the axis of rotation is fixed or in a fixed direction through the centre of mass, the sum of the moments of external impulses about the axis of rotation is equal to the change in the angular momentum, $I\dot{\theta}$, about the axis. See also page 188.

This is the rotational equivalent of 'impulse = change in momentum' for the linear motion of a particle.

EXAMPLE

A large yo-yo of mass 1.8 kg, moment of inertia 0.008 83 kg m², and outside radius 10 cm, is moving with a speed of 2.98 m s⁻¹ and an angular speed of 149 rad s⁻¹ when the thong is just about to reach its maximum length of 6 m. At this moment, it hits its target with a tangential blow which reduces both its linear and angular momentum to zero.

(i)   Find the impulsive reaction on the yo-yo and the angle the blow makes with the vertical.

(ii) Compare this with the impulse received when the yo-yo is dropped from a height of 6 m without rotation and is stopped by the target.

Solution

(i)   Suppose the impulse on the yo-yo is J N s and makes an angle α with the vertical as shown. The equation for linear momentum is

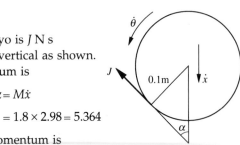

$$J\cos\alpha = M\dot{x}$$

$$= 1.8 \times 2.98 = 5.364$$

and the equation for angular momentum is

$$J \times 0.1 = I\dot{\theta}$$

$$\Rightarrow \qquad 0.1J = 0.008\,83 \times 149$$

$$\Rightarrow \qquad J = 13.16$$

Hence
$$\cos\alpha = \frac{5.364}{13.16} = 0.4077$$

$$\Rightarrow \qquad \alpha = 65.9°$$

The impulse is 13.2 N s delivered at an angle of 66° with the vertical.

(ii) When the yo-yo is dropped, its speed is $\sqrt{(2gh)} = 10.844$ m s⁻¹ and its momentum is $1.8 \times 10.844 = 19.5$ N s (to 3 significant figures). The impulse is 19.5 N s delivered vertically.

A yo-yo such as this would have been used for hunting or fighting. Although the speed of the yo-yo is much greater in the second case, the impulse it can deliver could be considered to be not so much greater that it negates the advantage of an implement which is able to return readily to the sender when it fails to hit its target.

Rolling and skidding

When a body rolls on a surface without slipping, any point of contact with the surface is instantaneously at rest. The speed, $\dot{x}$, of the centre of a circle of cross-section of radius r through a point of contact is then related to the angular speed $\dot{\theta}$, by the equation $\dot{x} = r\dot{\theta}$. When $\dot{x} \neq r\dot{\theta}$ the body skids rather than rolls. In this case, frictional forces act in such a way as to reduce the point of contact to rest by equalising $\dot{x}$ and $r\dot{\theta}$. Work is done against friction and mechanical energy is lost. You might have noticed, however, that snooker balls sometimes speed up after they are hit. The next example shows how this can happen without violating the principle of conservation of energy.

EXAMPLE

A snooker ball of mass M and radius r (which is initially at rest on a horizontal table) is hit horizontally by a cue at a point on the vertical plane through its centre which is at a height $\frac{1}{2}r$ above the centre. The impulse imparted by the cue is J and as a result the ball moves forward with a speed u and spins with an angular speed ω_0. The coefficient of friction between the ball and the table is μ.

(i) Find ω_0 in terms of u.

(ii) Find the time taken for the ball to stop skidding and start rolling.

(iii) Find the loss in kinetic energy of the ball.

(iv) Verify that this is equal to $\int Fv\,dt$, where v is the velocity of the point of contact relative to the table.

Solution

(i) Although the table cannot be assumed to be smooth, the maximum frictional force possible can be assumed to have a negligible impulse over the very short period of time that the ball is hit by the cue, so there is no impulsive reaction between the ball and the table. The first diagram shows the impulse and the linear and angular speeds just after the ball is hit.

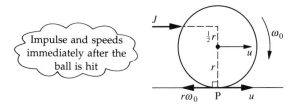

Impulse and speeds immediately after the ball is hit

The equation for the linear momentum of the ball is $J = Mu$. The moment of the impulse about the centre of the ball is $J \times \frac{1}{2}r$, so the equation for angular momentum is

$$J \times \tfrac{1}{2}r = I\omega_0$$

Assuming the ball is a uniform sphere $I = \frac{2}{5}Mr^2$ and, as $J = Mu$

$$\tfrac{1}{2}Mur = \tfrac{2}{5}Mr^2\omega_0$$

$$\Rightarrow \quad \omega_0 = \frac{5u}{4r}$$

(ii) The ball rolls when the point P in contact with the table is instantaneously at rest.

Immediately after the ball is struck, the forward speed of P is

$$u - r\omega_0 = u - \frac{5u}{4} = -\frac{u}{4}$$

so the ball skids, and because P is moving backwards there is a forward frictional force F. This serves to increase the speed $\dot{x}$ of the centre of mass and decrease the angular speed $\dot{\theta}$ until $\dot{x} = r\dot{\theta}$ and the ball begins to roll. While the ball skids, $F = \mu N$. Resolving vertically gives $N = Mg$.

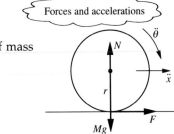

Forces and accelerations

$$\Rightarrow \quad F = \mu Mg$$

The equation of motion for the centre of mass is

$$F = M\ddot{x}$$

where $\ddot{x}$ is the acceleration.

$$\Rightarrow \quad \ddot{x} = \mu g$$

After time t the velocity of the centre of mass is

$$\dot{x} = u + \mu g t \qquad \text{①}$$

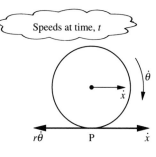

Speeds at time, t

The equation for rotation is

$$Fr = -I\ddot{\theta}$$

$$\Rightarrow \quad \mu Mgr = -\tfrac{2}{5} Mr^2 \ddot{\theta}$$

$$\Rightarrow \quad \ddot{\theta} = -\frac{5\mu g}{2r}$$

After time t the angular speed $\dot{\theta}$ is

$$\dot{\theta} = \omega_0 - \frac{5\mu g t}{2r}$$

$$\Rightarrow \quad r\dot{\theta} = \frac{5u}{4} - \frac{5\mu g t}{2} \qquad \text{②}$$

The ball stops skidding and starts rolling when the point in contact with the table has zero velocity, so rolling occurs when $\dot{x} = r\dot{\theta}$

$$\Rightarrow \quad u + \mu g t = \frac{5u}{4} - \frac{5\mu g t}{2} \quad \text{(from ① and ②)}$$

$$\Rightarrow \quad \frac{7\mu g t}{2} = \frac{u}{4}$$

$$\Rightarrow \quad \mu g t = \frac{u}{14}$$

The time taken to start rolling is $\dfrac{u}{14\mu g}$.

(iii) The kinetic energy of the ball just after being hit is

$$\tfrac{1}{2}Mu^2 + \tfrac{1}{2}\left(\tfrac{2}{5}Mr^2\right)\omega_0^2 = \tfrac{1}{2}Mu^2 + \tfrac{1}{5}M\left(\frac{5u}{4}\right)^2$$

This is $> u$, so the ball has 'speeded up'

$$= \tfrac{13}{16}Mu^2$$

When rolling begins, the speed of G is $u + \mu g t$ and $\mu g t = \dfrac{u}{14}$

$\Rightarrow$ the speed $= \dfrac{15u}{14}$

$\Rightarrow$ the angular speed $= \dfrac{15u}{14r}$

this is $> u$ as indicated on page 208

The kinetic energy when rolling begins is

$$\tfrac{1}{2}M\left(\frac{15u}{14}\right)^2 + \tfrac{1}{2}\left(\tfrac{2}{5}Mr^2\right)\left(\frac{15u}{14r}\right)^2 = \frac{7 \times 15^2\, Mu^2}{10 \times 14^2}$$

$$= \frac{45\, Mu^2}{56}$$

The kinetic energy *lost* is

$$Mu^2\left(\frac{13}{16} - \frac{45}{56}\right) = \frac{Mu^2}{112}$$

(iv) The friction is μMg and the velocity, v, of the point of contact relative to the table after time t is

$$v = \dot{x} - r\dot{\theta}$$

$$= u + \mu g t - r\left(\omega_0 - \frac{5\mu g t}{2r}\right) \quad \text{(from ① and ②)}$$

$$\Rightarrow \quad v = -\frac{u}{4} + \frac{7\mu g t}{2} \quad \left(r\omega_0 = \frac{5u}{4}\right)$$

So the work done by the friction is

$$\int Fv\, dt = \mu Mg\left(-\frac{ut}{4} + \frac{7\mu g t^2}{4}\right)$$

$$= \mu Mg\left[-\frac{u^2}{4 \times 14\mu g} + \frac{7\mu g\, u^2}{4(14\mu g)^2}\right]$$

$$= -\frac{Mu^2}{112}$$

The work done against the friction is $+\dfrac{Mu^2}{112}$ and is equal to the loss in kinetic energy of the ball.

N O T E

The work done in the foregoing example is not the product of the friction and the distance moved by the centre of the ball, which you can verify (using $s = ut + \tfrac{1}{2}at^2$) is equal to $\dfrac{29Mu^2}{392}$.

The work done against friction depends on the distance one surface has slid over the other rather than the actual distance moved. When the ball turns without moving forwards, it slides against the surface and the work done against the friction depends on the angle turned through. When the ball rolls without sliding, no work is done. The motion of a skidding ball is a combination of these.

Summary for rolling bodies

You now have a set of equations which can be used for the motion of rolling bodies. When $\mathbf{F}$, $\mathbf{J}$, and C represent the external forces, impulses and moments about the axis respectively,

$$\bullet \quad \sum \mathbf{F} = M\ddot{\mathbf{r}}_G$$
$$\left. \bullet \quad \sum \mathbf{J} = [M\dot{\mathbf{r}}_G] \right\} \quad \text{for the motion of the centre of mass, G}$$

$$\bullet \quad \sum \text{moment of } \mathbf{F} = C = I_G\ddot{\theta}$$
$$\left. \bullet \quad \sum \text{moment of } \mathbf{J} = [I_G\dot{\theta}] \right\} \quad \begin{array}{l} \text{for rotation about an axis through G} \\ \text{which remains in a constant direction.} \end{array}$$

• The work–energy equation

$$\int C \, d\theta + \sum \int \mathbf{F}.\dot{\mathbf{r}}_G \, dt = [\tfrac{1}{2} I_G\dot{\theta}^2 + \tfrac{1}{2} M|\dot{\mathbf{r}}_G|^2]$$

$$\text{or} \quad \int C \, d\theta + \sum \int \mathbf{F}.d\mathbf{r}_G = [\tfrac{1}{2} I_G\dot{\theta}^2 + \tfrac{1}{2} Mv_G^2]$$

where $\mathbf{v}_G = \dot{\mathbf{r}}_G$ is the velocity of G.

• When no external forces other than gravity do work, the total mechanical energy is conserved.

Notice in particular that the axis of rotation has been assumed to be through the centre of mass. You will remember that this is a condition for many of the equations of rotation when the axis is not fixed.

Other simplifying assumptions are
• The body rotates about an axis of rotational symmetry or is symmetrical about the plane through G perpendicular to the axis.
• The direction of the axis of rotation does not change.

Exercise 6B

1. (i) Find the kinetic energy of:
   (a) a uniform solid sphere of mass M and radius a rolling at a speed v m s^{-1}.
   (b) a uniform solid cylinder of mass m, radius r, rolling with angular speed ω rad s^{-1}.
   (ii) The sphere and cylinder, starting at rest, each roll a distance d down a plane inclined at an angle α to the horizontal. Find the speeds of their centres.

2. Alan wishes to stop a football of mass M and radius r which is rolling towards him along horizontal ground with angular speed ω, by applying an impulse J with his foot at a height h above the centre of the ball. Modelling the football as a *hollow* sphere,
   (i) use the change in momentum of the centre to find J in terms of M, r and ω;
   (ii) take moments about the centre to find h in terms of r.

3. A uniform solid ball, of mass m and radius a, is held at rest on a rough plane which is inclined to the horizontal at an angle α and then released. After a time t seconds the ball has rolled, without slipping, a distance x down the plane and turned through an angle θ.
   (i) Write x in terms of θ and hence write $\ddot{x}$ in terms of $\ddot{\theta}$.
   (ii) Using F as the frictional force between the ball and the table, write down the equations of motion for the rotation of the ball and the linear motion of its centre of mass.
   (iii) Hence find the acceleration of the ball and the time taken for it to roll a distance s down the plane.

4. Some students are trying to find g by rolling a uniform solid ball from rest down an inclined track. They assume no air resistance.
   (i) Use energy considerations to find the speed of the centre of the ball when it has dropped a vertical height y.
   (ii) The students assume the ball is an object sliding without friction. Find the speed of such an object after dropping the same vertical height, y.
   (iii) Hence show that, even when air resistance is negligible and all measurements made are accurate, the students' estimate of g will only be $\frac{5}{7}$ of the true value.

5. A cylindrical drum has mass M, radius r and moment of inertia I about its axis. The drum rolls without slipping down a ramp which is inclined at an angle α to the horizontal.
   (i) Write down the equations for the linear and angular motion of the drum and hence show that when it has rotated through an angle θ, the acceleration of its centre of mass is $g \sin \alpha - \dfrac{I\ddot{\theta}}{Mr}$.
   (ii) Use the condition for rolling without slipping to show that the acceleration of the axis of the drum is $\dfrac{Mgr^2 \sin \alpha}{(I + Mr^2)}$.

(iii) Use the principle of conservation of energy to find the speed of the axis when it has travelled a distance d from rest down the slope.

6. (i) You have two cylindrical objects of equal mass and size and you know that one is a uniform solid cylinder while the other is a closed cylindrical shell. If you roll them both from rest down an inclined plane, how can you decide which is the solid cylinder?
   (ii) The closed cylindrical shell is modelled as two uniform discs of mass m and radius R and a thin uniform hollow cylinder of mass M and radius R. Calculate its moment of inertia about its axis.
   (iii) Show that, if the plane is sufficiently rough to prevent slipping and is inclined at $30°$ to the horizontal, the acceleration of the centre of mass of the shell as it rolls down the plane is

$$\frac{(2m + M)g}{2(3m + 2M)}.$$

   (iv) By calculating the acceleration of the centre of mass of the solid cylinder, find which completes the journey down the plane in the shortest time.

[Oxford 1991]

7. Billiard and snooker balls vary in mass, but they are all of the same radius. The height of the cushion is such that any ball which rolls towards it without slipping and which hits the cushion at right angles will roll away from it without slipping. A ball, of mass m and radius a, rolls towards the cushion with angular speed ω_1 and receives a horizontal impulse J in the vertical plane containing its centre and at a height h above its centre, as shown in the diagram.

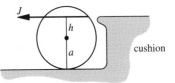

The ball then rolls away with angular speed ω_2 and without slipping.

(i) Draw a diagram to show the velocities of the centre of mass of the ball before and after the impact and hence obtain an expression for J in terms of ω_1 and ω_2.

(ii) By considering the change in angular momentum of the ball, show that

$$Jh = \tfrac{2}{5}ma^2(\omega_1 + \omega_2)$$

(iii) Show that the edge of the cushion is a height $\tfrac{7}{5}a$ above the table and hence that it is independent of the mass and angular speed of the ball.

8. In some preliminary investigations of road accidents a car is modelled by a uniform plane rectangular lamina with sides of length a and $2a$, and mass M. The car is to travel at a constant speed V in the direction parallel to the longer sides and to collide with a bollard. The collision is to be represented by an impulse of magnitude kMV at the front corner of the car which is nearer the kerb, in a direction opposite to the motion of the car before impact (see diagram).

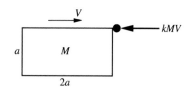

(i) Find the moment of inertia of the lamina about an axis through its centre of mass and perpendicular to its plane.

(ii) Determine the velocity of the centre of mass and the angular velocity of the car just after the collision.

(iii) Assuming no other forces are acting, find how far the centre of mass travels after impact before the car is facing in the opposite direction.

[MEI 1994]

9. A uniform rod, of mass m and length $2l$, has a small light ring attached to one end and the ring is free to slide on a smooth horizontal wire AB. When the rod is at rest in the vertical position, it receives a horizontal impulse at its bottom end of magnitude mV and in the direction AB.

(i) Find expressions for the angular speed and the speed of the centre of mass of the rod immediately after the impulse is applied.

(ii) Show that the ring starts to move in the direction BA with speed $2V$.

In the subsequent motion the rod makes an angle θ with the vertical and its centre is at a depth x below the wire. Find $\dot{x}$ in terms of $\dot{\theta}$.

(iii) Use the principle of conservation of energy to show that

$$(1 + 3\sin^2\theta)l^2\dot{\theta}^2 = 9V^2 - 6gl(1 - \cos\theta)$$

(iv) Find the minimum value of V^2 so that the rod may at some subsequent time be parallel to the wire.

[Oxford 1991 adapted]

10. A snooker ball, of mass m and radius r, is at rest on a rough horizontal table when it is hit in a horizontal direction by a cue. The impulse on the ball is of magnitude mu and is in a vertical plane through its centre at a height $3r/25$ above the centre. The coefficient of friction between the ball and the table is $\tfrac{1}{2}$. Assuming there is no instantaneous frictional impulse on the ball, find in terms of u

(i) the initial speed of the centre of mass;

(ii) the initial angular speed ω_0 of the ball;

(iii) the initial speed of the point in contact with the table.

Draw a diagram of the forces acting on the ball after it has been hit, paying particular attention to the direction of the friction.

(iv) Find the frictional force acting on the ball and hence the deceleration of the centre of mass and its speed after time t.

(v) Find the angular acceleration of the ball and hence its angular speed after time t.

(vi) By considering the speed of the point of contact between the ball and the table at time t, show that the ball starts rolling after a time $2u/5g$.

(vii) Find the distance travelled in this time.

11. A uniform solid sphere, of mass m and radius r, is projected down a line of greatest slope of a fixed rough plane which is inclined at an angle α to the horizontal. Initially the sphere has no angular speed and its centre has a speed u (>0).

(i) By considering the angular momentum of the sphere, show that while slipping takes place, its angular speed at time t is $(5\mu gt \cos \alpha)/2r$, where μ is the coefficient of friction between the sphere and the plane.

(ii) Find the speed of the centre of mass at time t and hence write down an equation satisfied by t when the sphere stops slipping and starts to roll.

(iii) Show that rolling without slipping cannot occur unless $\tan \alpha > 7\mu/2$.

12. A uniform circular hoop of mass m and radius a is projected along a rough horizontal plane so that it remains vertical. The coefficient of friction is μ and the initial speed of the centre of the hoop is $a\omega$. The hoop is also given an initial angular speed 2ω with the result that the speed of the point in contact with the plane is $3a\omega$.

(i) Show that when slipping ceases, the centre of the hoop is moving back towards its initial position with speed $\frac{1}{2} a\omega$.

(ii) Find the maximum distance of the hoop from its initial position before it returns.

(iii) Find also the kinetic energy lost by the hoop while it is slipping.

(iv) In the general case when the hoop is propelled forwards with a speed u and at the same time given a backwards angular speed ω, find the condition that the hoop will eventually roll back towards its original position.

13. A uniform rod of mass m and length $2a$ slides in a fixed vertical plane with one end against a smooth vertical wall and the other end on a smooth horizontal floor. The rod is released from rest in an almost vertical position. Show that at time t, when the rod has rotated through an angle θ, and the ends are still in contact with the wall and the floor, the mid-point of the rod has speed $a\dfrac{d\theta}{dt}$.

Write down in terms of m, a, and $\dfrac{d\theta}{dt}$, an expression for the total kinetic energy of translation and rotation of the rod. Hence show that

$$2a\left(\frac{d\theta}{dt}\right)^2 = 3g(1 - \cos \theta).$$

Find the inclination of the rod to the vertical at the instant when the upper end leaves the wall.

[AEB 1994]

(*Hint*: consider the motion of the centre of mass of ther rod)

14. A yo-yo is modelled as a circular disc of radius a with a narrow groove around its rim. A thin inextensible string is attached to a point in this groove and then wrapped round the rim. The mass of the yo-yo is m and its moment of inertia about an axis through its centre perpendicular to its plane is $\frac{2}{5} ma^2$. The yo-yo is released from rest and falls in a vertical plane, as shown in the diagram. The vertical distance dropped in time t is x and the angle turned through in this time is θ.

(i) Find the equations of motion and deduce that the downwards acceleration of the centre of the yo-yo is $\frac{5}{7}g$.

(ii) Find the angular speed of the yo-yo when it has dropped a distance l from rest.

Exercise 6B continued

The upper end of the string is now given an upward acceleration of constant magnitude *f*.

(iii) Explain why $\ddot{x} + f = a\ddot{\theta}$.

(iv) Find the acceleration of the centre of the yo-yo in terms of *f*.

[Oxford 1993]

15. A cylindrical roll of carpet is placed on the horizontal floor of a lorry so that its axis is perpendicular to the direction of motion of the lorry. The lorry starts to move with constant acceleration *f*. By modelling the roll as a uniform solid cylinder of radius *r* and assuming that it rolls without slipping relative to the lorry,

(i) show that the axis of the roll has acceleration $\frac{2}{3}f$ relative to the lorry;

(ii) find the distance the lorry moves for the axis of the roll to travel a distance *d* relative to the lorry;

(iii) determine the condition satisfied by the coefficient of friction between the lorry and the roll for the above motion to be possible.

[Oxford 1994]

Investigations

1. Do wheels save energy?

   In icy conditions sledges and skis are useful means of transport, but wheels are preferable under normal conditions.

   Use energy considerations to compare the speeds achieved when a block slides down a slope against friction and when a solid disc of the same mass rolls down the same slope. Under what conditions does the wheel save energy?

   Consider the motion of a block with wheels.

2. Murphy's Law, otherwise known as Sod's Law, states that if anything can go wrong, it will. One version of this is that if you drop a piece of buttered toast it will land buttered-side down. Model this by considering the angle a lamina falling off the edge of a table will turn through before hitting the ground.

KEY POINTS

Motion of the centre of mass

- The centre of mass G moves as though it is a particle with all the external forces concentrated at G.

$$\sum F_{ext} = M\ddot{r}_G$$

$$\sum J_{ext} = [M\dot{r}_G]$$

Rotation about a fixed axis

- Angular momentum: $L = I\dot\theta$

- Moment of force about axis: $C = \dfrac{dL}{dt} = I\ddot\theta$

- When there is no resultant moment about the axis, angular momentum is conserved

- Moment of impulse $= \displaystyle\int C\,dt =$ change in angular momentum

- Kinetic energy $= \frac{1}{2}I\dot\theta^2$

- Work $= \displaystyle\int C\,d\theta$

- Mechanical energy is conserved if no external forces other than gravity do work

Rotation about an axis in a fixed direction through the centre of mass

- Angular momentum about axis through G: $L_G = I_G\dot\theta$

- Moment of force about axis through G: $C_G = \dfrac{dL_G}{dt} = I_G\ddot\theta$

- Moment of impulse about axis $=$ change in angular momentum

- Kinetic energy and work both have two parts. The first part is to do with rotation about the centre of mass, G. The second part can be found by assuming the body is a particle with all its mass and external forces concentrated at G.
  * Kinetic energy $= \frac{1}{2}I_G\dot\theta^2 + \frac{1}{2}Mv_G^2$

  * Work $= \displaystyle\int C\,d\theta + \int \mathbf{F}\cdot\dot{\mathbf{r}}_G\,dt = \int C\,d\theta + \int \mathbf{F}\cdot d\mathbf{r}_G$

- Mechanical energy is conserved if no external forces other than gravity do work

- Rolling occurs when $\dot{x}_G = r\dot\theta$

- The frictional force between a rolling body and the surface on which it rolls does no work

Stability and oscillations

A government so situated is in the condition called in mechanics 'unstable equilibrium' like a thing balancing on its smaller end.

John Stuart Mill

What happens when each of these objects is slightly displaced?

The train *can* balance at the high point of a roller coaster, a coin *can* sit on its edge. But in both cases the slightest disturbance causes them to move rapidly away from their equilibrium positions. The train and coin are in *unstable equilibrium*. Compare this with the swing or the bunjee jumper suspended on an elastic string. When these are disturbed, the resulting net forces will tend to restore the object to its equilibrium position. The swing and bunjee jumper are in *stable equilibrium*. When an up-and-over garage door is fully open, you would like it to be in stable equilibrium so that a breath of wind does not bring it crashing on to your car. But if you pull it sufficiently far from this position, to close it, you would like the door to take up a new state of stable equilibrium, in its closed position.

This chapter deals with stability of equilibrium and the oscillations about positions of stable equilibrium which are caused by small displacements. The approach is to apply *energy principles*, not only to analyse the oscillations about equilibrium positions, but also to find those positions. It may be surprising to learn that energy methods often provide an easier way to find equilibrium positions than analysing resultant forces and moments.

Potential energy at equilibrium

Consider a body modelled as a particle, such as the ball hanging on the end of an elastic string, able to move in one dimension, its displacement from some fixed point denoted by x. Assume also that *all forces on it are conservative*, so that the energy equation applies:

$$\text{kinetic energy} + \text{potential energy} = \text{constant}$$

or $$\tfrac{1}{2}mv^2 + U(x) = E$$

where the potential energy $U(x)$ is the total of *all* types of energy due to position x, in this case the total of gravitational and elastic energy. It is denoted by $U(x)$ to indicate that it is simply a function of the position of the body. Differentiating the energy equation with respect to x gives

$$mv\frac{dv}{dx} + \frac{dU}{dx} = 0$$

Now $mv\dfrac{dv}{dx} = ma = F$, by Newton's Second Law. So

$$F + \frac{dU}{dx} = 0 \qquad \Rightarrow \qquad F = -\frac{dU}{dx}$$

But a particle is in equilibrium if and only if the resultant force is zero. Hence $\dfrac{dU}{dx} = 0$ in an equilibrium position.

The potential energy has a stationary value if and only if the body is in an equilibrium position.

Sometimes the position of a body is specified not by its displacement from a given position but by some other variable. For example, the position of a pendulum is usually expressed by giving the angle θ it makes with the vertical (figure 7.1). Then the derivative with respect to *this* variable is still zero at equilibrium, since by the chain rule

$$\frac{dU}{d\theta} = \frac{dU}{dx}\frac{dx}{d\theta} = 0 \quad \text{when} \quad \frac{dU}{dx} = 0$$

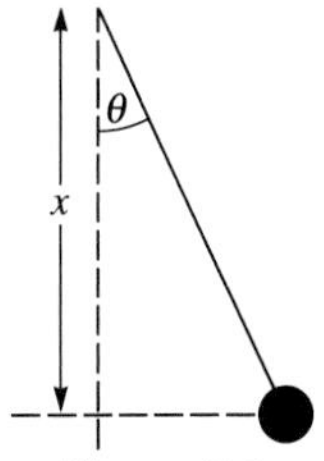

This argument can be applied not just to a particle but to a more general system, provided that it is constrained to move in such a way that its position, and therefore its potential energy, can be specified by the value of a single variable: normally an angle or a displacement. The conclusion is

Figure 7.1

> when a body acted on by conservative forces is free to move so that its potential energy can be given as a function $U(x)$ of a single variable x, its equilibrium positions are given by the stationary values of U with respect to this variable.

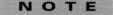

A rigid body is in equilibrium only if the resultant force and torque are both zero (see Chapter 9). To show that a stationary potential energy value implies an equilibrium position needs an extension to the argument above, not given here.

EXAMPLE

One end of an elastic string of natural length l and modulus λ is fixed to the ceiling and a ball of mass m is attached to the other end so that the string hangs vertically. Show, *using energy methods*, that in equilibrium the string is stretched an amount mgl/λ.

Solution

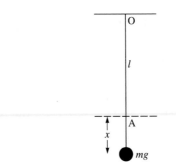

In the diagram, OA represents the position of the unstretched string. Consider the potential energy of the system when the mass is displaced x below A. The elastic energy is $\lambda x^2/2l$. The gravitational energy, relative to the level of point A, is $-mgx$: negative because the ball is below A. So the total potential energy is

$$U = \frac{\lambda x^2}{2l} - mgx \qquad \frac{dU}{dx} = \frac{\lambda x}{l} - mg$$

When $\dfrac{dU}{dx} = 0$, $\dfrac{\lambda x}{l} - mg = 0$

$$\Rightarrow \qquad x = \frac{mgl}{\lambda} \qquad\qquad\qquad ①$$

 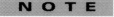

The choice of the horizontal through A as the zero level of gravitational potential energy is arbitrary. Any fixed point will do. Changing U by a constant does not affect the result when dU/dx is found.

This familiar example shows the energy method works, even though in this case simply balancing the forces ($mg = T = \lambda x/l$) gives equation ① immediately and is rather easier. The following examples show that energy methods can often be a quicker way of locating equilibrium positions than considering forces.

EXAMPLE

The diagram shows a uniform ladder AB of mass m and length $2l$ resting in equilibrium with its upper end A against a smooth vertical wall and its lower end B on a smooth inclined plane. The inclined plane makes an angle θ with the horizontal and the ladder makes an angle ϕ with the wall. What is the relationship between θ and ϕ?

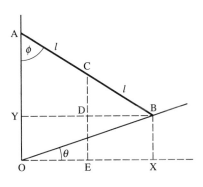

Solution

There is no friction at A and B, so the reaction forces do no work when the ladder moves. The equilibrium position can thus be found using the energy method. The only contribution to the potential energy of the system is the gravitational force, which may be considered to act at the centre of the ladder, C. Note that ϕ varies as the ladder moves, but θ, the inclination of the plane to the ground, is constant. So the energy can be expressed in terms of a single variable, ϕ.

Take the ground level OX as the zero level of gravitational energy. The height of C above this level is

$$CE = CD + DE = CD + BX$$

$$= l\cos\phi + OX\tan\theta$$

$$= l\cos\phi + 2l\sin\phi\tan\theta \quad (OX = YB = 2l\sin\phi)$$

Hence the potential energy is

$$U = mg(l\cos\phi + 2l\sin\phi\tan\theta)$$

$$\frac{dU}{d\phi} = mgl(-\sin\phi + 2\cos\phi\tan\theta)$$

Equilibrium occurs when $\dfrac{dU}{d\phi} = 0$, that is

$$-\sin\phi + 2\cos\phi\tan\theta = 0$$

Hence $\tan\phi = 2\tan\theta$ at equilibrium.

NOTE

Any fixed level can be taken as the reference for zero energy; OX is the most convenient here. It would be wrong to choose the zero energy level through A or B because A and B move.

The next example shows a situation with more than one equilibrium position.

EXAMPLE

A uniform rod OQ of mass m and length a is smoothly jointed to a fixed point at O, so that it can rotate in a vertical plane. P is a fixed point vertically above O so that $OP = a$. The ends of an elastic string of length a and modulus $2mg$ are connected to P and Q. Find the potential energy of the system when the rod makes an angle θ with the upward vertical, assuming the string is stretched. Hence find positions of equilibrium of the rod. Are there any other equilibrium positions when the string is unstretched? (You may assume that the string is attached in such a way that the rod can rotate completely in the vertical plane without being caught up by the string.)

Solution

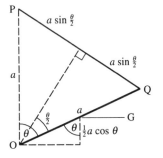

The diagram shows the position when the rod makes a general angle θ clockwise from the upward vertical. The string is just taut when OPQ is equilateral, that is $\theta = \frac{\pi}{3}$ (or $-\frac{\pi}{3}$). Consider the rod when in the general position *with the string taut*, that is $|\theta| > \frac{\pi}{3}$. Then

$$PQ = 2a \sin \frac{\theta}{2}$$

so the extension is $x = 2a \sin \dfrac{\theta}{2} - a$. The elastic energy is then

$$\frac{\lambda x^2}{2a} = \frac{2mg \left(2a \sin \dfrac{\theta}{2} - a \right)^2}{2a}$$

$$= mga \left(2 \sin \frac{\theta}{2} - 1 \right)^2$$

The centre of mass of the rod is at the mid-point G, so the gravitational potential energy, referred to zero level through O, is $mg \times$ height of G above O, that is $mg(\frac{1}{2}a \cos \theta)$. This remains correct when $\theta > \frac{\pi}{2}$, as $\cos \theta$ is then negative *and* G is *below* O.

The total potential energy is $U = mga(\frac{1}{2} \cos \theta) + mga \left(2 \sin \dfrac{\theta}{2} - 1 \right)^2$

Differentiating to find the equilibrium points gives

$$\frac{dU}{d\theta} = -mga(\tfrac{1}{2} \sin \theta) + 2mga \left(2 \sin \frac{\theta}{2} - 1 \right) \cos \frac{\theta}{2}$$

$$= -mga \left(\sin \frac{\theta}{2} \cos \frac{\theta}{2} \right) + 2mga \left(2 \sin \frac{\theta}{2} - 1 \right) \cos \frac{\theta}{2}$$

$$= mga \cos \frac{\theta}{2} \left(-\sin \frac{\theta}{2} + 4 \sin \frac{\theta}{2} - 2 \right)$$

$$= mga \cos \frac{\theta}{2} \left(3 \sin \frac{\theta}{2} - 2 \right)$$

Equilibrium points are given where $\dfrac{dU}{d\theta} = 0$:

$$\cos\dfrac{\theta}{2} = 0 \quad \text{or} \quad \sin\dfrac{\theta}{2} = \dfrac{2}{3}$$

$$\dfrac{\theta}{2} = \dfrac{\pi}{2} \qquad \dfrac{\theta}{2} = \arcsin\dfrac{2}{3} = 0.730 \ \text{(3 significant figures)} \quad \text{or} \quad \pi - 0.730$$

$$\theta = \pi \qquad \theta = 1.46 \quad \text{or} \quad 2\pi - 1.46$$

The three equilibrium positions with the string taut, are shown in the diagram.

It remains to look at the case when the string is *unstretched*, $|\theta| < \frac{\pi}{3}$. In this case, the equilibrium position is the one where the rod is balanced vertically upwards. The energy method confirms this: only the gravitational term $mga(\frac{1}{2}\cos\theta)$ contributes to the potential energy and this has a stationary value at $\theta = 0$.

Stability of equilibrium

For Discussion

In the example at the end of the last section, there are four equilibrium positions: $\theta = 0,\ \pi,\ \pm1.46$. But not all of these are stable. Which do you think are stable and which unstable?

The potential energy can be written

$$U = mga\left[\tfrac{1}{2}\cos\theta + \left(2\sin\dfrac{\theta}{2} - 1\right)^2\right] \quad \text{when } |\theta| > \tfrac{\pi}{3}$$

$$U = mga(\tfrac{1}{2}\cos\theta) \quad \text{when } |\theta| \leqslant \tfrac{\pi}{3} \text{ (string not taut)}$$

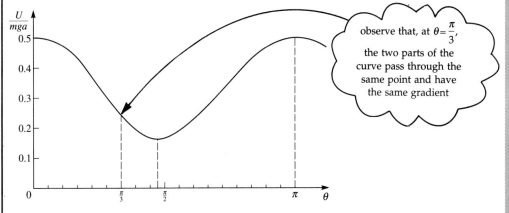

observe that, at $\theta = \dfrac{\pi}{3}$, the two parts of the curve pass through the same point and have the same gradient

The graph shows how the potential energy varies with θ. Can you infer any relationship between the type of stationary point of a potential energy curve and the stability of the equilibrium position? The following argument shows that there is indeed such a relationship.

Condition for stability of equilibrium

Suppose the potential energy U has a *minimum*, U_0, at the equilibrium position x_0. Displacing the body from its equilibrium position is equivalent to supplying a small amount of kinetic energy K_0 in the form of a velocity taking it away from the equilibrium position. Assume this is in the direction of increasing x (see figure 7.2).

During the subsequent motion

$$U + K = \text{constant} \quad E = U_0 + K_0$$

where K denotes the kinetic energy.

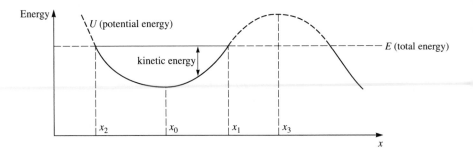

Figure 7.2

The graph shows the potential energy $U(x)$ and total energy E of the body at points near x_0. The difference $E - U(x)$ is the kinetic energy. You can see that at the point x_1, the kinetic energy is zero. It cannot become negative so the body cannot move past x_1 (dotted region of curve). Furthermore, since x_1 is *not* a position of equilibrium, the body begins to return to its equilibrium position. When it returns to x_0, it will have some velocity so will continue to a point x_2 and oscillate about the position of equilibrium (the motion is considered in the next section.) This is what is meant by a position of *stable equilibrium*: the body returns after a small displacement.

Note, if K_0 is greater leading to a *large* displacement past the point x_3, this would take the body into a new energy region and it would not return to x_0.

Activity

Show that when U is a maximum at the equilibrium position, and the body is given a small displacement, the kinetic energy *increases* and the body moves further from the equilibrium point. Draw an equivalent graph to that above (remember that $E \geqslant U(x)$). The equilibrium position is then unstable.

Summary of stability criteria

The results can be summarised as follows.

- The potential energy is a minimum $\Leftrightarrow$ stable equilibrium point:
  $U'(x) = 0$, $U''(x) > 0$
- The potential energy is a maximum $\Leftrightarrow$ unstable equilibrium point:
  $U'(x) = 0$, $U''(x) < 0$

A way to remember this is to imagine the graph of potential energy as a surface on which a ball is placed; the ball oscillates in the hollows (minima) which correspond to stable equilibrium and rolls away from the hilltops (maxima) which correspond to unstable equilibrium.

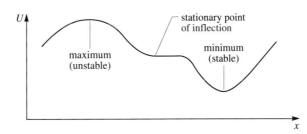

For Discussion

There are other types of equilibrium position. For example, imagine a ball on a horizontal surface. $U = $ constant, $\dfrac{dU}{dx} = 0$ and a displacement does not result in any force. This is known as neutral equilibrium. Suppose the potential energy graph has a point of inflexion as shown in the diagram. What could this represent? Would you describe it as stable or unstable?

EXAMPLE

A radio mast is to be erected and supported by four light cables. During the design stage, the stability is to be investigated by considering the motion of the mast and just *two* of the supporting cables. For this design the model is represented by a uniform rod of length a and mass m, freely pivoted at its base. The two cables extend from the top of the mast to anchoring points on the ground on opposite sides of the mast and at equal distances a from the base of the mast. The mast and the two supporting cables all lie in the same vertical plane and motion takes place in this plane. Each cable has a natural length a and elastic modulus λ.

(i) Find an expression for the potential energy of the model system in terms of a, m, g, λ and θ, where 2θ is the acute angle the mast makes with the horizontal when tilted towards one of the ground anchoring points. It may be assumed that neither cable becomes slack.

(ii) Use this potential energy expression to investigate the stability of the vertical position of the mast. Determine the condition on λ which makes this position one of stable equilibrium. Show that for such λ no other positions of equilibrium exist.

[MEI 1987]

Solution

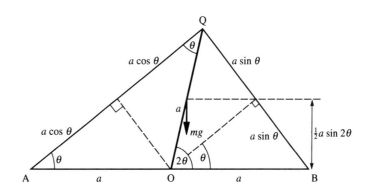

(i) The diagram shows the mast OQ in a tilted position, with anchoring cables AQ and BQ. $AO = BO = OQ = a$, so triangles AOQ and BOQ are both isoceles. Then, from the geometry,

$$BQ = 2a \sin \theta$$

$$\angle OAQ = \angle AQO = \theta \quad \Rightarrow \quad AQ = 2a \cos \theta$$

The extension of BQ from its natural length is thus

$$2a \sin \theta - a = a(2 \sin \theta - 1)$$

and of AQ is

$$2a \cos \theta - a = a(2 \cos \theta - 1)$$

The combined elastic energy is thus

$$\frac{\lambda}{2a}[a^2(2 \sin \theta - 1)^2] + \frac{\lambda}{2a}[a^2(2 \cos \theta - 1)^2]$$

$$= \tfrac{1}{2}\lambda a(4 \sin^2\theta - 4 \sin \theta + 1 + 4 \cos^2\theta - 4 \cos \theta + 1)$$

$$= \lambda a(3 - 2 \sin \theta - 2 \cos \theta) \quad (\sin^2\theta + \cos^2\theta = 1)$$

The gravitational energy, relative to ground level, is $\tfrac{1}{2}mga \sin 2\theta$. Hence the total potential energy is

$$U = \tfrac{1}{2}mga \sin 2\theta + \lambda a(3 - 2 \sin \theta - 2 \cos \theta)$$

(ii) Differentiate to investigate equilibrium positions:

$$\frac{dU}{d\theta} = mga \cos 2\theta + \lambda a(-2 \cos \theta + 2 \sin \theta) \qquad \text{①}$$

$$= mga (\cos^2\theta - \sin^2\theta) - 2\lambda a(\cos \theta - \sin \theta)$$

At equilibrium positions $\dfrac{dU}{d\theta} = 0$

$$\Rightarrow \quad mga(\cos \theta - \sin \theta)(\cos \theta + \sin \theta) - 2\lambda a(\cos \theta - \sin \theta) = 0$$

$$\Rightarrow \quad \cos \theta - \sin \theta = 0 \quad \text{or} \quad mga(\cos \theta + \sin \theta) = 2\lambda a$$

$$\theta = \tfrac{\pi}{4} \quad \text{or} \quad \cos \theta + \sin \theta = \frac{2\lambda}{mg}$$

Using the standard method of writing $a \cos \theta + b \sin \theta$ in the form $r \cos (\theta - \alpha)$

$$2\theta = \tfrac{\pi}{2} \quad \text{or} \quad \sqrt{2} \cos (\theta - \tfrac{\pi}{4}) = \frac{2\lambda}{mg}$$

That is, the mast is vertical, or

$$\cos (\theta - \tfrac{\pi}{4}) = \frac{\sqrt{2}\lambda}{mg}$$

$$\Rightarrow \quad \theta - \tfrac{\pi}{4} = \pm \arccos\left(\frac{\sqrt{2}\lambda}{mg}\right)$$

$$\Rightarrow \quad 2\theta = \tfrac{\pi}{2} \pm 2 \arccos\left(\frac{\sqrt{2}\lambda}{mg}\right)$$

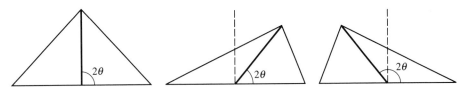

Thus, apart from the vertical position, there may be an equilibrium position each side of the vertical, where $\cos(\theta - \frac{\pi}{4}) = (\sqrt{2}\lambda)/mg$. The other solutions exists if $(\sqrt{2}\lambda)/mg < 1$.

Investigate stability of equilibrium by differentiating ①:

$$\frac{d^2U}{d\theta^2} = -2mga\sin 2\theta + \lambda a(2\sin\theta + 2\cos\theta)$$

$$= -2mga + \lambda a(\sqrt{2} + \sqrt{2}) \quad \text{when } \theta = \frac{\pi}{4} \text{ (vertical)}$$

For stable equilibrium, $\frac{d^2U}{d\theta^2} > 0$ (minimum):

$$-2mga + 2\sqrt{2}\lambda a > 0 \qquad \Rightarrow \qquad \lambda > \frac{mg}{\sqrt{2}}$$

As shown above, this is just the condition that there is no other position of equilibrium.

N O T E

Potential energy as a function of several variables
The discussion of stationary values of potential energy U has been confined to the case when U can be expressed as a function of a single variable, e.g. distance or angle. In this case, the system is said to have 'one degree of freedom'. In fact, the principle that equilibrium positions coincide with stationary values of potential energy extends to systems with more than one degree of freedom, i.e. where U is a function of several variables. Handling stationary values of functions of several variables is beyond the scope of this course. However, the following investigation demonstrates the energy principle in use in such a case. Although the problem has two degrees of freedom (the potential energy can be expressed as a function of two variables), this complication does not have to be considered in the investigation.

Investigation

Using potential energy to find the shortest path connecting points
Mechanics can be used to solve *geometrical* problems. You are given three points, representing villages perhaps, and you want to find the shortest length of path which connects all three.

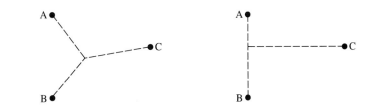

Investigation continued

As you will see, this can be solved by modelling an experiment in mechanics. It is even more interesting to *do* the experiment, if you have the equipment.

The theoretical model
Imagine a smooth horizontal plane lamina with three holes in it at positions A, B, C, representing the positions of three villages. Three strings threaded as shown (they need not be of equal length), tied together at P and passing smoothly through the holes. At the end of each is a weight W. The system is allowed to rest in equilibrium.

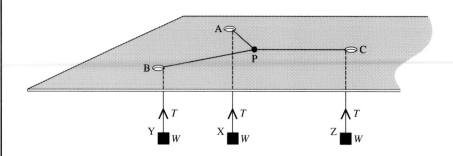

(i)   Write down the potential energy of the system in terms of string lengths, and deduce that the total length PA + PB + PC must have a minimum value in the equilibrium position.

(ii) Show that when the triangle ABC has no angle $> 120°$, P takes up a position where $\angle BPA = \angle APC = \angle CPA = 120°$. This configuration represents the solution to the problem.

(iii) Show that when $\angle BAC$ (say) in the triangle ABC is greater than $120°$, you could not have an equilibrium position with P inside the triangle. What would then happen if you performed the experiment? What is the solution to the problem in this case?

(iv) Does the experiment extend to four or more points? How could you then state the problem and solution?

The actual experiment
The experiment works very well if you use small pulleys at A, B and C and adjust their angles to the final directions of the strings. Try it with various positions of A, B and C, including a case where $\angle BAC > 120°$.

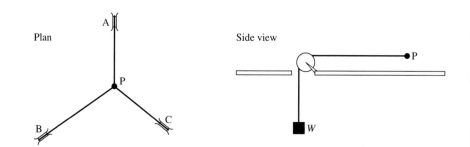

Small oscillations about stable equilibrium

You have already met numerous situations in which an oscillation about a stable equilibrium position can be modelled by simple harmonic motion.

- Spring mass oscillator: $\ddot{x} = -\left(\dfrac{\lambda}{ml}\right)x$

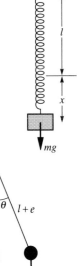

- Simple pendulum: $\ddot{\theta} = -\left(\dfrac{g}{l}\right)\theta$ when θ is small

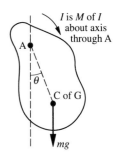

- Compound pendulum: $\ddot{\theta} = -\left(\dfrac{mgh}{I}\right)\theta$ when θ is small

I is *M* of *I* about axis through A

In this chapter, you have seen that a system displaced from a position of stable equilibrium will oscillate. This oscillation can often be modelled by simple harmonic motion when the amplitude of the oscillation is small, that is, the initial displacement is small. When the energy method is used to find such equilibrium positions, it is normally easier to differentiate the energy equation to obtain the equation of motion, rather than to consider forces explicitly. This is demonstrated in the following example.

EXAMPLE

A small ring A of mass m is free to slide on a smooth vertical wire. The ring is attached to one end of an elastic string of modulus $\lambda = mg$ and natural length a. The string passes through a smooth fixed ring P, distance a from the wire, and is attached to a fixed point at B, where PB is horizontal and PB $= a$ (see diagram).

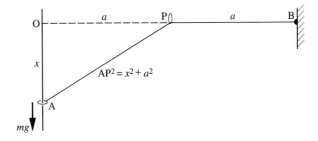

(i) Write down an expression for the potential energy $U(x)$ of the system when A is at a distance x below the level of P. Hence show that the ring will rest in a position of stable equilibrium at $x = a$.

(ii) When the ring is displaced from equilibrium and released, it oscillates about its equilibrium position. Write down an equation representing conservation of energy. Differentiate it and hence show that the motion of the ring can be modelled by simple harmonic motion with a period $2\pi\sqrt{(a/g)}$.

Solution

(i) Take OP as the zero level of gravitational energy. The gravitational PE is $-mgx$ (negative since A is *below* O). The total string length is

$$AP + PB = \sqrt{(x^2 + a^2)} + a$$

and the elastic energy is thus

$$\frac{\lambda(x^2 + a^2)}{2a} \quad \text{(initial length is } a \text{, so extension is } \sqrt{(x^2 + a^2)})$$

The total potential energy is

$$U = \frac{\lambda(x^2 + a^2)}{2a} - mgx$$

Differentiate to find the equilibrium position:

$$\frac{dU}{dx} = \frac{2x\lambda}{2a} - mg = \frac{x\lambda}{a} - mg \qquad \text{①}$$

At the equilibrium position, $\dfrac{dU}{dx}$ is zero

$$\Rightarrow \quad x\lambda = mga$$

$$\Rightarrow \quad x = a \quad \text{(since } \lambda = mg)$$

So $x = a$ is an equilibrium position.

$$\frac{d^2U}{dx^2} = \frac{\lambda}{a} > 0$$

hence U is a minimum $\Rightarrow$ the equilibrium is stable.

(ii) Energy is conserved so

$$\tfrac{1}{2}m\dot{x}^2 + U(x) = E \quad \text{(the total energy)}$$

Differentiating with respect to time t gives

$$m\dot{x}\ddot{x} + \left(\frac{dU}{dx}\right)\dot{x} = 0 \quad \left(\text{by the chain rule: } \frac{dU}{dt} = \frac{dU}{dx}\frac{dx}{dt}\right)$$

$$\Rightarrow \quad \ddot{x} = -\frac{1}{m}\frac{dU}{dx} = -\frac{x\lambda}{ma} + g \quad \text{from ①}$$

$$= -\frac{xg}{a} + g$$

$$= -\frac{g}{a}(x-a)$$

Compare this with the equation $\ddot{x} = -\omega^2(x-x_0)$, which you know represents simple harmonic motion about the position $x = x_0$. The ring thus oscillates about the equilibrium position at $x = a$, with period $2\pi\sqrt{(a/g)}$.

It is normally best to express the position of the body as a displacement *from the equilibrium position* in order to demonstrate that the motion is simple harmonic. In the above example, this means translating the origin by putting $y = x - a$. The displacement y is then zero at equilibrium. The equation becomes $\ddot{y} = -(g/a)y$.

EXAMPLE

Two positively charged spheres, their centres a distance d apart on a horizontal line, have charges of respectively 4 and 1 microcoulombs (or μC). A positively charged particle is free to move on the line joining their centres. When the particle is at a distance r m from the centre of a sphere with positive charge q μC, the potential energy associated with the electrical force between them is qk/r J.
(i) Use the energy method to find a position of equilibrium.
(ii) Show that a *small* displacement will lead to simple harmonic motion about the equilibrium position. Find the period.

Solution

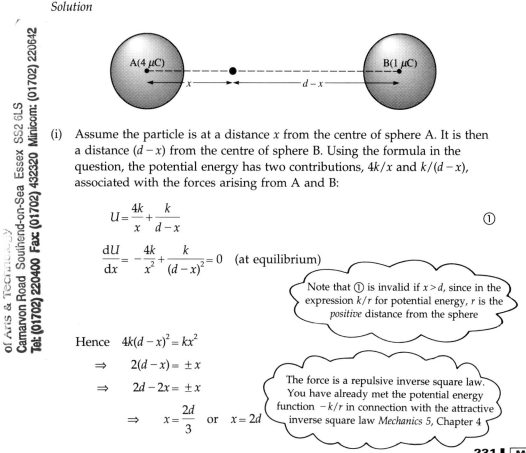

(i) Assume the particle is at a distance x from the centre of sphere A. It is then a distance $(d-x)$ from the centre of sphere B. Using the formula in the question, the potential energy has two contributions, $4k/x$ and $k/(d-x)$, associated with the forces arising from A and B:

$$U = \frac{4k}{x} + \frac{k}{d-x} \qquad \text{①}$$

$$\frac{dU}{dx} = -\frac{4k}{x^2} + \frac{k}{(d-x)^2} = 0 \quad \text{(at equilibrium)}$$

> Note that ① is invalid if $x > d$, since in the expression k/r for potential energy, r is the *positive* distance from the sphere

Hence $4k(d-x)^2 = kx^2$

$\Rightarrow \quad 2(d-x) = \pm x$

$\Rightarrow \quad 2d - 2x = \pm x$

> The force is a repulsive inverse square law. You have already met the potential energy function $-k/r$ in connection with the attractive inverse square law *Mechanics 5*, Chapter 4

$\Rightarrow \quad x = \frac{2d}{3} \quad \text{or} \quad x = 2d$

$$\Rightarrow \quad x = \frac{2d}{3} \quad \text{(since } 0 < x < d)$$

(ii) Conservation of energy gives

$$\tfrac{1}{2}m\dot{x}^2 + U(x) = E \quad \text{(the total energy)}$$

Differentiating with respect to time t gives

$$m\dot{x}\ddot{x} + \left(\frac{dU}{dx}\right)\dot{x} = 0$$

$$\Rightarrow \quad m\ddot{x} = -\frac{dU}{dx} = \frac{4k}{x^2} - \frac{k}{(d-x)^2} \quad \text{from } \textcircled{1}$$

Put $z = \left(x - \dfrac{2d}{3}\right)$, so that z is the (small) displacement from the known

equilibrium position. Hence $x = \left(z + \dfrac{2d}{3}\right)$ and $\ddot{z} = \ddot{x}$, giving

$$m\ddot{z} = 4k\left(\frac{2d}{3} + z\right)^{-2} - k\left(\frac{d}{3} - z\right)^{-2}$$

$$= 4k \frac{9}{4d^2}\left(1 + \frac{3z}{2d}\right)^{-2} - k \times \frac{9}{d^2}\left(1 - \frac{3z}{d}\right)^{-2}$$

$$\approx \frac{9k}{d^2}\left(1 - \frac{3z}{d}\right) - \frac{9k}{d^2}\left(1 + \frac{6z}{d}\right)$$

> The equation is put in terms of z, the displacement from the equilibrium position, so that the binomial theorem can be used to expand the expressions in brackets in powers of z. For a sufficiently small displacement, terms in z^2 and higher powers can be neglected. Then the motion can be modelled by simple harmonic motion.

$$= -\left(\frac{81k}{d^3}\right)z$$

$$\Rightarrow \quad \ddot{z} = -\left(\frac{81k}{md^3}\right)z$$

This is simple harmonic motion with period $2\pi\sqrt{\left(\dfrac{md^3}{81k}\right)} = \dfrac{2\pi d}{9k}\sqrt{(mkd)}$.

Use of the second derivative of potential energy

The analysis of the oscillations about equilibrium positions can be assisted by use of the second derivative of the potential energy. As before, assume all forces are conservative so that the equation of motion is

$$m\ddot{x} = -\frac{dU}{dx} = F(x)$$

where F is the total force and U is the potential energy.

Assume an equilibrium position at $x = x_0$, that is $F(x_0) = 0$. Then, when x is near x_0, you can see from figure 7.3 that

$$F'(x_0) \simeq \text{slope of PQ}$$

$$= \frac{F(x)}{(x - x_0)}$$

$$\Rightarrow \quad F(x) \simeq (x - x_0)F'(x_0)$$

(Compare with the Newton–Raphson formula.)

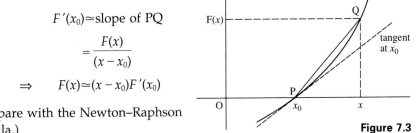

Figure 7.3

Hence for a sufficiently small displacement from equilibrium

$$m\ddot{x} = (x - x_0)F'(x_0)$$

$$= -(x - x_0)U''(x_0) \quad (\text{since } F(x) = -U'(x))$$

This is the standard form of simple harmonic motion, $\ddot{x} = -\omega^2(x - x_0)$, when

$$m\omega^2 = U''(x_0)$$

which is the value of the second derivative of potential energy at the equilibrium point. Since x_0 is a *stable* equilibrium point, $U''(x_0)$ cannot be negative. Unless it is zero (which can happen), the motion can be modelled by simple harmonic motion for small displacements.

This method can be applied to the last example. It was shown that the derivative of potential energy is

$$U' = -\frac{4k}{x^2} + \frac{k}{(d - x)^2}$$

Differentiating again with respect to x gives

$$U'' = \frac{8k}{x^3} + \frac{2k}{(d - x)^3}$$

The value of the second derivative at equilibrium, $x = 2d/3$, is

$$U''\left(\frac{2d}{3}\right) = \frac{27 \times 8k}{8d^3} + \frac{27 \times 2k}{d^3} = \frac{81k}{d^3}$$

Then using the above result, $m\omega^2 = \dfrac{81k}{d^3}$ and the equation of motion can be written

$$m\ddot{x} = -\left(x - \frac{2d}{3}\right)U''\left(\frac{2d}{3}\right)$$

$$= -\frac{81k}{d^3}\left(x - \frac{2d}{3}\right)$$

Oscillations involving rotation

In some of the previous examples, the equilibrium points were specified by the value of an *angle* rather than a linear displacement. The energy method can still be used to find the oscillation period, provided that you can find an expression for kinetic energy in terms of the angle.

EXAMPLE

The diagram shows a spring controlled flap. The flap can be modelled as a uniform rectangular lamina length $AB = 2l$, smoothly hinged about an axis through B. The spring BD is horizontal and is connected by a cord passing over a smooth pulley C and attached to the flap at A. $BC = 2l$. When the flap is horizontal (A is at C), the spring has its natural length l. The modulus λ of the spring is $\frac{1}{4}mg$.

(i)   Write down the potential energy of the system when the flap makes an angle θ with the horizontal. Hence show that $\theta = \frac{\pi}{4}$ is an equilibrium position.

(ii) Differentiate the energy equation to give the equation of motion and express this in terms of $\phi = \theta - \frac{\pi}{4}$. Hence show that the period of small oscillations about the equilibrium position is $2\pi/\omega$, where $\omega^2 = (3\sqrt{2}g)/4l$.

Solution

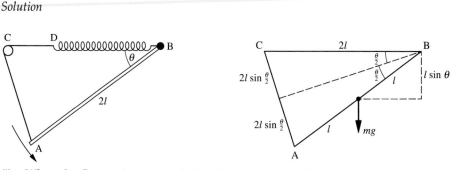

(i)   When the flap makes an angle θ with the horizontal, the spring is extended from its natural length by $CA = 4l\sin\left(\dfrac{\theta}{2}\right)$. Hence the elastic potential energy is

$$\lambda\left[\frac{\left(4l\sin\dfrac{\theta}{2}\right)^2}{2l}\right] = 2mgl\sin^2\frac{\theta}{2} \quad (\text{since } \lambda = \tfrac{1}{4}mg)$$

The gravitational potential energy (taking BC as zero level) is $-mgl\sin\theta$. Therefore, the total potential energy is

$$U = 2mgl\sin^2\frac{\theta}{2} - mgl\sin\theta$$

Differentiate to find the equilibrium position:

$$\frac{dU}{d\theta} = 2mgl\sin\frac{\theta}{2}\cos\frac{\theta}{2} - mgl\cos\theta$$

$$= mgl\sin\theta - mgl\cos\theta$$

$$\frac{d}{d\theta}\left(\sin^2\frac{\theta}{2}\right) = \left(2\sin\frac{\theta}{2}\cos\frac{\theta}{2}\right) \times \frac{1}{2}$$

$$= \sin\frac{\theta}{2}\cos\frac{\theta}{2}$$

When $dU/d\theta = 0$, $\sin\theta = \cos\theta \Rightarrow \theta = \frac{\pi}{4}$.

(ii) When the flap is moving, PE + KE = constant.

The moment of inertia of the flap, rotating about a pivot through one end is $\frac{4}{3}ml^2$. So the kinetic energy of the flap is

$$\frac{1}{2}\left(\frac{4}{3}ml^2\right)\dot\theta^2 = \frac{2}{3}ml^2\dot\theta^2$$

Therefore

$$\frac{2}{3}ml^2\dot\theta^2 + U = \text{constant}$$

Differentiate with respect to t to give the equation of motion:

$$\left(\frac{4ml^2}{3}\right)\dot\theta\ddot\theta + \left(\frac{dU}{d\theta}\right)\dot\theta = 0$$

$$\Rightarrow \quad \left(\frac{4ml^2}{3}\right)\ddot\theta = -\frac{dU}{d\theta} \qquad ①$$

$$= -mgl(\sin\theta - \cos\theta)$$

Put $\phi = \theta - \frac{\pi}{4}$ in order to investigate motion about the equilibrium position.

$$\left(\frac{4ml^2}{3}\right)\ddot\phi = -mgl[\sin(\phi+\tfrac{\pi}{4}) - \cos(\phi+\tfrac{\pi}{4})]$$

$$= -mgl\left(\frac{1}{\sqrt2}\sin\phi + \frac{1}{\sqrt2}\cos\phi - \frac{1}{\sqrt2}\cos\phi + \frac{1}{\sqrt2}\sin\phi\right)$$

$$= -\sqrt2\,mgl\sin\phi$$

$$\Rightarrow \quad \ddot\phi = -\left(\frac{3\sqrt2 g}{4l}\right)\sin\phi$$

When ϕ remains small, $\sin\phi \approx \phi$:

$$\Rightarrow \quad \ddot\phi = -\left(\frac{3\sqrt2 g}{4l}\right)\phi$$

This is simple harmonic motion, where $\omega^2 = (3\sqrt2 g/4l)$.

NOTE

Use of second derivative in rotational problems
The method of using the second derivative of energy to find the equation of motion near equilibrium can be extended to rotational problems such as part (ii) of the last example. In part (i) it was shown that

$$\frac{dU}{d\theta} = mgl(\sin\theta - \cos\theta)$$

and this is zero at $\theta = \frac{\pi}{4}$.

Then the second derivative is

$$\frac{d^2U}{d\theta^2} = mgl(\cos\theta + \sin\theta)$$

$$= \sqrt2 mgl \quad \text{at } \theta = \tfrac{\pi}{4}, \text{ the equilibrium point}$$

Now the first derivative near its zero point $(\theta = \frac{\pi}{4})$ can be written

$$\frac{dU}{d\theta} \approx \left[\frac{dU}{d\theta}\right]_{\text{at } \theta = \pi/4} + (\theta - \tfrac{\pi}{4}) \times \left[\frac{d^2U}{d\theta^2}\right]_{\text{at } \theta = \pi/4}$$

$$= 0 + \sqrt{2}mgl\,(\theta - \tfrac{\pi}{4})$$

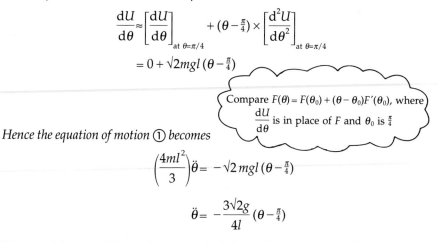

Compare $F(\theta) = F(\theta_0) + (\theta - \theta_0)F'(\theta_0)$, where $\dfrac{dU}{d\theta}$ is in place of F and θ_0 is $\frac{\pi}{4}$

Hence the equation of motion ① becomes

$$\left(\frac{4ml^2}{3}\right)\ddot{\theta} = -\sqrt{2}\,mgl\,(\theta - \tfrac{\pi}{4})$$

$$\ddot{\theta} = -\frac{3\sqrt{2}g}{4l}(\theta - \tfrac{\pi}{4})$$

The use of the second derivative has avoided the explicit translation of the origin and the trigonometric manipulation in the previous solution.

Exercise 7A

1. A particle P of mass m moves along the line of greatest slope of a smooth plane which makes an angle θ with the horizontal. The particle is attached to one end of a light elastic string, the other end of which is fixed to a point O on the plane. The natural length of the string is a and its modulus is $2mg$.

(i) Write down the gravitational potential energy of the particle when it is at a distance x down the plane from O. Take O as the zero energy level.

(ii) Write down the elastic energy, assuming $x \geqslant a$ (i.e. the string is not loose)

(iii) Denoting the total potential energy by $U(x)$, find $\dfrac{dU}{dx}$ and hence find e, the value of x when the particle is in equilibrium.

(iv) Write down an energy equation (kinetic + potential = E, a constant) and differentiate with respect to t to give the equation of motion. Show that this can be written in the form

$$\ddot{x} = -\frac{2g}{a}(x - e)$$

Hence deduce the period of oscillations about the equilibrium position, assuming the string remains taut.

2. A spring of modulus λ and natural length a is attached to a point P on the ceiling and an identical spring is attached to a point Q on the floor a distance $2a$ vertically below P. The other end of each spring is attached to a ball of mass m, so that it is free to oscillate in a vertical line.

(i) Write down the gravitational potential energy and the elastic energy when the ball is displaced x below O, the mid-point of PQ. Hence deduce there is an equilibrium position at $x = mga/2\lambda$.

(ii) Differentiate the energy equation to show the equation of motion is

$$\ddot{x} = -\frac{2\lambda}{ma}\left(x - \frac{mga}{2\lambda}\right)$$

Hence write down the period of the oscillation about the equilibrium position.

3. In the diagram, ABC represents a light spring of natural length $2a$ and modulus of elasticity λ, which is coiled round a smooth horizontal rod. B is the mid-point of AC. The two ends of a light inelastic string of

length 2*a* are attached to the spring at A
and C. A particle of mass *m* is fixed to the
spring at its mid-point D. Thus, as the
particle descends vertically, the spring is
compressed.

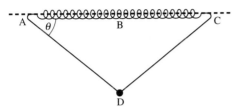

(i) Obtain an expression for the total
potential energy of the system in terms
of θ, the angle DAB.

(ii) Hence show that, if $\theta = \frac{\pi}{6}$ is an
equilibrium position, $mg = \lambda\left(\frac{2}{\sqrt{3}} - 1\right)$.

[O&C 1990 adapted]

4. A uniform rod OA of length 2*a* and mass *m*
is freely pivoted to a fixed point at O. The
end A is attached by a light elastic string of
natural length *a* and modulus λ to a point B
a distance 2*a* vertically above O. When the
rod is in a general position, the angle AOB
is denoted by 2θ.

(i) Find the gravitational potential energy
with respect to O, and the elastic
energy, as functions of θ.

(ii) Show there is always one position of
equilibrium with the string stretched
and determine the condition of λ for
there to be a second.

(iii) Discuss the nature of the stability in
each case.

5. A small lamp of mass *m* is at the end A of
a light rod AB of length 2*a* attached at B to
a vertical wall in such way that the rod can
rotate freely about B in a vertical plane
perpendicular to the wall. A spring CD of

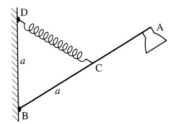

natural length *a* and modulus of elasticity λ
is joined to the rod at its mid-point C and
to the wall at a point D a distance *a*
vertically above B.

Show that if λ > 4*mg* the lamp can hang in
equilibrium away from the wall and
calculate the angle DBA.

[O&C 1993]

6. The diagram shows a smooth wire bent
into the form of a circle in a vertical plane.
A ring P is threaded on the circle and tied
to a light inextensible string which passes
over a pulley O at the highest point of the
circle. A particle of the same mass as the
ring hangs at the other end of the string.
Use the energy method to find two
positions of equilibrium, one stable and
one unstable.

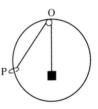

7. A smooth thin wire in the form of a
semicircle of diameter *d* is fastened in
a vertical plane with its ends A and B at
the same horizontal level and uppermost.
A smooth ring P of mass *m* is threaded on
the wire and small smooth rings are fixed
at A and B. An endless elastic band of
natural length *d* and stiffness *kmg/d* passes
through the three rings but is otherwise
unrestricted.

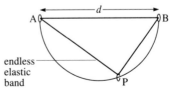

(i) Write down the total gravitational and
elastic energy of the system in terms of
θ = angle PAB.

(ii) Show that there is a value of k for which P can be in equilibrium at every point of the wire between A and B.

(iii) Find out the range of values of k for which there is just one stable equilibrium point.

8. The rod PQ, whose length is $2a$ and whose mid-point is at O, is rigidly attached to another rod OY perpendicular to PQ. At Y is a concentrated mass m. $OY = 2a$. The system is pivoted about the fixed point O so that it can rock about O in the vertical plane of PQY. Springs with stiffness k are attached at P and Q and have their natural lengths when OY is vertical. Masses apart from m may be neglected.

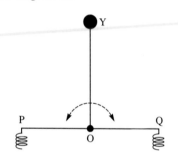

(i) Derive an expression for the gravitational and elastic potential energy of the system when OY makes a small angle θ with the vertical.

(ii) Derive the equation of motion from the energy equation. Hence show that the system will oscillate in simple harmonic motion provided that $k > mg/a$. Confirm that this is just the condition that the potential energy is minimum when OY is vertical.

9. In a diatomic molecule, one atom is very much heavier than the other and the lighter atom is free to move on a straight line through the centre of the heavier atom. The potential energy for the force between two atoms in a diatomic molecule can be represented approximately by

$$V(x) = -ax^{-6} + bx^{-12}$$

where x is the atomic separation and a and b are positive constants.

(i) Locate any positions of stable equilibrium of the lighter atom.

(ii) Find the period of small oscillations about such positions, assuming the mass of the lighter atom is m.

(iii) Sketch the potential energy as a function of x. How does the force vary as the atoms approach each other?

[MEI 1985 adapted]

10.

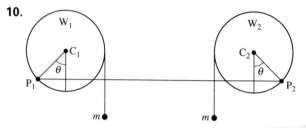

In the figure, W_1 and W_2 are wheels, both of radius r. Their centres C_1 and C_2 are fixed at the same height, a distance d apart, and each wheel is free to rotate, without friction, about its centre. Both wheels are in the same vertical plane. Particles of mass m are suspended from W_1 and W_2 as shown, by light inextensible strings wound round the wheels. A light elastic string of natural length d and modulus of elasticity λ is fixed to the rims of the wheels at the points P_1 and P_2. The lines joining C_1 to P_1 and C_2 to P_2 both make an angle θ with the vertical. The system is in equilibrium. Show that

$$\sin 2\theta = \frac{mgd}{\lambda r}$$

For what value or values of λ (in terms of m, d, r and g) are there

(i) no equilibrium positions;

(ii) just one equilibrium position;

(iii) exactly two equilibrium positions;

(iv) more than two equilibrium positions?

[O&C 1992]

11. Three points are defined by the vertices of an equilateral triangle which has medians of length $3a/2$ and lies in a vertical plane. One vertex is above the other two which lie at the same horizontal level. Three light springs of natural lengths a each have one end attached to a different vertex. Their

free ends are joined together and lie at the intersection of the medians. A particle of mass m is fixed to this junction which falls under the influence of gravity. The configuration is shown in the diagram.

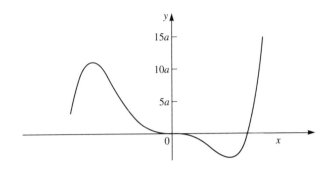

The spring attached to the top vertex has a stiffness $2mg/a$ and those attached to the lower vertices have five times this stiffness.

(i) Show that the potential energy, V, of the system is given by

$$V = \tfrac{1}{4}mga(33\sec^2\theta - 40\sqrt{3}\sec\theta + 38)$$

where θ is the angle that one of the lower springs makes with the horizontal and the gravitational potential energy is taken relative to the horizontal line AB.

(ii) Show that one point of equilibrium occurs at $\theta = 0$ and locate any other equilibrium points.

(iii) By considering the behaviour of $\dfrac{dV}{d\theta}$ for small values of θ determine the stability of equilibrium at $\theta = 0$.

[MEI 1993]

12. (a) Explain what is meant by a system being
(i) in equilibrium;
(ii) in stable equilibrium.

Give an example of a system in which a particle is in
(iii) stable equilibrium;
(iv) unstable equilibrium.

(b) A smooth wire is fixed in a vertical plane and is defined by

$$y = \frac{1}{10a^4}\left(\tfrac{1}{5}x^5 + \tfrac{1}{4}ax^4 - 4a^2x^3\right) \qquad -5a \leqslant x \leqslant 4.7a$$

where y is measured vertically upwards and x is a horizontal co-ordinate. The form of the wire is shown in the diagram.

Copy the diagram and, given that there is a small bead threaded on the wire, mark on your diagram the equilibrium positions of the bead, noting whether each position is stable (S) or unstable (US). Also calculate the exact locations of the equilibrium positions.

Describe the motion (no calculations are necessary) when the bead is released, in turn from
(i) the point $(-4.5a, 9.8a)$;
(ii) the point $(2a, -2.16a)$;
(iii) the point on the wire where $y = 15a$.

[Oxford 1992]

13. A uniform circular cylinder of mass M and radius a lies at rest inside a fixed hollow cylinder of radius $3a$. The axes of both cylinders are horizontal. A particle of mass m is fixed to a point at the top of the inner cylinder which can roll without slipping inside the hollow cylinder. The system is subject to a vertical gravitational field g.

Show that $\phi = \tfrac{1}{2}\theta$, where ϕ and θ are as shown in the diagram (overleaf).

Find an expression for the gravitational potential energy of the system as a function of θ.

Locate the positions of equilibrium and determine the relationship between M and m for the initial position to be one of stable

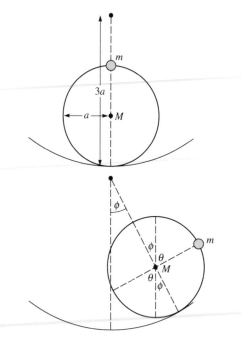

equilibrium. Show that in this case it is the only equilibrium position.

[MEI 1988]

14. The force F of repulsion between two particles with positive charges Q and Q' is given by $F = kQQ'/r^2$, where k is a positive constant and r is the distance between the particles. Two small beads P_1 and P_2 are fixed to a straight horizontal smooth wire, a distance d apart. A third bead P_3 of mass m is free to move along the wire between P_1 and P_2. The beads carry positive electrical charges Q_1, Q_2 and Q_3. If P_3 is in equilibrium at a distance a from P_1, show that

$$a = \frac{d\sqrt{Q_1}}{\sqrt{Q_1} + \sqrt{Q_2}}$$

Suppose that P_3 is displaced slightly from its equilibrium position and released from rest. Show that it performs approximate simple harmonic motion with period

$$\frac{\pi d}{(\sqrt{Q_1} + \sqrt{Q_2})^2} \sqrt{\left[\frac{2md\sqrt{(Q_1 Q_2)}}{kQ_3}\right]}$$

[You may use the fact that $\dfrac{1}{(a+y)^2} \approx \dfrac{1}{a^2} - \dfrac{2y}{a^3}$ for small y.]

[O&C 1993]

8

8

Variable mass

Plus ça change, plus c'est la même chose

Alphonse Karr

For Discussion

1. What is causing the rocket to move?
2. A wagon is rolling freely along a level track at constant speed when it passes under a hopper which dumps coal into it. A short way further on, the bottom of the wagon opens up to allow the coal to drop out. Assuming friction can be neglected, will the speed of the wagon at the end be greater than, less than or the same as at the beginning?

A rocket loses mass as it expels fuel. For example, the first stage of Ariane IV uses 250 kg of fuel a second. A falling raindrop may shrink because of evaporation or, when falling in mist, grow because of condensation. The motion of the rocket and that of the raindrop are affected not only by external forces but by the fact that their mass is changing. This chapter deals with such situations. The topic is usually called *variable mass* because you form an equation of motion referring to a system (e.g. the rocket) whose mass is changing. Of course, mass is not (normally) created or destroyed. The mass of the rocket indeed decreases but the lost mass is expelled into space as fuel. As you will see, the velocity of this fuel is a crucial factor in the motion of the rocket.

Conservation of linear momentum

You have already met problems involving conservation of linear momentum, for example when a gun is fired or somebody jumps off the back of a sledge. In the absence of external forces, the *total linear momentum* of gun plus bullet or sledge plus person is conserved. Both these involve variable mass, although the change of mass occurs in discrete lumps. This chapter covers situations where the mass can be regarded as changing continuously. The technique in such problems, illustrated in the following examples, is to consider the momentum–impulse relationship over a small interval of time from t to $t + \delta t$.

EXAMPLE

A hopper truck containing gravel is moving along a horizontal railway line. The gravel is dropping out of the bottom at a constant rate of k kg s^{-1}. By considering the total linear momentum before and after a small interval of time from t to $t + \delta t$,

(i) show that the truck does not accelerate if there are no horizontal forces on it.

(ii) The truck is actually being pulled by a variable force F. Derive the relationship between F and the acceleration of the truck when the mass of the truck plus gravel is m.

Solution

(i) Denote by v the velocity of the truck at a time t when the mass of the truck and remaining gravel is m. The total linear momentum is thus mv.

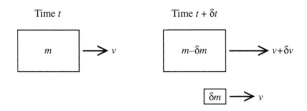

During a short time δt, an amount of gravel of mass $\delta m = k\delta t$ drops out. When this gravel begins dropping, its horizontal velocity is the same as that of the truck, i.e. v. At the end of the period δt, the velocity of the truck will have increased to $v + \delta v$. The total momentum is now

$$(m - \delta m)(v + \delta v) + v\delta m = mv + m\delta v - \delta m\delta v$$

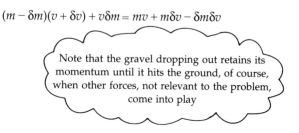

Note that the gravel dropping out retains its momentum until it hits the ground, of course, when other forces, not relevant to the problem, come into play

The *increase* in linear momentum of the whole system is thus $m\delta v - \delta m\delta v$.

When there is no force, there is no change in momentum, so

$$m\delta v - \delta m\delta v = 0$$

Dividing through by the time interval δt gives

$$\frac{m\delta v}{\delta t} = \frac{\delta m}{\delta t} \times \delta v$$

$$= k\delta v \quad \left(\text{since } \frac{\delta m}{\delta t} = k\right)$$

In the limit as $\delta t \to 0$ and $\delta v \to 0$, the equation becomes $m\dfrac{dv}{dt} = 0$. The acceleration is zero.

(ii) When the truck is pulled by a force F, the impulse of the force over this period, $F\delta t$, is equal to the increase of momentum over the period:

$$F\delta t = m\delta v - \delta m\delta v$$

$$\Rightarrow \quad F = m\frac{\delta v}{\delta t} - \frac{\delta m}{\delta t} \times \delta v$$

$$= m\frac{\delta v}{\delta t} - k\delta v \quad \left(\text{since } \frac{\delta m}{\delta t} = k\right) \qquad \text{①}$$

In the limit as $\delta t \to 0$, and $\delta v \to 0$, the equation becomes $F = m\dfrac{dv}{dt}$.

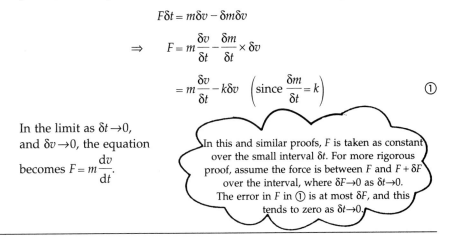

In this and similar proofs, F is taken as constant over the small interval δt. For more rigorous proof, assume the force is between F and $F + \delta F$ over the interval, where $\delta F \to 0$ as $\delta t \to 0$. The error in F in ① is at most δF, and this tends to zero as $\delta t \to 0$.

This is just Newton's Second Law $F = ma$, where m happens to be varying. You may think that this example was rather trivial and you could have written down the answers straight away. The fact that gravel is dropping out of the bottom of the truck does not affect the instantaneous relation between force and acceleration on the truck. This is correct, but you will see in the next example why it is useful to go through the momentum calculation.

EXAMPLE

A canal barge is drifting without power under a hopper which is filling it with coal at a rate of r kg s^{-1}. The barge was initially moving with a velocity u m s^{-1} and its mass without the coal is M kg. The resistance of the water may be neglected.

(i) Denoting by m and v respectively the mass and velocity of the barge plus coal at time t, show that

$$m\frac{dv}{dt} = -v\frac{dm}{dt}$$

(ii) Show that the velocity of the barge t s after the coal begins to drop into the barge is given by

$$v = \frac{Mu}{(M + rt)}$$

(iii) Show that the distance travelled in this time is

$$S = \left(\frac{Mu}{r}\right)\ln\left(1 + \frac{r}{M}t\right)$$

Solution

(i) Consider an instant at time t when the barge plus coal has mass m moving with velocity v. At a time δt later, a mass δm (whose initial velocity is zero) has been added, and the whole is now moving with a velocity $v + \delta v$. The change in momentum is

$$(m + \delta m)(v + \delta v) - mv = mv + v\delta m + m\delta v + \delta m\delta v - mv$$

$$= v\delta m + m\delta v + \delta m\delta v$$

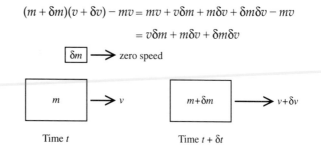

There is no net horizontal force, so there is no change in the total linear momentum of the coal plus barge:

$$v\delta m + m\delta v + \delta m\delta v = 0$$

Dividing through by δt and letting $\delta t \to 0$ gives

$$v\frac{dm}{dt} + m\frac{dv}{dt} = 0$$

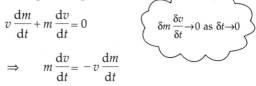

$$\Rightarrow \quad m\frac{dv}{dt} = -v\frac{dm}{dt}$$

As each δm of coal is added, its momentum increases from zero to $v\delta m$ so $v\dfrac{dm}{dt}$ is the rate of increase in momentum of the added mass. Hence the equation can be interpreted as

mass × acceleration = – (rate of increase of momentum from added mass)

Compare with $ma = -F$ and you can see that the effect of the coal being added is the same as a *resisting force of magnitude equal to the rate of increase in momentum of the new coal.*

(ii) This equation can be solved by separating the variables, but it is easier to go back to

$$m\frac{dv}{dt} + v\frac{dm}{dt} = 0$$

$$\Rightarrow \quad \frac{d}{dt}(mv) = 0$$

Integrating gives $mv = $ constant.

After time t, a mass rt of coal has been added. The mass is then $m = M + rt$. Hence

$$(M + rt)v = \text{constant} = Mu \quad \text{(since when } t = 0,\ v = u\text{)}$$

$$v = \frac{Mu}{(M + rt)}$$

(iii) To work out distance travelled, v must be written as $\dfrac{ds}{dt}$:

$$\frac{ds}{dt} = \frac{Mu}{(M + rt)}$$

$$\Rightarrow \quad s = \left(\frac{Mu}{r}\right) \ln(M + rt) + k$$

where k is the constant of integration. When $t = 0$, $s = 0$, giving

$$k = -\left(\frac{Mu}{r}\right) \ln M$$

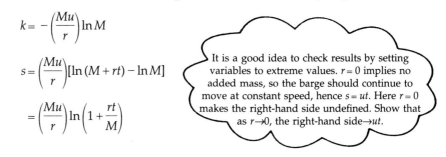

It is a good idea to check results by setting variables to extreme values. $r = 0$ implies no added mass, so the barge should continue to move at constant speed, hence $s = ut$. Here $r = 0$ makes the right-hand side undefined. Show that as $r \to 0$, the right-hand side $\to ut$.

$$s = \left(\frac{Mu}{r}\right)[\ln(M + rt) - \ln M]$$

$$= \left(\frac{Mu}{r}\right) \ln\left(1 + \frac{rt}{M}\right)$$

Newton's Second Law revisited

In the last example, the equation of motion was written as $\dfrac{d}{dt}(mv) = 0$. This is saying that the *rate of change of the total momentum in the system* is zero. If an external force F is included in this example, then

$$F = \frac{d}{dt}(mv)$$

In other words:

$$\text{force} = \text{rate of change of momentum}$$

This is the *full form* of Newton's Second Law of motion. When m is constant, it reduces to the familiar

$$F = \frac{d}{dt}(mv) = m\frac{dv}{dt} = ma$$

where a is acceleration.

In variable mass problems, you have to be careful when using Newton's Second Law in the form $F = \dfrac{d}{dt}(mv)$. The momentum mv refers to the *total momentum of the system.* In the last example, the coal has no linear momentum (in the direction of the barge motion) before it drops on to the barge, so that the momentum of the barge plus contents is equal to the momentum of the whole system. So to solve part (i) of this problem, you could simply have written

$$0 = \frac{d}{dt}(mv) = m\frac{dv}{dt} + v\frac{dm}{dt}$$

But in the first example, the gravel dropping off the truck retained its velocity, so applying $F = \frac{d}{dt}(mv)$ *only to the truck* would have been wrong.

The lost mass still had its momentum. Of course, the gravel eventually hits the ground and loses its velocity, but then other forces are involved.

The following is another example where it is safe to apply the full form of Newton's Second Law. The mass being added to a moving object has no initial velocity and therefore $F = \frac{d}{dt}(mv)$ can be applied simply to the moving object.

EXAMPLE

A spherical droplet, initially of radius a, falls from rest under the influence of a gravitational field g, through a stationary light mist. The mass of the droplet increases due to the condensation of mist on its surface. Air resistance may be ignored.

(i)   Denoting the mass and velocity of the droplet at any time t by m and v respectively, write down Newton's Second Law for the motion. Deduce that the equation is consistent with an initial acceleration of g.

(ii) Given that the mass increases at a rate proportional to the instantaneous surface area of the droplet, show that its radius r increases linearly with time t, i.e. can be expressed in the form $r = ct + a$, where c is a constant.

(iii) Derive v in terms of r, a, c and g.

(iv) Deduce that when $r \gg a$, the acceleration is approximately $g/4$.

Solution

(i)   The extra mass is at rest before condensing, therefore force = rate of change of momentum.

$$mg = \frac{d}{dt}(mv) = m\frac{dv}{dt} + v\frac{dm}{dt}$$

When $v = 0$, $mg = m\frac{dv}{dt} \Rightarrow \frac{dv}{dt} = g$.

Initial acceleration is g.

(ii) Mass of droplet = volume × density

$$m = \frac{4\pi r^3}{3}\rho$$

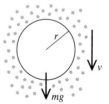

$$\frac{dm}{dt} = (4\pi r^2 \rho)\frac{dr}{dt}$$

The rate of increase of m is proportional to the surface area. Therefore

$$\frac{dm}{dt} = k \times 4\pi r^2$$

for some constant k. Hence

$$4\pi r^2 k = (4\pi r^2 \rho)\frac{dr}{dt} \ .$$

$$\Rightarrow \quad \frac{dr}{dt} = \frac{k}{\rho} = c \quad \text{(say)}$$

Integrating gives

$$r = ct + a \qquad \text{(constant of integration } a \text{, since } r = a \text{ when } t = 0\text{)}$$

Therefore r increases linearly with t.

(iii) $\dfrac{d}{dt}(mv) = mg = \tfrac{4}{3}\pi\rho g r^3 = \tfrac{4}{3}\pi\rho g\,(ct + a)^3$

Integrating gives

$$mv = \frac{4\pi\rho g}{3}(ct + a)^4 \times \frac{1}{4c} + d \qquad (d \text{ is the constant of integration})$$

$$= \left(\frac{\pi\rho g}{3c}\right)r^4 + d \qquad (r = ct + a)$$

When $t = 0$, $r = a$ and $v = 0$ giving

$$d = -\left(\frac{\pi\rho g}{3c}\right)a^4$$

Hence

$$mv = \frac{\pi\rho g}{3c}(r^4 - a^4)$$

But $m = \tfrac{4}{3}\pi r^3 \rho$, thus

$$\frac{4\pi r^3 \rho v}{3} = \frac{\pi\rho g}{3c}(r^4 - a^4)$$

$$\Rightarrow \quad v = \frac{g}{4c}\left(r - \frac{a^4}{r^3}\right)$$

(iv) Acceleration $\dfrac{dv}{dt} = \dfrac{dv}{dr} \times \dfrac{dr}{dt}$

$$= \frac{g}{4c}\left(1 + \frac{3a^4}{r^4}\right)\frac{dr}{dt}$$

$$= \frac{g}{4} + \frac{3g}{4}\left(\frac{a}{r}\right)^4 \qquad \text{since } \frac{dr}{dt} = c$$

when $r = a$, $\dfrac{dv}{dt} = g$ as expected (initial acceleration).

When $r \gg a$, the second term is negligible, thus the acceleration is approximately $g/4$.

Exercise 8A

1. A charged droplet of mass M, initially at rest, is under the influence of a constant horizontal electrical force of magnitude F. The droplet evaporates so that at time t its mass is Me^{-kt}, where k is a positive constant.
   (i) Show that the equation of motion in the direction of the electrical force is

   $$F = Me^{-kt}\frac{dv}{dt}$$

   where v is the speed of the droplet in this direction.
   (ii) Solve this equation to express v in terms of t.

2. A wagon containing sand moves under a constant horizontal force F while the sand drops out at a constant rate k. The wagon is initially at rest with a total mass M of which the mass of sand is m_0. What velocity does it have when all the sand has gone?

3. As it moves forward, a dumper truck is releasing topsoil at a rate of r kg s^{-1}. The net forward force on the truck works at a constant rate of P kW. Initially the truck is at rest and the total load (truck plus soil) is M_0 kg. The soil drops out of the truck with no horizontal velocity relative to the truck.
   (i) Show that the equation of motion may be written

   $$1000P = (M_0 - rt)v\frac{dv}{dt}$$

   (ii) Solve the equation to express v^2 in terms of t. Hence show that when the initial load is 25 tonnes, $r = 50$ and $P = 20$, the truck would be moving at just over 7 m s^{-1} after half a minute.

4. A trailer releasing fertilizer is being pulled across a field by a constant force F. In order to give an even spread, the releasing mechanism ensures that the rate of release, r kg s^{-1} is proportional to the speed of the trailer: $r = kv$, where k is constant. Fertilizer release starts when the mass of the trailer plus load is M and its speed is u.
   (i) Show that the total mass of the trailer after it has moved a distance s is $M - ks$.
   (ii) Write the equation of motion in terms of v and s.
   (iii) Show that

   $$v^2 = u^2 - \frac{2F}{k}\ln\left(1 - \frac{ks}{M}\right)$$

   (iv) Confirm that as $k \to 0$, the formula becomes $v^2 = u^2 + 2(F/M)s$. (Use the series expansion of $\ln(1+x)$.)

5. An oil sheikh decides to tow icebergs to the Middle East for fresh water. The iceberg is modelled as a homogeneous sphere of radius r, which melts at a rate proportional to its surface area. An iceberg has an initial mass M and one eighth of its mass remains after a time T.
   (i) Show that after time t, its mass is

   $$M\left(1 - \frac{t}{2T}\right)^3.$$

   (ii) Assuming a net towing force which works at a constant rate P, write down the equation of motion.
   (iii) Assuming the iceberg starts from rest and has a velocity V at time T, show that

   $$V^2 = \frac{6TP}{M}$$

6. A raindrop falls vertically from rest through a mist under the influence of gravity. It

grows by condensation of mist droplets, initially at rest, on its surface. After falling for a time t, the mass of the raindrop is Me^{kt}, where M and k are positive constants. After a time T the mass has increased to $2M$.

(i)   Determine the value of k in terms of T.
(ii) Formulate the equation of motion of the raindrop.
(iii) Show that the speed of the raindrop at time T is $gT/(2\ln 2)$.
(iv) Determine how far it has fallen in time T in terms of g and T.

[MEI 1993]

7. A raindrop falling through mist may be modelled as a uniform sphere whose radius r increases according to the law $\dfrac{dr}{dt} = kr$, where k is constant. Assuming the only force is gravity, show that the raindrop approaches a limiting velocity of $g/3k$.

8. The mass m of a raindrop falling under gravity through a stationary cloud increases at a constant rate k. The resistance to motion may be modelled as kv, where v is the velocity of the raindrop.
(i)   Show that the dimensions of k are consistent in the preceding two statements.
(ii) Show that the equation of motion can be put in the form

$$m\frac{dv}{dm} + 2v = \frac{mg}{k}$$

(iii) Assuming the raindrop is initially at rest, find v in terms of g, k, m and the initial mass M.

9. A train of mass M moving under a force with constant power P picks up stationary water at a rate k. The initial velocity is v_0. Show that when the total mass (train plus water) is m, its velocity v satisfies the equation

$$k(m^2v^2 - M^2v_0^2) = P(m^2 - M^2)$$

10. In heavy mist a particle is projected horizontally with a speed U. Its initial mass is M but it increases at a constant rate k due to condensation.
(i)   Denoting its velocity at time t by $\mathbf{v}$, show that

$$\frac{d\mathbf{v}}{dt} + \frac{k\mathbf{v}}{(M+kt)} = \mathbf{g}$$

where $\mathbf{g}$ is the gravitational acceleration vector. (Use the vector form of Newton's Second Law $\mathbf{F} = \dfrac{d}{dt}(m\mathbf{v})$)

(ii) By resolving $\mathbf{v}$ into its horizontal component v_x and (downward) vertical component v_y show that

$$\frac{dv_x}{dt} + \frac{kv_x}{M+kt} = 0$$

and

$$\frac{dv_y}{dt} + \frac{kv_y}{M+kt} = g$$

(iii) Hence show that at time t

$$v_x = \frac{UM}{M+kt}$$

and

$$v_y = \tfrac{1}{2}gt\left(\frac{2M+kt}{M+kt}\right)$$

11. An electron of mass m_0 is initially at rest when an electric field is applied to it. This field subjects the electron to a force of magnitude E in a constant direction. When the electron is moving very fast, its mass as measured by a stationary observer increases to

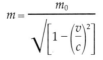

$$m = \frac{m_0}{\sqrt{\left[1 - \left(\dfrac{v}{c}\right)^2\right]}}$$

where the constant c is the velocity of light. (This is a consequence of Einstein's Special Theory of Relativity.)
(i)   Assuming Newton's Second Law applies, that force is rate of change of

Exercise 8A continued

momentum, show that

$$v^2 = \frac{c^2 k^2 t^2}{1 + k^2 t^2}$$

where $k = E/cm_0$.

Deduce that the velocity of the electron never reaches the velocity of light.

(ii) Find the distance travelled by the electron after time T. Show that when $kt \ll 1$, the distance travelled is approximately what it would have been if the mass had remained constant at m_0.

Rockets

One of the most interesting applications of variable mass methods is the motion of space rockets. The rocket stores a large amount of fuel which is burnt and ejected at very high speed. As you see in the example below, the rocket gains forward momentum to compensate for the backward momentum of the expelled fuel. Rockets are normally designed so that the fuel leaves the rocket at a constant mass rate and at a constant velocity *relative to the rocket*. This is more complicated than the first example of this chapter, where the lost mass simply had the *same* velocity as the truck.

The following example demonstrates the essential mathematical principles of rocket motion.

EXAMPLE

At time t, a rocket in one-dimensional motion has mass m, is travelling with speed v and is burning fuel at a constant mass rate k. The burnt fuel is ejected with a constant exhaust speed u relative to the rocket. Such a rocket has initial mass m_0, of which a fraction α is fuel, and is fired vertically upwards from rest in a constant gravitational field g.

(i) By considering the momentum over a small period, derive the equation of motion:

$$(m_0 - kt) \frac{dv}{dt} = uk - (m_0 - kt)g$$

(ii) Write down a condition for the rocket to start rising when its motor is fired.

(iii) For how long will the motor burn?

(iv) Assuming that the rocket does start to rise immediately, find its final speed when all the fuel has burnt.

(v) Use the expression derived in part (iv) to suggest ways in which the final speed might be increased, assuming that the initial mass m_0 is fixed.

[MEI 1992]

Solution

(i) At time t, the mass of the rocket is $m = m_0 - kt$. The momentum at time t is mv.

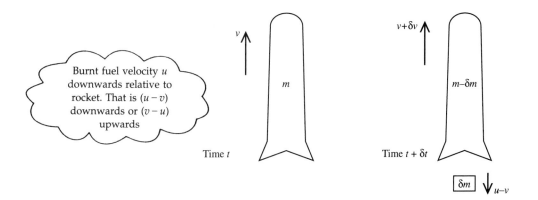

At a small time δt later, an amount of mass δm has been expelled at a speed of u *relative to the rocket*. The velocity of the fuel is thus $(v - u)$ upwards.

At time $t + \delta t$, the velocity of the rocket has increased to $v + \delta v$. The momentum at time $t + \delta t$ is

$$(m - \delta m)(v + \delta v) + \delta m(v - u) = mv + m\delta v - u\delta m - \delta m\delta v$$

The *increase* in momentum over this period is thus approximately $m\delta v - u\delta m - \delta m\delta v$. This is equal to the impulse of the external force which is $-mg\delta t$:

$$-mg\delta t = m\delta v - u\delta m - \delta m\delta v$$

$$\Rightarrow \quad -mg = m\frac{\delta v}{\delta t} - u\frac{\delta m}{\delta t} - \frac{\delta m\delta v}{\delta t}$$

251   M5

As $\delta t \to 0$, $\dfrac{\delta v}{\delta t} \to \dfrac{dv}{dt}$, $\dfrac{\delta m}{\delta t} \to k$, the rate of *loss* of mass. The other term tends to zero, so

$$-mg = m\dfrac{dv}{dt} - uk$$

> Compare with $ma = F$. The left-hand side is mass × acceleration of the rocket. The right-hand side, in addition to the external force of gravity, includes a term uk which can be interpreted as the forward force (thrust) of the engine; it is the *rate of change of momentum of the expelled mass*

$$\Rightarrow \quad m\dfrac{dv}{dt} = uk - mg$$

Substituting $m = m_0 - kt$, gives the required result:

$$(m_0 - kt)\dfrac{dv}{dt} = uk - (m_0 - kt)g$$

NOTE

In this proof δm, the mass of the fuel expelled, has been taken as a positive quantity to fit the mechanics of the situation. The rate of change of mass of the rocket $\dfrac{dm}{dt}$ is actually negative (since the mass is decreasing) and so $\dfrac{dm}{dt} = -k$.

(ii) $(m_0 - kt)\dfrac{dv}{dt} = uk - (m_0 - kt)g$

When the rocket is fired, $t = 0$. For the rocket to start rising, the acceleration must be positive, i.e. $\dfrac{dv}{dt} > 0$. Hence

$$uk - m_0 g > 0$$

$$\Rightarrow \quad uk > m_0 g$$

The thrust of the engine, uk, must be greater than the gravitational force $m_0 g$.

(iii) The motor burns as long as there is fuel. Since the total fuel is αm_0, then when all the fuel is burnt

$$\alpha m_0 = kt \quad \Rightarrow \quad t = \dfrac{\alpha m_0}{k}$$

(iv) From part (i)

$$\dfrac{dv}{dt} = \dfrac{uk}{m_0 - kt} - g$$

Integrating gives

$$v = -u\ln(m_0 - kt) - gt + c$$

where c is the constant of integration. When $t = 0$, $v = 0$, so $c = u\ln m_0$.

Substituting $t = \alpha m_0/k$ gives the final velocity

$$v_f = u\ln\left(\dfrac{m_0}{m_0 - \alpha m_0}\right) - \dfrac{g\alpha m_0}{k}$$

$$= u\ln\left(\dfrac{1}{1-\alpha}\right) - \dfrac{g\alpha m_0}{k}$$

(v) Increasing v_f can be achieved by increasing the first term or decreasing the second. The possibilities are
- increase u: use a fuel with a high exhaust rate;
- make α near 1: increase the proportion of the total mass that is fuel;
- make k larger: increase the burn rate.

Rocket with no external force

If the rocket is fired in space where there is no appreciable external force, the rocket equation is simply

$$uk = m\frac{dv}{dt}$$

where, using the notation of the above example, u is the relative exhaust speed, m is the rocket mass at a given time and $k = -\dfrac{dm}{dt}$ is the rate at which fuel is burned.

The effective thrust of the engine is the relative speed of the fuel × rate of expulsion of its mass, relative to the rocket.

The solution from the above example is

$$v_f = u \ln\left(\frac{1}{1-\alpha}\right)$$

$$\Rightarrow \quad v_f = u \ln\left(\frac{m_0}{m_0 - \alpha m_0}\right)$$

$$= u \ln\left(\frac{\text{initial mass}}{\text{final mass}}\right)$$

The final velocity depends only on u, the relative velocity of the fuel, and on the proportion of the total mass that is fuel. Note that k does not appear in this expression. When there is no external force, the rate at which mass is burned does not affect the *final* velocity, although it obviously determines how long the rocket takes to reach this velocity.

If the rocket begins with some initial velocity v_0, then the result is

$$v_f - v_0 = u \ln\left(\frac{\text{initial mass}}{\text{final mass}}\right)$$

$(v_f - v_0)$ is the *velocity gained* over the burn period.

EXAMPLE

Captain Kirk's spaceship *Enterprise* is pointing towards a newly discovered planet. Unfortunately, his hyperdrive is not functioning and he has to depend on conventional rocket propulsion technology to reach the planet. The initial total mass of the ship is M_0 kg and the fuel is ejected at a rate of k kg s^{-1} with a speed u m s^{-1} relative to the spaceship. The ship starts at rest relative to the planet, and its speed, when all the fuel is exhausted, is $2u$ m s^{-1}. However, it is still so far from the planet that the gravitational force can be neglected.

(i) From the equation of motion derive an expression for v in terms of the remaining mass m at any time. Hence show that the total mass at the end is M_0/e^2 kg.

(ii) Find the velocity of the *ejected fuel* relative to the planet just before the end of the period. Comment on this result.

(iii) Show that the distance travelled during the period of the rocket firing is

$$\left(1 - \frac{3}{e^2}\right)\frac{u}{k}M_0$$

Solution

(i) After time t, the mass of the ship is $m = M_0 - kt$. Since there are no external forces, the equation of motion is

$$m\frac{dv}{dt} = uk$$

$$\Rightarrow \quad \frac{dv}{dt} = \frac{uk}{(M_0 - kt)}$$

Integrating gives

$$v = -u\ln(M_0 - kt) + c$$

When $t = 0$, $v = 0$, so $c = u\ln(M_0)$. This gives

$$v = u\ln\left(\frac{M_0}{M_0 - kt}\right) = u\ln\left(\frac{M_0}{m}\right) \qquad \text{①}$$

At the end of the burn, $v = 2u$, giving

$$\ln\left(\frac{M_0}{m}\right) = 2$$

$$\frac{M_0}{m} = e^2$$

$$\Rightarrow \quad m = \frac{M_0}{e^2}$$

(ii) At the end of the burn period, the velocity of the fuel relative to the planet is $2u - u = u$ m s^{-1}. The expelled fuel is actually travelling *towards* the planet, although not as fast as the rocket! There is no paradox here; all velocities are relative.

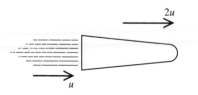

(iii) From ①

$$\frac{ds}{dt} = u\ln\left(\frac{M_0}{M_0 - kt}\right)$$

where s is the displacement from the start. Integrating gives

$$s = u \int \ln\left(\frac{M_0}{M_0 - kt}\right) dt$$

$$= u \int \ln M_0 \, dt - u \int \ln(M_0 - kt) \, dt \qquad \text{②}$$

The second integral is easier if you make the substitution $m = M_0 - kt$.

$$\int \ln(M_0 - kt) \, dt = -\frac{1}{k} \int \ln m \, dm$$

$$= -\frac{1}{k}(m \ln m - m) + \text{constant}$$

The standard result $\int \ln y \, dy = y \ln y - y$ is easily proved by integration by parts (write as $\int 1 \times \ln y \, dy$) or by differentiating the right-hand side

Substituting this in ② gives

$$s = ut \ln M_0 + \frac{mu}{k}(\ln m - 1) + c$$

When $s = 0$, $t = 0$, $m = M_0$. So

$$c = -\frac{M_0 u}{k}(\ln M_0 - 1)$$

Hence

$$s = ut \ln M_0 + \frac{mu}{k}(\ln m - 1) - \frac{M_0 u}{k}(\ln M_0 - 1) \qquad \text{③}$$

At burn out, $m = \dfrac{M_0}{e^2}$ (proved in part (i))

$$\Rightarrow \qquad M_0 - kT = \frac{M_0}{e^2}$$

where T is time of burn out

$$\Rightarrow \qquad T = \frac{M_0}{ke^2}(e^2 - 1)$$

Substituting $m = \dfrac{M_0}{e^2}$ and $t = T = \dfrac{M_0}{ke^2}(e^2 - 1)$ in ③ gives:

$$s = \frac{uM_0}{ke^2}(e^2 - 1)\ln M_0 + \frac{M_0 u}{ke^2}\left(\ln\left(\frac{M_0}{e^2}\right) - 1\right) - \frac{M_0 u}{k}(\ln M_0 - 1)$$

$$= \frac{uM_0}{ke^2}(e^2 \ln M_0 - \ln M_0 + \ln M_0 - 2 - 1 - e^2 \ln M_0 + e^2)$$

$$= \frac{uM_0}{ke^2}(e^2 - 3)$$

$\ln e^2 = 2$

The first rockets were probably bamboo tubes packed with 'black powder',
a precursor of gunpowder. These were used by the Chinese in the 10th century.
The technology spread both for warfare and entertainment; from the 16th century
onwards, firework displays with rockets were used widely in Europe. Rockets were
successfully used against the British by Indian forces during the battles of
Seringapatam in 1792 and 1799. As a result, the British army adopted rockets for
military purposes and used them for most of the next century. A typical military
rocket of that time was described as an iron cylinder, 200 mm long, 40 mm wide
with a 3 m guiding stick. By the end of the 19th century, better artillery superseded
the rocket and the military lost interest. A few enthusiasts maintained development
in rocket technology during the following decades until World War 2 when
Germany developed the V2. This could reach a target some 300 km away in 5
minutes. Despite many technical problems, it was used to devastating effect, with
about 1500 landing in southern England. With the end of the war, and the
beginning of the Cold War, most work on the development of rockets moved to the
USA and USSR.

In 1957, the space age began, with the USSR putting the first artificial satellite
Sputnik into orbit. This led to manned space launches, the most spectacular of
which was the moon landing in 1969. Huge rockets were built: the first stage of the
enormous Saturn V rocket which put the first men on the moon had a thrust of
3.4×10^7 *N.*

The prohibitive expense of space exploration has curtailed development in recent
years. However, the huge growth of the need for satellites for communication
purposes has meant that rocket technology has continued to develop and the space
rocket is now a standard part of contemporary technology.

Jet aircraft

Figure 8.1

A jet aircraft is propelled in a similar way to a rocket, in the sense that
propulsion is achieved by expelling mass. However, the engine uses the
surrounding medium in that it takes in air at the front, compresses it and
ejects it at high speed (figure 8.1). The only change in mass is the
expenditure of some fuel also expelled as exhaust gases, but this is small
compared with the mass of air. Assuming air is entering and leaving at a
mass rate k, the aircraft is moving at a velocity v relative to the air, and
expels it at a rate u relative to the aircraft, the effective thrust is the rate of
change of momentum of this air, which is $k(u - v)$.

Unlike a rocket, it is impossible for the aircraft, in still air, to achieve a speed
greater than the relative speed of ejection of the fuel.

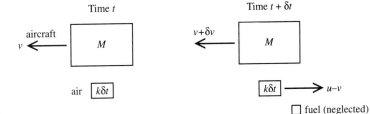

Figure 8.2

A mass of still air $k\delta t$ is picked up and expelled in a time period δt. Equating linear momentum before and after this period (figure 8.2):

$$Mv + k\delta t \times 0 = M(v + \delta v) - k\delta t(u - v)$$
$$\Rightarrow \quad M\delta v = k\delta t(u - v)$$
$$\Rightarrow \quad M\frac{\delta v}{\delta t} = (u - v)k$$

In the limit, the effective thrust $M\dfrac{\mathrm{d}v}{\mathrm{d}t}$ is $(u - v)k$.

Investigations

Multistage rockets

This investigation will use the previous result which gives the gain in velocity of a rocket when there are no external forces, that is gravitational forces are ignored, as

$$u \ln \left(\frac{\text{initial mass}}{\text{final mass}} \right)$$

The increase in velocity is proportional to u, the exhaust speed of the fuel, so fuels are chosen which would burn with a high exhaust speed.

Note also that the larger the proportion of the initial mass which is fuel, the greater will be the gain in velocity. However, there is a limit here. The rocket is being used to move some payload—whether a communications satellite, or three astronauts and their dog—and that is not part of the fuel. Also the fuel must be contained in some casing. If the amount of fuel is increased, the size and therefore the mass of this casing is increased.

Suppose the payload mass is m kg and the total mass (rocket plus payload) is Nm kg. Therefore, the fuel plus casing has a mass $(N-1)m$ kg. Assume that only a fraction α of this is fuel.

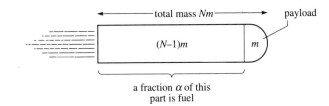

Investigation continued

(i) Given a fuel ejection velocity of u km s^{-1}, what gain in velocity is achieved by the rocket?

(ii) The velocity gain can be increased by building a bigger rocket: increasing N. But assuming α is constant, there is an upper limit to what can be achieved. What is the limit?

(iii) Taking $u = 2000$, and $\alpha = 0.9$, plot the gain in velocity against N.

Two-stage rocket
The way round this limit is to use multistage rockets. A sequence of rockets is used and once one has burned all its fuel, the casing etc of this rocket is ejected.

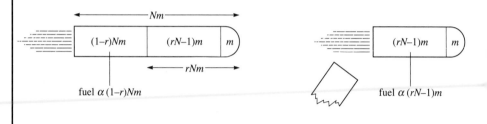

fuel $\alpha (1-r)Nm$ fuel $\alpha (rN-1)m$

At end of first stage, casing for first stage rocket ejected

Suppose the whole system has a *total mass* Nm, but is divided into two stages (see diagram). The second stage carries the payload and has a total mass of rNm, where $0 < r < 1$. The first stage, simply a rocket with fuel and casing has mass $(1-r)Nm$ and so contains $\alpha(1-r)Nm$ fuel. In the first part of the trip, stage 1 fuel is all burnt, giving a gain in velocity v_1. The casing from stage 1 is then ejected. This leaves the stage 2 and the payload: a mass of fuel $\alpha(rN-1)m$. This is then burned, giving another gain in velocity v_2.

For some value of N investigate graphically how the total gain in velocity varies for different values of r and hence find when the gain is a maximum. Use the values of α and u from part (iii). If you do this for different values of N, you should spot a relationship between r and N when the gain is a maximum. There is also a simple relationship between v_1 and v_2 for maximum gain.

Practical rocket experiments
Some model shops sell rockets where burning rates and thrusts are supplied (for examples, see Question 1 in Exercise 8B). Devise simple experiments you could do with such rockets which would enable you to verify the principles of rocket motion given in this chapter.

1. A rocket expels fuel at a constant relative velocity u. The mass of the rocket at time t is m and the fuel is used at a constant rate k, so that $\dfrac{dm}{dt} = -k$.

(i) Assuming no external forces act on the rocket, use conservation of linear momentum to show that

$$m\frac{dv}{dt} = uk$$

where v is the velocity of the rocket at time t.

(ii) The following data have been published for a range of model rockets.

Model	Time of burn	Thrust	Mass of Fuel
R1	0.216 s	5.80 N	2.03 g
R2	0.325 s	7.70 N	4.16 g
R3	1.20 s	4.15 N	8.33 g
R4	0.862 s	5.80 N	6.23 g
R5	0.625 s	8.00 N	6.23 g
R6	1.72 s	5.80 N	12.5 g
R7	1.70 s	11.8 N	24.9 g

Work out the relative expulsion speed of the fuel (m s^{-1}) in each case. How many distinct fuels do you think are used in this range of rockets?

2. A rocket expels fuel at a constant relative velocity u. The mass of the rocket at time t is m and the fuel is used at a constant rate k

(i) Show that the motion of a rocket launched upwards under gravity is governed by the equation

$$m\frac{dv}{dt} = uk - mg$$

and hence prove that the velocity after time t is given by

$$v = u \ln\left(\frac{m_0}{m_0 - kt}\right) - gt$$

where m_0 is the initial total mass.

(ii) A model rocket, R5, has the following characteristics.

Time of burn: 0.625 s
Thrust: 8.00 N
Mass of fuel: 6.23 g

Assume that the mass of fuel represents 80% of the total rocket mass. What will be the velocity of R5 at burnout if it is launched upwards from rest?

3. A rocket of initial total mass M burns fuel at a constant rate rM and expels the fuel with a constant relative velocity u. When the rocket is launched vertically under gravity, the thrust is just sufficient for the rocket to rise.

(i) Show that $ru = g$.

(ii) Show that when a mass M' remains, the rocket's velocity is

$$u\ln\left(\frac{M}{M'}\right) - u\left(1 - \frac{M'}{M}\right)$$

4. A rocket under no external forces has a mass m and velocity v at time t and expels fuel backwards at a mass rate k and a relative speed u.

(i) Write down the equation of motion

(ii) By considering the change in *total* kinetic energy (rocket and expelled fuel) over a small interval δt, show that the rate of gain of total kinetic energy is $\frac{1}{2}ku^2$.

5. A rocket-propelled vehicle starts from rest with mass M and ejects fuel at a constant mass rate r per unit time with a constant relative speed u. When travelling horizontally with speed v the total resistance to motion is kv.

(i) Show that its acceleration at time t is $(ru - kv)/m$, where $m = M - rt$.

(ii) Show that its speed at time t is

$$\frac{ru}{k}\left[1 - \left(\frac{m}{M}\right)^{k/r}\right]$$

6. At time t a rocket in one-dimensional motion has mass M, is travelling with speed v and is burning fuel at a constant mass rate k. The burnt fuel is ejected with a constant exhaust speed c relative to the rocket. The rocket has an initial mass of M_0, of which a fraction α is fuel, and is fired from rest in a gravity free environment.

(i) Write down the mass of the rocket after time t and find the time for which the fuel burns.

(ii) Derive the equation of motion

$$M\frac{dv}{dt} = ck$$

The mass ratio R of a rocket is defined to be the mass of the rocket when fuelled divided by the mass when empty.

(iii) Show that the final speed V of the rocket is given by

$$V = -c\ln(1-\alpha) = c\ln R.$$

A two-stage rocket consists of a first stage of mass M_1 which carries a second stage of mass M_2. Each stage has a fraction α of fuel and $M_1 + M_2 = M_0$. The first stage is fired from rest and carries the second stage until all of the first-stage fuel is burnt. At this point the second stage separates and its motor is fired until all of its fuel is burnt.

(iv) Find the mass ratios R_1 and R_2 of the first stage (carrying the second) and the second stage of the rocket.

(v) Determine the speed of the two-stage system at separation in terms of c and R_1.

(vi) Find the final speed of the second stage in terms of c, R_1 and R_2.

[MEI 1994]

7. At time t a rocket has a mass m and is in one-dimensional motion with speed v. The exhaust of burnt fuel is expelled with a constant speed c relative to the rocket. Derive the differential equation

$$m\frac{dv}{dt} + c\frac{dm}{dt} = F$$

where F is the total external force.

A rocket initially of mass M, of which a fraction α is fuel, $0 < \alpha < 1$, burns fuel at a constant mass rate k. The burnt fuel is expelled with constant exhaust speed c relative to the rocket. The rocket is fired vertically upwards in a constant gravitational field g. Air resistance is neglected. If the rocket starts to rise immediately upon ignition, find its speed and altitude when all the fuel has burnt.

(The result $\displaystyle\int \ln x\, dx = x(\ln x - 1) + C$

may be assumed.)

[MEI 1987]

KEY POINTS

In the following, F is the external force on a body whose mass is changing, m is the mass of the body at time t, and v is its velocity.

The first key point is the crucial one, from which the others follow.

- The standard approach to variable mass problems is to apply the principle

  change in momentum = impulse of force

  over a small time interval t to $t + \delta t$. The change in the force over this interval may be neglected.

- Newton's Second Law can be used directly in the form

    external force = rate of change of *total* momentum

    However, you must ensure that you have taken into account the *total* momentum change in the system.

- For a body losing mass, where the lost mass maintains its momentum as it leaves, (e.g. coal dropping from a hopper truck), the equation becomes

$$F = m \frac{dv}{dt}$$

- For a body gaining mass, where the extra mass previously has no velocity (e.g. moisture from still air condensing onto a raindrop), the equation is

$$F = \frac{d}{dt}(mv) = m\frac{dv}{dt} + v\frac{dm}{dt}$$

- For a rocket expelling mass at a constant rate k with a relative velocity u opposite to the direction of rocket motion

$$m\frac{dv}{dt} = uk + F$$

    where uk may be regarded as the thrust on the rocket due to the burning of fuel.

    * A rocket under no external force will have a velocity gain given by

$$v_f - v_0 = u \ln \left(\frac{\text{initial mass}}{\text{final mass}} \right)$$

9

The vector product in mechanics

The great man would get bored and quietly fade away, leaving me to solve Timmy's problem by some method a little simpler than vector analysis.

Aldous Huxley

For Discussion

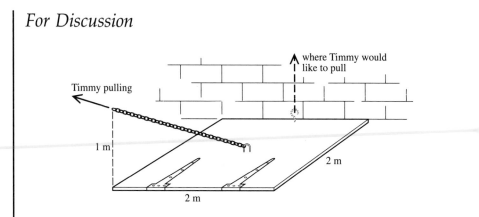

Timmy finds a square trapdoor (side 2 m) set into the floor by a wall. A chain is attached to a hook in the middle of door. He wants to pull open the trapdoor, but walls and other obstructions mean that he can stand only at a corner by the hinged end of the trapdoor. He pulls the chain at a height of 1 m as shown. The trap door only just begins to lift. If he could pull vertically upwards, and on a hook attached at the opposite edge from the hinges it would take only a fraction of the effort. Why? Exactly what fraction?

Thinking in three dimensions is not easy. The diagrams are difficult to draw and you cannot always visualise what is happening. This chapter describes the use of *vectors* which are, despite the impression given by the opening quotation to this chapter, an invaluable aid to modelling and solving three-dimensional problems in mechanics.

The important tool introduced here is the *vector product*, used particularly in situations involving the turning effect of forces (as in the case above) and rotation in general.

The vector product

Mechanics 5 described the use of the scalar product **a.b** of two vectors in finding scalars such as the component of a vector in a given direction, the work done by a force etc. There is another important operation, the vector product (or cross product), where the result of combining two vectors is also a vector. This is described here in sufficient detail for your needs in mechanics, but you should refer to *Pure Mathematics 4* for a fuller treatment.

Representing an area by a vector

Consider a rectangular box, with sides x, y and z as shown in figure 9.1, containing a gas under a pressure p. Pressure means force per unit area, so the total force of the gas on the top face of the box is

$$\text{force} = \text{pressure} \times \text{area} = pxy$$

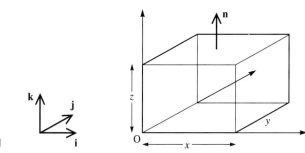

Figure 9.1

The force is at right angles to the face, so the total force on the face is, as a vector,

$$\mathbf{F} = pxy\mathbf{k} \quad \text{where } \mathbf{k} \text{ is the unit vector in direction } Oz$$
$$= p\mathbf{n} \quad \text{where } \mathbf{n} = xy\mathbf{k}$$

The *magnitude* of the vector $\mathbf{n}$ is the area of the top face, and its *direction* is normal to the face, and directed outwards from the box. Thus $\mathbf{n}$ may be thought of as a 'vector area'.

Activity

Write down the 'vector area' of each of the other faces in terms of x, y, z and the standard unit vectors.

Apply this notion of a vector area to a parallelepiped: this is a box whose six faces are all parallelograms, as shown in figure 9.2. Every edge can be represented by one of the three vectors $\mathbf{a}(\overrightarrow{OA})$, $\mathbf{b}(\overrightarrow{OB})$ and $\mathbf{c}(\overrightarrow{OC})$.

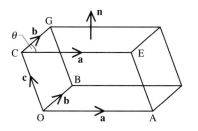

Figure 9.2

The force on, say, the top face can again be written $\mathbf{F} = p\mathbf{n}$, where $\mathbf{n}$ is a vector whose magnitude is the area of this face and whose direction is perpendicular to it, pointing outwards. So

$$|\mathbf{n}| = |\mathbf{a}||\mathbf{b}|\sin\theta$$

where θ is the angle ECG. The vector $\mathbf{n}$ depends simply on $\mathbf{a}$ and $\mathbf{b}$, and is known as the *vector product* of $\mathbf{a}$ and $\mathbf{b}$, written

$$\mathbf{n} = \mathbf{a} \times \mathbf{b}$$

Introducing the vector product as a 'vector area' may help to illuminate the normal definition which is given below.

Definition and properties of the vector product

Given two vectors $\mathbf{a}$ and $\mathbf{b}$, inclined at an angle θ (where $\theta < 180°$), the vector product $\mathbf{a} \times \mathbf{b}$ (you say 'a cross b') is defined as a vector which
- is perpendicular to both $\mathbf{a}$ and $\mathbf{b}$;
- has a magnitude $|\mathbf{a}||\mathbf{b}|\sin\theta$;
- has a direction such that $\mathbf{a}$, $\mathbf{b}$ and $\mathbf{a} \times \mathbf{b}$ in that order form a right-handed set of vectors.

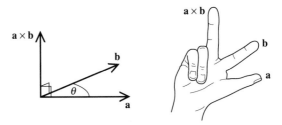

Figure 9.3

This is illustrated in figure 9.3. Without the last condition, there would be two possible candidates, pointing up or down from the plane containing $\mathbf{a}$ and $\mathbf{b}$. The right-hand rule means that if you point your right-hand thumb in the direction of $\mathbf{a}$ and your index finger in the direction of $\mathbf{b}$, then your second finger gives the direction of $\mathbf{a} \times \mathbf{b}$ (assuming it is perpendicular to the other two).

Thus $\mathbf{b} \times \mathbf{a}$ has the same magnitude but the opposite direction (figure 9.4):

$$\mathbf{a} \times \mathbf{b} = -(\mathbf{b} \times \mathbf{a})$$

The vector product is *not commutative*, unlike the scalar product $(\mathbf{a} . \mathbf{b} = \mathbf{b} . \mathbf{a})$

Figure 9.4

Parallel and perpendicular vectors
When two vectors are parallel, θ is zero, so the vector product is zero. In particular,

$$\mathbf{a} \times \mathbf{a} = \mathbf{0}$$

Note that the bold $\mathbf{0}$ is written for the zero vector.

When two vectors are perpendicular, $\sin \theta$ is 1, so in this case

$$|\mathbf{a} \times \mathbf{b}| = |\mathbf{a}||\mathbf{b}|$$

Then $\mathbf{a}$, $\mathbf{b}$ and $\mathbf{a} \times \mathbf{b}$ form a set of mutually perpendicular vectors.

Note the following relationships between the standard unit vectors $\mathbf{i}$, $\mathbf{j}$ and $\mathbf{k}$:

$$\mathbf{i} \times \mathbf{j} = \mathbf{k} \qquad \mathbf{j} \times \mathbf{k} = \mathbf{i} \qquad \mathbf{k} \times \mathbf{i} = \mathbf{j}$$

$$\mathbf{j} \times \mathbf{i} = -\mathbf{k} \qquad \mathbf{k} \times \mathbf{j} = -\mathbf{i} \qquad \mathbf{i} \times \mathbf{k} = -\mathbf{j}$$

$$\mathbf{i} \times \mathbf{i} = 0 \qquad \mathbf{j} \times \mathbf{j} = 0 \qquad \mathbf{k} \times \mathbf{k} = 0$$

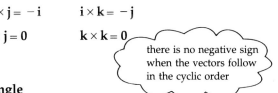

there is no negative sign when the vectors follow in the cyclic order

Area of a parallelogram and triangle

You saw from the preceding section that $|\mathbf{a} \times \mathbf{b}|$ is the area of the parallelogram OAPB whose sides OA and OB can be represented by vectors $\mathbf{a}$ and $\mathbf{b}$. Similarly, the area of the triangle OAB is $\frac{1}{2}|\mathbf{a} \times \mathbf{b}|$ (figure 9.5).

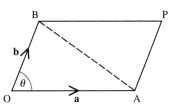

Figure 9.5

Volume of parellelepiped

The volume of a parallelepiped is given by (area of base) × (perpendicular height).

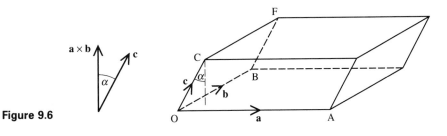

Figure 9.6

Consider $(\mathbf{a} \times \mathbf{b}).\mathbf{c}$ (figure 9.6). By the definition of the scalar product, this has a value $|\mathbf{a} \times \mathbf{b}||\mathbf{c}|\cos \alpha$. But $|\mathbf{a} \times \mathbf{b}|$ is the area of the base parallelogram and $|\mathbf{c}| \cos \alpha$ is the height. So volume $= (\mathbf{a} \times \mathbf{b}).\mathbf{c}$.

Distributive law for the vector product

Like the scalar product, the vector product is *distributive* over addition:

$$\mathbf{a} \times (\mathbf{b}_1 + \mathbf{b}_2 + \ldots + \mathbf{b}_n) = (\mathbf{a} \times \mathbf{b}_1) + (\mathbf{a} \times \mathbf{b}_2) + \ldots + (\mathbf{a} \times \mathbf{b}_n)$$

This useful property is not easy to see from the geometrical definition. A proof is given in *Pure Mathematics 4*.

Triple scalar product

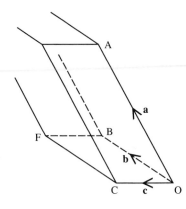

Figure 9.7

The same argument as above, but using OCFB (figure 9.7) as the base of the parallelepiped, would lead to the formula for the volume as $(\mathbf{b} \times \mathbf{c}) \cdot \mathbf{a}$. Similarly starting from a side face, you would get $(\mathbf{c} \times \mathbf{a}) \cdot \mathbf{b}$. Thus

$$\text{volume} = (\mathbf{a} \times \mathbf{b}) \cdot \mathbf{c} = (\mathbf{b} \times \mathbf{c}) \cdot \mathbf{a} = (\mathbf{c} \times \mathbf{a}) \cdot \mathbf{b}$$

The expression $(\mathbf{a} \times \mathbf{b}) \cdot \mathbf{c}$ is called a *triple scalar product*. It is positive when $\mathbf{a}$, $\mathbf{b}$, $\mathbf{c}$ form a right-handed set: for example, $(\mathbf{i} \times \mathbf{j}) \cdot \mathbf{k} = \mathbf{k} \cdot \mathbf{k} = +1$. Otherwise it will give the negative of the volume enclosed by the parallelepiped formed by the vectors. This illustrates the fact that the triple scalar product is the same if the vectors are permuted, but they must retain the same cyclic order. For example:

$$(\mathbf{b} \times \mathbf{a}) \cdot \mathbf{c} = -(\mathbf{a} \times \mathbf{b}) \cdot \mathbf{c}$$

When two of the vectors are the same (or parallel), the triple scalar product is zero:

$$\mathbf{a} \cdot (\mathbf{a} \times \mathbf{b}) = (\mathbf{a} \times \mathbf{a}) \cdot \mathbf{b} = 0 \quad \text{since } \mathbf{a} \times \mathbf{a} \text{ is zero}$$

NOTE

Other notations for the triple scalar product are: $\mathbf{a} \times \mathbf{b} \cdot \mathbf{c}$ (without any brackets) or $[\mathbf{a}, \mathbf{b}, \mathbf{c}]$.

Components of the vector product

Given two vectors in component form

$$\mathbf{a} = a_1\mathbf{i} + a_2\mathbf{j} + a_3\mathbf{k}$$
$$\mathbf{b} = b_1\mathbf{i} + b_2\mathbf{j} + b_3\mathbf{k}$$

then the vector product is

$$\mathbf{a} \times \mathbf{b} = (a_1\mathbf{i} + a_2\mathbf{j} + a_3\mathbf{k}) \times (b_1\mathbf{i} + b_2\mathbf{j} + b_3\mathbf{k})$$

Since the vector product is distributive, this can be worked out term by term in the normal algebraic way. Three of the terms disappear since $\mathbf{i} \times \mathbf{i} = \mathbf{j} \times \mathbf{j} = \mathbf{k} \times \mathbf{k} = 0$. This leaves

$$a_1b_2(\mathbf{i} \times \mathbf{j}) + a_1b_3(\mathbf{i} \times \mathbf{k}) + a_2b_1(\mathbf{j} \times \mathbf{i}) + a_2b_3(\mathbf{j} \times \mathbf{k}) + a_3b_1(\mathbf{k} \times \mathbf{i}) + a_3b_2(\mathbf{k} \times \mathbf{j})$$

Using the relationships $\mathbf{i} \times \mathbf{j} = \mathbf{k}$ etc. given above, this reduces to

$$\mathbf{a} \times \mathbf{b} = (a_2b_3 - a_3b_2)\mathbf{i} + (a_3b_1 - a_1b_3)\mathbf{j} + (a_1b_2 - a_2b_1)\mathbf{k}$$

This important formula is used to work out the components of a vector product when you know the components of its constituents $\mathbf{a}$ and $\mathbf{b}$. The $\mathbf{i}$ component of the result involves only the $\mathbf{j}$ and $\mathbf{k}$ components of $\mathbf{a}$ and $\mathbf{b}$, similarly with the other two components of the vector product. Notice how the corresponding subscripts in the brackets follow in cyclic order. For example, the first terms in the brackets start with a_2, a_3, a_1.

Determinant notation for vector products
You will often see the calculation of a vector product written in the following way.

Given $\mathbf{a} = a_1\mathbf{i} + a_2\mathbf{j} + a_3\mathbf{k}$ and $\mathbf{b} = b_1\mathbf{i} + b_2\mathbf{j} + b_3\mathbf{k}$:

$$\mathbf{a} \times \mathbf{b} = \begin{vmatrix} \mathbf{i} & a_1 & b_1 \\ \mathbf{j} & a_2 & b_2 \\ \mathbf{k} & a_3 & b_3 \end{vmatrix} = \mathbf{i}(a_2b_3 - a_3b_2) + \mathbf{j}(a_3b_1 - a_1b_3) + \mathbf{k}(a_1b_2 - a_2b_1)$$

The components are written vertically alongside the $\mathbf{ijk}$ symbols, all enclosed in vertical lines. If you are familiar with determinants, you will understand how appropriate this notation is; the calculation of the components of the vector product is exactly the calculation you make when working out a determinant.

To find the $\mathbf{i}$ component, cover up the $\mathbf{i}$ row, and look at the remaining components:

$$\begin{vmatrix} a_2 & b_2 \\ a_3 & b_3 \end{vmatrix}$$

The $\mathbf{i}$ component is the 2×2 determinant $a_2b_3 - a_3b_2$.

For the $\mathbf{j}$ component, cover up the $\mathbf{j}$ row, leaving

$$\begin{vmatrix} a_1 & b_1 \\ a_3 & b_3 \end{vmatrix}$$

This time you want the negative of the 2×2 determinant $-(a_1b_3 - a_3b_1)$.

Finally, for the $\mathbf{k}$ component cover up the $\mathbf{k}$ row and take the positive determinant $a_1b_2 - a_2b_1$.

EXAMPLE

(i) Work out $\mathbf{a} \times \mathbf{b}$, where $\mathbf{a} = 4\mathbf{i} + \mathbf{j} + 2\mathbf{k}$ and $\mathbf{b} = 3\mathbf{i} - 3\mathbf{k}$.
(ii) Confirm that the result is perpendicular to $\mathbf{a}$ and $\mathbf{b}$.

Solution

(i) Using the determinant notation:

$$\mathbf{a} \times \mathbf{b} = \begin{vmatrix} \mathbf{i} & 4 & 3 \\ \mathbf{j} & 1 & 0 \\ \mathbf{k} & 2 & -3 \end{vmatrix} = [1 \times (-3) - 2 \times 0]\mathbf{i} - [4 \times (-3) - 2 \times 3]\mathbf{j} + (4 \times 0 - 1 \times 3)\mathbf{k}$$

write down the minus and work out the determinant $\begin{vmatrix} 4 & 3 \\ 2 & -3 \end{vmatrix}$

$$\Rightarrow \quad \mathbf{a} \times \mathbf{b} = -3\mathbf{i} + 18\mathbf{j} - 3\mathbf{k}$$

(ii) To confirm this is perpendicular to **a** and **b**, take the scalar product with each in turn.

$$\mathbf{a}\cdot(\mathbf{a}\times\mathbf{b})=\begin{pmatrix}4\\1\\2\end{pmatrix}\cdot\begin{pmatrix}-3\\18\\-3\end{pmatrix}=(4)(-3)+(1)(18)+(2)(-3)=0$$

$$\mathbf{b}\cdot(\mathbf{a}\times\mathbf{b})=\begin{pmatrix}3\\0\\-3\end{pmatrix}\cdot\begin{pmatrix}-3\\18\\-3\end{pmatrix}=(3)(-3)+(0)(18)+(-3)(-3)=0$$

This is as expected and confirms the answer to part (i).

Once you get used to working out vector products, you will do much of the working in your head. *But always work out the scalar product with at least one of the original vectors to check the answer is zero.* It is very easy to make mistakes when evaluating numerical vector products.

EXAMPLE

The diagram shows a unit cube ABCDEFGH. The line CB is extended to X so that BX = 1. Show that $\overrightarrow{AX}=\overrightarrow{AC}\times\overrightarrow{AG}$

(i) from the geometrical definition of the vector product;

(ii) by working out the components in a Cartesian frame of reference.

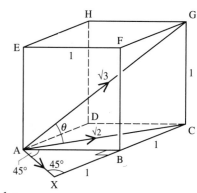

Solution

(i) The direction of $\overrightarrow{AC}\times\overrightarrow{AG}$ is at right angles to both AC and AG, i.e. to the plane ACG. The right-hand rule says that $\overrightarrow{AC}\times\overrightarrow{AG}$ is out of the page in the diagram. Thus it is in the direction $\overrightarrow{AX}$.

By definition $|\overrightarrow{AC}\times\overrightarrow{AG}|=|\overrightarrow{AC}||\overrightarrow{AG}|\sin\theta$
  $=\sqrt{2}\times\sqrt{3}\times(1/\sqrt{3})$
  $=\sqrt{2}=|\overrightarrow{AX}|.$

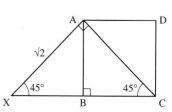

So $\overrightarrow{AX}=\overrightarrow{AC}\times\overrightarrow{AG}$.

(ii) Taking **i** along AB, **j** along AD, **k** along AE, then

$$\overrightarrow{AC}=\mathbf{i}+\mathbf{j}\qquad\overrightarrow{AG}=\mathbf{i}+\mathbf{j}+\mathbf{k}$$

$$\overrightarrow{AC}\times\overrightarrow{AG}=\begin{vmatrix}\mathbf{i}&1&1\\\mathbf{j}&1&1\\\mathbf{k}&0&1\end{vmatrix}=(1-0)\mathbf{i}+(0-1)\mathbf{j}+(1-1)\mathbf{k}$$

$$=\mathbf{i}-\mathbf{j}$$

This is the position vector of X with respect to A.

EXAMPLE

Find a unit vector at right angles to the plane containing $\mathbf{a} = 2\mathbf{i} - 6\mathbf{j} - 3\mathbf{k}$ and $\mathbf{b} = 4\mathbf{i} + 3\mathbf{j} - \mathbf{k}$.

Solution

The vector product $\mathbf{a} \times \mathbf{b}$ is perpendicular to both $\mathbf{a}$ and $\mathbf{b}$ and therefore to the plane containing them.

$$\mathbf{a} \times \mathbf{b} = \begin{vmatrix} \mathbf{i} & 2 & 4 \\ \mathbf{j} & -6 & 3 \\ \mathbf{k} & -3 & -1 \end{vmatrix} = \mathbf{i}(6+9) + \mathbf{j}(-12+2) + \mathbf{k}(6+24)$$

$$= 15\mathbf{i} - 10\mathbf{j} + 30\mathbf{k}$$

(Check: $\mathbf{a} \cdot (\mathbf{a} \times \mathbf{b}) = (2\mathbf{i} - 6\mathbf{j} - 3\mathbf{k}) \cdot (15\mathbf{i} - 10\mathbf{j} + 30\mathbf{k}) = 30 + 60 - 90 = 0$, as expected.)

$$|\mathbf{a} \times \mathbf{b}| = \sqrt{(15^2 + 10^2 + 30^2)} = 35$$

So the unit vector in the direction of $\mathbf{a} \times \mathbf{b}$ is

$$\frac{15\mathbf{i} - 10\mathbf{j} + 30\mathbf{k}}{35} = \tfrac{3}{7}\mathbf{i} - \tfrac{2}{7}\mathbf{j} + \tfrac{6}{7}\mathbf{k}$$

Notice that $-\tfrac{3}{7}\mathbf{i} + \tfrac{2}{7}\mathbf{j} - \tfrac{6}{7}\mathbf{k}$ would also be at right angles to the plane.

Differentiation of scalar and vector products

When modelling in mechanics, the value of a vector, for instance a force or a velocity, often depends on a variable, such as time t. You saw in *Mechanics 5* that a vector can be differentiated by differentiating its Cartesian components:

$$\frac{d\mathbf{F}}{dt} = \frac{dF_x}{dt}\mathbf{i} + \frac{dF_y}{dt}\mathbf{j} + \frac{dF_z}{dt}\mathbf{k}$$

where $\mathbf{i}$, $\mathbf{j}$ and $\mathbf{k}$ are the usual unit vectors in the direction of the axes.

When two vectors, $\mathbf{F}$ and $\mathbf{G}$ depend on t, their scalar and vector products can be differentiated using the familiar product rule of calculus applied to the scalar and vector product operations:

$$\frac{d}{dt}(\mathbf{F} \cdot \mathbf{G}) = \frac{d\mathbf{F}}{dt} \cdot \mathbf{G} + \mathbf{F} \cdot \frac{d\mathbf{G}}{dt}$$

$$\frac{d}{dt}(\mathbf{F} \times \mathbf{G}) = \frac{d\mathbf{F}}{dt} \times \mathbf{G} + \mathbf{F} \times \frac{d\mathbf{G}}{dt}$$

These can be shown by resolving the vectors into components in the $\mathbf{i}$, $\mathbf{j}$ and $\mathbf{k}$ directions and differentiating each component. It is recommended that you work through one of these for yourself.

EXAMPLE

The position vector $\mathbf{r}$ of a particle relative to an origin satisfies the vector differential equation $\dot{\mathbf{r}} = \boldsymbol{\omega} \times \mathbf{r}$, where $\boldsymbol{\omega}$ is a constant vector.

(i) By taking the scalar product of this equation with $\mathbf{r}$, show that the path of the particle lies on a sphere.

(ii) By taking the scalar product of the equation with the unit vector $\hat{\boldsymbol{\omega}}$ show that the path must also lie in a plane.

(iii) Hence deduce that the path is a circle and show that the particle moves with constant speed.

[MEI 1992]

Solution

(i) $$\dot{\mathbf{r}} = \boldsymbol{\omega} \times \mathbf{r}$$

$$\Rightarrow \quad \mathbf{r} \cdot \dot{\mathbf{r}} = \mathbf{r} \cdot (\boldsymbol{\omega} \times \mathbf{r}) = 0 \quad \text{(two equal vectors in a triple scalar product)}$$

The left-hand side is $\dfrac{1}{2}\dfrac{d}{dt}(\mathbf{r} \cdot \mathbf{r})$ by the product rule for differentiating dot products. So $\mathbf{r} \cdot \mathbf{r} =$ constant, i.e. $|\mathbf{r}|$ is constant, so the particle lies on a sphere.

(ii) $$\hat{\boldsymbol{\omega}} \cdot \dot{\mathbf{r}} = \hat{\boldsymbol{\omega}} \cdot (\boldsymbol{\omega} \times \mathbf{r}) = 0$$

$\hat{\boldsymbol{\omega}}$ and $\boldsymbol{\omega}$ are parallel, so the triple scalar product is zero. The left-hand side is $\dfrac{d}{dt}(\hat{\boldsymbol{\omega}} \cdot \mathbf{r})$, again by the product rule ($\hat{\boldsymbol{\omega}}$ is constant). So $\hat{\boldsymbol{\omega}} \cdot \mathbf{r} =$ constant.

This is the standard vector equation of a plane which is normal to $\hat{\boldsymbol{\omega}}$.

(iii) The particle lies on both a sphere and a plane, and must therefore lie on the intersection of these, i.e. a circle.

Differentiating $\dot{\mathbf{r}} = \boldsymbol{\omega} \times \mathbf{r}$ gives

$$\ddot{\mathbf{r}} = \boldsymbol{\omega} \times \dot{\mathbf{r}} \quad \text{(by the product rule for vector products: } \boldsymbol{\omega} \text{ is constant)}$$

$$\Rightarrow \quad \dot{\mathbf{r}} \cdot \ddot{\mathbf{r}} = \dot{\mathbf{r}} \cdot (\boldsymbol{\omega} \times \dot{\mathbf{r}}) = 0$$

The left-hand side is $\dfrac{1}{2}\dfrac{d}{dt}(\dot{\mathbf{r}} \cdot \dot{\mathbf{r}})$. Hence $\dot{\mathbf{r}} \cdot \dot{\mathbf{r}} =$ constant $= |\dot{\mathbf{r}}|^2$ so the speed $|\dot{\mathbf{r}}|$ is constant.

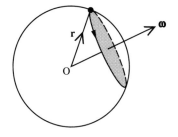

This example shows the power of vector manipulations, although you may find it difficult to see the physical situation modelled by this problem. The vector $\boldsymbol{\omega}$ is the *angular velocity* vector of the particle, through the origin. It is described later in this chapter.

Exercise 9A

1. Given $\mathbf{a} = \mathbf{i} + 2\mathbf{j} - 3\mathbf{k}$ and $\mathbf{b} = 2\mathbf{i} + \mathbf{j} - 4\mathbf{k}$, work out

   (i) $\mathbf{a} \times \mathbf{b}$ (ii) $\mathbf{b} \times \mathbf{a}$ (iii) $(\mathbf{a} + \mathbf{b}) \times (\mathbf{a} - \mathbf{b})$

   Verify that part (iii) is equal to $2\mathbf{b} \times \mathbf{a}$ and prove this result holds for any $\mathbf{a}$ and $\mathbf{b}$.

2. Given $\mathbf{a} = 3\mathbf{i} - \mathbf{j} + 2\mathbf{k}$, $\mathbf{b} = 2\mathbf{i} + \mathbf{j} - \mathbf{k}$, $\mathbf{c} = \mathbf{i} - 2\mathbf{j} + 2\mathbf{k}$, verify that
   (i) $(\mathbf{a} \times \mathbf{b}) . \mathbf{c} = \mathbf{a} . (\mathbf{b} \times \mathbf{c})$
   (ii) $\mathbf{a} \times (\mathbf{b} + \mathbf{c}) = (\mathbf{a} \times \mathbf{b}) + (\mathbf{a} \times \mathbf{c})$
   (iii) $(\mathbf{a} \times \mathbf{b}) \times \mathbf{c}$ is not equal to $\mathbf{a} \times (\mathbf{b} \times \mathbf{c})$

3. Find a vector perpendicular to the plane containing $\mathbf{a} = 6\mathbf{i} + 3\mathbf{j} - 4\mathbf{k}$, $\mathbf{b} = -2\mathbf{i} - 9\mathbf{j} - 2\mathbf{k}$.

4. Find a vector of magnitude 3 perpendicular to both $2\mathbf{i} + \mathbf{j} - 3\mathbf{k}$ and $\mathbf{i} - 2\mathbf{j} + \mathbf{k}$.

5. The three sides of a triangle are represented by vectors $\mathbf{p}$, $\mathbf{q}$, $\mathbf{r}$ (see diagram). Show that

   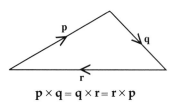

   $$\mathbf{p} \times \mathbf{q} = \mathbf{q} \times \mathbf{r} = \mathbf{r} \times \mathbf{p}$$

   What well-known geometrical formula is represented by this result?

6. Given points $A(1, 2, 1)$, $B(1, 0, 3)$, $C(-1, 2, -1)$,

   (i) write down the vectors $\overrightarrow{AB}$, $\overrightarrow{BC}$, $\overrightarrow{CA}$;

   (ii) use the formula $\frac{1}{2}|\overrightarrow{AB} \times \overrightarrow{AC}|$ to work out the area of the triangle ABC;

   (iii) confirm the result is the same as $\frac{1}{2}|\overrightarrow{BC} \times \overrightarrow{BA}|$.

7. Prove that the area of a triangle whose vertices have position vectors $\mathbf{a}$, $\mathbf{b}$, $\mathbf{c}$ is

   $$\tfrac{1}{2}|\mathbf{a} \times \mathbf{b} + \mathbf{b} \times \mathbf{c} + \mathbf{c} \times \mathbf{a}|$$

8. By writing $\mathbf{a} = a_1\mathbf{i} + a_2\mathbf{j} + a_3\mathbf{k}$, and similarly for $\mathbf{b}$ and $\mathbf{c}$, show that

   $$\mathbf{a} \times (\mathbf{b} \times \mathbf{c}) = (\mathbf{a} . \mathbf{c})\mathbf{b} - (\mathbf{a} . \mathbf{b})\mathbf{c}$$

9. Given $\mathbf{a} = \sin t\mathbf{i} - \cos t\mathbf{j}$, $\mathbf{b} = \sin t\mathbf{i} + \cos t\mathbf{j} + t\mathbf{k}$

   (i) work out $\mathbf{a} . \mathbf{b}$ and hence $\dfrac{\mathrm{d}}{\mathrm{d}t}(\mathbf{a} . \mathbf{b})$

   (ii) work out $\mathbf{a} \times \mathbf{b}$ and hence $\dfrac{\mathrm{d}}{\mathrm{d}t}(\mathbf{a} \times \mathbf{b})$

   In each case confirm that the same result is given when the product rule is used for differentiation.

10. A charged particle of mass m moving in a constant magnetic field has, at time t, position vector

    $$\mathbf{r} = \cos \omega t\mathbf{i} + \sin \omega t\mathbf{j}$$

    (i) Show that $\mathbf{r}$ is always perpendicular to the velocity $\mathbf{v}$ of the particle.
    (ii) Show that $\mathbf{r} \times \mathbf{v}$ is a constant vector.
    (iii) Show that the force on the particle is $m\omega \mathbf{v} \times \mathbf{H}$, where $\mathbf{H} = -\mathbf{k}$
    ($\mathbf{H}$ is the magnetic field).

Vector moments

Torque

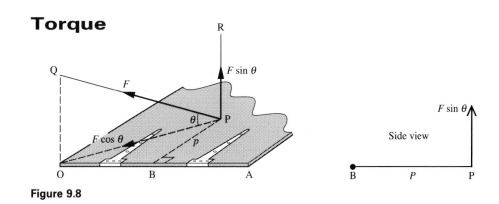

Figure 9.8

Consider again the situation depicted at the opening of this chapter (see figure 9.8).

How can you work out the turning effect of the force about the hinged edge of the trapdoor? You might argue as follows. The tension **F** in the chain pulling at P can be resolved into two components: a horizontal component along PO which has no turning effect about the axis OA, and a vertical component with magnitude $F \sin \theta$ along PR. The moment of this vertical component about the axis OA is $pF \sin \theta$.

Suppose now that there are no hinges in the trapdoor: they are broken, it is simply resting in its frame. You pull along PQ, but prevent (with your foot) the corner O of the door from leaving the frame (figure 9.9). The force is strong enough to pull the rest of the door up, rotating about the corner O. You can probably see that it will begin to rotate about an axis OX *perpendicular to the plane POQ* (figure 9.10). This is the axis through O about which the turning effect of **F** is largest.

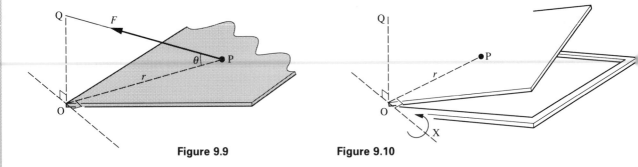

Figure 9.9   **Figure 9.10**

Now the turning effect of **F** about OX is $rF \sin \theta = |\overrightarrow{OP} \times \mathbf{F}|$ and $\overrightarrow{OP} \times \mathbf{F}$ is parallel to OX.

This suggests a *vector* definition for the turning effect of a force about a *point*. In this example, the vector in the direction OX with magnitude $rF \sin \theta$ is known as the *torque* (or *vector moment*) of the force **F** about the point O (figure 9.11). Its *direction* is perpendicular to the plane containing the force and the point O; this is the direction through O along which the turning effect is greatest. Its *magnitude* is the moment of the force about that axis. The vector moment of the force **F** about the point O is thus $\overrightarrow{OP} \times \mathbf{F}$.

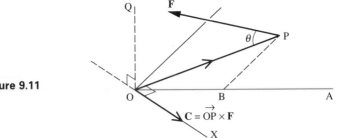

Figure 9.11

Definition of torque

The vector moment of torque, **C**, of a force **F** about a point O is defined as

$$\mathbf{C} = \mathbf{r} \times \mathbf{F}$$

where **r** is the position vector of *any* point P
on the line of action of the force.

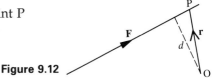

Figure 9.12

In figure 9.12 **F** and **r** are in the plane of the page. Thus $\mathbf{r} \times \mathbf{F}$ is directed into
the page and its magnitude is $rF\sin\theta = Fd$, i.e. the moment of **F** about the
perpendicular axis through O. *Notice that you get the same answer wherever P is
on the line of action of* **F**. The direction of **C** is perpendicular to both **F** and **r**.
It is in the same direction as the axis through O about which the turning
effect of the force is greatest.

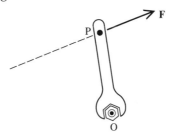

Figure 9.13

The direction (into the page) is that in which a right-handed bolt at O would
be driven if OP were a spanner (figure 9.13). Do not make the mistake of
using $\mathbf{F} \times \mathbf{r}$, which would give a vector in the reverse direction.

The torque of a force about a point thus embodies both the magnitude of the
turning effect *and* the axis along which this turning effect is directed.

Relationship between torque about a point and moment about an axis

There is a close relationship between the familiar concept of the *moment of a
force about an axis* (a scalar quantity) and *torque* (a vector). The moment of a
force **F** about an axis through O is the component of the torque of **F** about O
in the direction of the axis.

You can verify this with the foregoing trapdoor problem. With reference to
figure 9.14, the component of **C** in the direction OB is

$$|\mathbf{C}|\cos\phi = rF\sin\theta\cos\phi$$

$$= pF\sin\phi \quad (\text{since } r\cos\phi = p)$$

This is the scalar moment of the force **F** about the axis OB as expected.

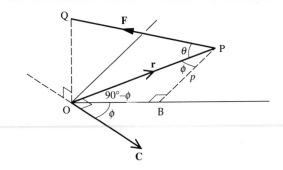

Figure 9.14

A vector derivation of the moment of a force about an axis

A force **F** acts through a point P (figure 9.15) and you wish to find its moment about an axis in the direction of a unit vector $\hat{\mathbf{n}}$ through a point O. The torque of **F** about O is $\overrightarrow{OP} \times \mathbf{F}$. The component of this in the direction $\hat{\mathbf{n}}$ is the scalar product $\hat{\mathbf{n}} \cdot (\overrightarrow{OP} \times \mathbf{F})$.

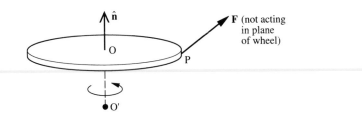

Figure 9.15

This can be useful when analysing problems of rotation about a fixed axis. Figure 9.15 represents a wheel which can rotate about an axis OO' with a force **F** acting at a point P on the wheel. In *Mechanics 5* you saw that $C = I\ddot{\theta}$, where I is the moment of inertia of the wheel and C is the moment of the force *about the axis*.

In previous examples, the lines of action of all forces were perpendicular to the axis, i.e. parallel to the plane of the wheel. When this is *not* the case, as in the figure 9.15, the geometry is complicated and the vector approach to finding the scalar moment C is useful.

Activity

In figure 9.15, show that C is the same whichever origin is chosen on the axis. That is

$$\hat{\mathbf{n}} \cdot (\overrightarrow{OP} \times \mathbf{F}) = \hat{\mathbf{n}} \cdot (\overrightarrow{O'P} \times \mathbf{F})$$

(Use the fact that $\overrightarrow{O'P} = \overrightarrow{OP} + \lambda\hat{\mathbf{n}}$ for some λ.)

EXAMPLE

A force $\mathbf{F} = F_x\mathbf{i} + F_y\mathbf{j}$ acts at the point $(x, y, 0)$.

(i) Show that the torque of **F** about the origin is in the **k** direction and that its magnitude is the same as that given by taking 'moments about O' in the xy plane (see diagram).

(ii) What is the moment of **F** about an axis through the origin in the direction $2\mathbf{i} + 2\mathbf{j} + \mathbf{k}$?

Solution

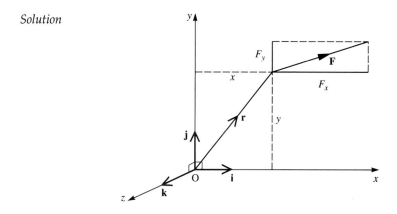

The diagram shows the xy plane with the z axis out of the page.

(i) The torque about O is

$$\mathbf{C} = \mathbf{r} \times \mathbf{F} = \begin{vmatrix} \mathbf{i} & x & F_x \\ \mathbf{j} & y & F_y \\ \mathbf{k} & 0 & 0 \end{vmatrix} = 0\mathbf{i} + 0\mathbf{j} + (xF_y - yF_x)\mathbf{k}$$

$$\Rightarrow \quad \mathbf{C} = (xF_y - yF_x)\mathbf{k}$$

Taking moments about O gives anticlockwise moment $= xF_y - yF_x$

(ii) $\mathbf{n} = 2\mathbf{i} + 2\mathbf{j} + \mathbf{k}$

$$\Rightarrow \quad |\mathbf{n}| = \sqrt{(4 + 4 + 1)} = 3$$

Therefore, unit vector $\hat{\mathbf{n}} = \frac{2}{3}\mathbf{i} + \frac{2}{3}\mathbf{j} + \frac{1}{3}\mathbf{k}$

Moment about axis is $\mathbf{C}.\hat{\mathbf{n}} = \frac{1}{3}(xF_y - yF_x)$

Adding torques

Just as moments can be added together, torques about the same point can be combined by usual vector addition.

EXAMPLE

Forces $\mathbf{i} - 2\mathbf{j}$ and $-\mathbf{i} + 2\mathbf{j}$ act respectively through points with position vectors $\mathbf{k}$ and $3\mathbf{k}$. Find the vector sum of their torques about O.

Solution

The sum of the torques is

$$\begin{vmatrix} \mathbf{i} & 0 & 1 \\ \mathbf{j} & 0 & -2 \\ \mathbf{k} & 1 & 0 \end{vmatrix} + \begin{vmatrix} \mathbf{i} & 0 & -1 \\ \mathbf{j} & 0 & 2 \\ \mathbf{k} & 3 & 0 \end{vmatrix}$$

$$= \quad 2\mathbf{i} + \mathbf{j} + -6\mathbf{i} - 3\mathbf{j}$$

Therefore, the total torque is $-4\mathbf{i} - 2\mathbf{j}$.

Torque of a pair of equal and opposite forces

In *Mechanics 5*, you were reminded that a pair of equal and opposite forces (which have no resultant but have a net turning effect) is known as a *couple*. It was shown that the moment about any axis perpendicular to the plane of the forces was the same. This generalises to the fact that the *torque* about *any point* is the same.

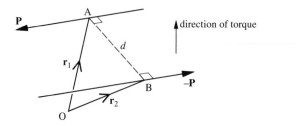

Figure 9.16

Figure 9.16 shows two equal and opposite forces $\mathbf{P}$ and $-\mathbf{P}$, with a distance d between their lines of action. Taking the total torque, relative to an origin O, gives

$$\mathbf{C} = \mathbf{r}_1 \times \mathbf{P} + \mathbf{r}_2 \times (-\mathbf{P})$$
$$= (\mathbf{r}_1 - \mathbf{r}_2) \times \mathbf{P}$$
$$= \overrightarrow{BA} \times \mathbf{P}$$

Since A and B can be chosen anywhere on the lines of action of the forces, take AB perpendicular to the lines of action. Since the vectors are perpendicular, the magnitude of the vector product is Pd, and its direction is perpendicular to the plane of the forces (upwards, in figure 9.16). Hence the torque $\mathbf{C}$ is independent of the position of O.

In general, a couple is any set of forces which have no resultant but do have a net non-zero torque. It is shown in the next section (Systems of forces) that the total torque of a couple is the same about any point.

Torques about different points

There is a relationship between the torques of a force about different points. The torque about a point O (figure 9.17) is

$$\mathbf{C}_O = \mathbf{r} \times \mathbf{F}$$

The torque $\mathbf{C}_P$ about some other point P, with position vector $\mathbf{p}$ relative to O is, by definition, $\overrightarrow{PA} \times \mathbf{F}$.

$$\mathbf{C}_P = \overrightarrow{PA} \times \mathbf{F}$$
$$= (\mathbf{r} - \mathbf{p}) \times \mathbf{F}$$
$$= (\mathbf{r} \times \mathbf{F}) - (\mathbf{p} \times \mathbf{F})$$
$$= \mathbf{C}_O - (\mathbf{p} \times \mathbf{F})$$

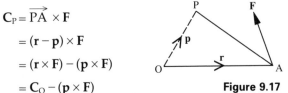

Figure 9.17

Activity

Introduce two equal and opposite forces $\mathbf{F}$ and $-\mathbf{F}$ at P. This does not change the system so the torque about O is still C_O. However, the system can be regarded as (i) a new force $\mathbf{F}$ through P together with (ii) a couple formed by the original $\mathbf{F}$ and the new $-\mathbf{F}$. Work out the torques about O due to (i) and (ii) and hence show that

$$C_O = C_P + (\mathbf{p} \times \mathbf{F})$$

EXAMPLE

A force $\mathbf{F}$ of magnitude 6 N acts along the line with vector equation

$$\mathbf{r} = \mathbf{i} + \mathbf{j} + \mathbf{k} + \lambda(\mathbf{i} + 2\mathbf{j} - 2\mathbf{k})$$

in the direction of increasing λ.

(i)   Work out the torque C_O of $\mathbf{F}$ about the origin.
(ii) Work out the torque C_A of $\mathbf{F}$ about the point A with position vector $\mathbf{a} = \mathbf{i} + 2\mathbf{j} + 3\mathbf{k}$.
(iii) Confirm $C_A = C_O - (\mathbf{a} \times \mathbf{F})$.

Solution

(i)   The equation of the line shows that the force acts through the point Q with position vector $\mathbf{q} = \mathbf{i} + \mathbf{j} + \mathbf{k}$ and that the direction of the force is $\mathbf{i} + 2\mathbf{j} - 2\mathbf{k}$. The magnitude of this latter vector is $\sqrt{(1^2 + 2^2 + 2^2)} = 3$. Since the magnitude of $\mathbf{F}$ is 6,

$$\mathbf{F} = 6(\mathbf{i} + 2\mathbf{j} - 2\mathbf{k})/3 = 2\mathbf{i} + 4\mathbf{j} - 4\mathbf{k}$$

The torque of $\mathbf{F}$ about the origin is thus

$$\mathbf{q} \times \mathbf{F} = \begin{vmatrix} \mathbf{i} & 1 & 2 \\ \mathbf{j} & 1 & 4 \\ \mathbf{k} & 1 & -4 \end{vmatrix} = -8\mathbf{i} + 6\mathbf{j} + 2\mathbf{k}$$

That is, $C_O = -8\mathbf{i} + 6\mathbf{j} + 2\mathbf{k}$ N m.

(ii) The torque of $\mathbf{F}$ about A is

$$(\mathbf{q} - \mathbf{a}) \times \mathbf{F} = \begin{vmatrix} \mathbf{i} & 0 & 2 \\ \mathbf{j} & -1 & 4 \\ \mathbf{k} & -2 & -4 \end{vmatrix} = 12\mathbf{i} - 4\mathbf{j} + 2\mathbf{k}$$

That is, $C_A = 12\mathbf{i} - 4\mathbf{j} + 2\mathbf{k}$ N m

(iii)

$$\mathbf{a} \times \mathbf{F} = \begin{vmatrix} \mathbf{i} & 1 & 2 \\ \mathbf{j} & 2 & 4 \\ \mathbf{k} & 3 & -4 \end{vmatrix} = -20\mathbf{i} + 10\mathbf{j} + 0\mathbf{k}$$

$$\Rightarrow \quad C_O - (\mathbf{a} \times \mathbf{F}) = -8\mathbf{i} + 6\mathbf{j} + 2\mathbf{k} - (-20\mathbf{i} + 10\mathbf{j})$$

$$= 12\mathbf{i} - 4\mathbf{j} + 2\mathbf{k}$$

which is C_A, as expected.

1. A force $\mathbf{F} = \mathbf{i} - \mathbf{j} + 2\mathbf{k}$ acts at the point P with position vector $\mathbf{i} - \mathbf{j} + \mathbf{k}$. Find the torque (vector moment) of $\mathbf{F}$ about the origin.

2. Forces $(\mathbf{i} - 2\mathbf{j})$ and $(\mathbf{i} + 2\mathbf{j})$ act through points with position vectors $\mathbf{k}$ and $2\mathbf{k}$ relative to the origin O. Find the torque of each force about O, and show that the sum of these torques is $-2\mathbf{i} + 3\mathbf{j}$.

3. A force $\mathbf{F} = \mathbf{i} + 2\mathbf{j}$ acts at the point $A(0, 1, 1)$. B is the point $(1, 3, 2)$.
   (i) Find the torque, $\mathbf{C}_O$ of $\mathbf{F}$ about the origin.
   (ii) Write down the vector $\overrightarrow{BA}$ and hence work out $\mathbf{C}_B$, the torque of $\mathbf{F}$ about B.
   (iii) Confirm that $\mathbf{C}_B = \mathbf{C}_O - (\overrightarrow{OB} \times \mathbf{F})$.

4. OABCPQRS is a unit cube (see diagram). A force $\mathbf{F}$, acting along the diagonal PB, has a magnitude equal to PB. What is its torque about (i) O, (ii) C? Give the answers in terms of unit vectors $\mathbf{i}, \mathbf{j}, \mathbf{k}$ in directions OA, OP, OC respectively.

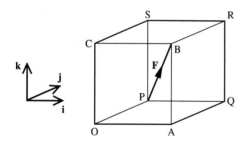

5. A force $\mathbf{F}$ of magnitude 9 N acts along a line with vector equation

   $$\mathbf{r} = \mathbf{i} + \mathbf{j} - \mathbf{k} + \lambda(\mathbf{i} + 2\mathbf{j} - 3\mathbf{k})$$

   in the direction of increasing λ.
   (i) Find the torque of $\mathbf{F}$ about the origin.
   (ii) Find the torque about the point $(-1, -3, 5)$. Explain the implication of the result.

6. A force $\mathbf{i} + \mathbf{j}$ acts through the point $(0, 1, 1)$ and an equal and opposite force acts through the point $(1, 0, 2)$. Find the torque of each force about the point $P(p, q, r)$ and show that the sum of these torques is

independent of the position of P. Explain why this is the case.

7. A force $\mathbf{F} = 5\mathbf{i} + 2\mathbf{j}$ acts through the point P with position vector $\mathbf{i} + \mathbf{j} + 2\mathbf{k}$ with respect to the origin O
   (i) Find the torque (vector moment) of $\mathbf{F}$ about O.
   (ii) Hence find the (scalar) moment of $\mathbf{F}$ about an *axis* through O in the direction $(3\mathbf{i} + 4\mathbf{j} + 2\mathbf{k})$.

8. Timmy tries to open a horizontal square trapdoor ABCD which is hinged along the side AB. Each side of the door is 2 m. One end of a chain is attached to the mid-point P of the trapdoor, and he holds the other end at a point Q 1 m above A. He pulls along the direction PQ with a force $\mathbf{F}$ of magnitude 30 N. Taking an origin at A and unit vectors $\mathbf{i}, \mathbf{j}, \mathbf{k}$ in directions AB, AD and vertical,
   (i) find the unit vector in the direction PQ, and hence write down the force $\mathbf{F}$ in terms of $\mathbf{i}, \mathbf{j}, \mathbf{k}$;
   (ii) find the torque of $\mathbf{F}$ about A;
   (iii) find the moment of $\mathbf{F}$ about the axis AB;
   (iv) If he were to pull with the same strength vertically upwards on a chain attached to a point on the side opposite the hinges, what would then be the moment of his force about AB?

9.

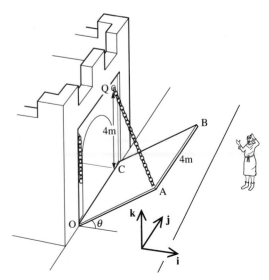

The diagram shows the drawbridge on Ethelred's castle. The chains which raise it had become detached and Ethelred, his mind on weightier matters, made an incompetent repair, with just one chain attached, and this to the wrong corner of the bridge. The drawbridge OABC is square with sides of 4 m and is hinged along OC. The chain attached to A enters the wall at Q, 4 m vertically above C. Take an origin O at one end of the hinged side and unit vectors $\mathbf{i}, \mathbf{j}$, horizontal and $\mathbf{k}$ vertical, as

shown. The tension in the chain is T. The drawbridge is raised from its horizontal position at an angle θ from the horizontal.

(i) Find the unit vector in the direction AQ and hence write down the vector representing the force in the chain holding up the drawbridge.

(ii) Find the torque of this force about O.

(iii) Show that the moment of the force about the pivot OC has magnitude

$$\frac{4T\cos\theta}{\sqrt{(3 - 2\sin\theta)}}$$

Systems of forces

You know how to find the effect of a set of forces in two dimensions by resolving and taking moments. In three dimensions the methods are similar. Forces are resolved in three directions $\mathbf{i}, \mathbf{j}$ and $\mathbf{k}$. Instead of moments you deal with torques (vector moments), which can also be resolved into components. When it is difficult to draw or imagine exactly what is going on, you have to rely on vector manipulations.

EXAMPLE

A crate, suspended from a crane, is being manoeuvred on to a lorry. There are four forces: its weight $\mathbf{W}$, the tension in the suspension $\mathbf{T}$, and forces in two ropes $\mathbf{R}_1$ and $\mathbf{R}_2$. In a certain position, these may be modelled as follows (relative to the origin and axes shown in the diagram).

Force		Acting at point with position vector
$\mathbf{W}$	$-1000\mathbf{k}$	$\mathbf{i}+\mathbf{j}+\mathbf{k}$
$\mathbf{T}$	$1000\mathbf{k}+100\mathbf{i}$	$\mathbf{i}+\mathbf{j}+2\mathbf{k}$
$\mathbf{R}_1$	$-50\mathbf{i}+20\mathbf{j}$	$\mathbf{k}$
$\mathbf{R}_2$	to be found	$\mathbf{j}+\mathbf{k}$

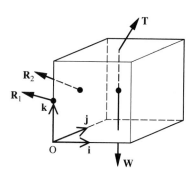

(i) Given that the resultant force on the crate is zero, find $\mathbf{R}_2$.

(ii) By finding the torque about the origin O, show that the crate is not in equilibrium.

(iii) Verify that the torque about O is the same as the torque about the point with position vector $\mathbf{i} + \mathbf{j} + 2\mathbf{k}$.

Solution

(i) The resultant force is zero, so

$$\mathbf{W} + \mathbf{T} + \mathbf{R}_1 + \mathbf{R}_2 = 0$$

$$\mathbf{R}_2 = -\mathbf{W} - \mathbf{T} - \mathbf{R}_1$$

$$= 1000\mathbf{k} - (1000\mathbf{k} + 100\mathbf{i}) - (-50\mathbf{i} + 20\mathbf{j})$$

$$= -50\mathbf{i} - 20\mathbf{j}$$

(ii) To find the torque, calculate $\mathbf{r} \times \mathbf{F}$ for each force and then find the vector sum.

Torque of $\mathbf{W}$ about O: $\begin{vmatrix} \mathbf{i} & 1 & 0 \\ \mathbf{j} & 1 & 0 \\ \mathbf{k} & 1 & -1000 \end{vmatrix} = -1000\mathbf{i} + 1000\mathbf{j}$

Torque of $\mathbf{T}$ about O: $\begin{vmatrix} \mathbf{i} & 1 & 100 \\ \mathbf{j} & 1 & 0 \\ \mathbf{k} & 2 & 1000 \end{vmatrix} = 1000\mathbf{i} - 800\mathbf{j} - 100\mathbf{k}$

Torque of $\mathbf{R}_1$ about O: $\begin{vmatrix} \mathbf{i} & 0 & -50 \\ \mathbf{j} & 0 & 20 \\ \mathbf{k} & 1 & 0 \end{vmatrix} = -20\mathbf{i} - 50\mathbf{j}$

Torque of $\mathbf{R}_2$ about O: $\begin{vmatrix} \mathbf{i} & 0 & -50 \\ \mathbf{j} & 1 & -20 \\ \mathbf{k} & 1 & 0 \end{vmatrix} = 20\mathbf{i} - 50\mathbf{j} + 50\mathbf{k}$

The total torque is thus

$$(-1000 + 1000 - 20 + 20)\mathbf{i} + (1000 - 800 - 50 - 50)\mathbf{j} + (-100 + 50)\mathbf{k}$$

which is

$$100\mathbf{j} - 50\mathbf{k}.$$

(iii) Note that the lines of action of both $\mathbf{T}$ and $\mathbf{W}$ pass through the point $(1, 1, 2)$ so have no torque about this point. If the torque about O is $\mathbf{r} \times \mathbf{F}$, say, then the torque about the point with position vector $\mathbf{p}$ is $(\mathbf{r} - \mathbf{p}) \times \mathbf{F}$, as was shown earlier. Thus the other two forces contribute:

Torque due to $\mathbf{R}_1$: $(\mathbf{k} - (\mathbf{i} + \mathbf{j} + 2\mathbf{k})) \times \mathbf{R}_1 = \begin{vmatrix} \mathbf{i} & -1 & -50 \\ \mathbf{j} & -1 & 20 \\ \mathbf{k} & -1 & 0 \end{vmatrix}$

$$= 20\mathbf{i} + 50\mathbf{j} - 70\mathbf{k}$$

Torque due to $\mathbf{R}_2$: $(\mathbf{j} + \mathbf{k} - (\mathbf{i} + \mathbf{j} + 2\mathbf{k})) \times \mathbf{R}_2 = \begin{vmatrix} \mathbf{i} & -1 & -50 \\ \mathbf{j} & 0 & -20 \\ \mathbf{k} & -1 & 0 \end{vmatrix}$

$$= -20\mathbf{i} + 50\mathbf{j} + 20\mathbf{k}$$

The total torque is thus $100\mathbf{j} - 50\mathbf{k}$, as before.

Reduction to a couple

The last example illustrated the general result that when the resultant of all the forces is zero, the torque about different points is the same. This can be shown as follows.

Denote the forces on the body by $\mathbf{F}_1$, $\mathbf{F}_2$, $\mathbf{F}_3$... (figure 9.18). The resultant force $\mathbf{F}$ is the vector sum of these:

$$\mathbf{F} = \sum \mathbf{F}_n = 0$$

since the resultant is zero.

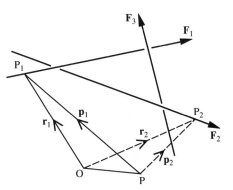

Figure 9.18

The total torque about O is

$$\mathbf{C}_O = \sum (\mathbf{r}_n \times \mathbf{F}_n)$$

The torque about another point P with position vector $\mathbf{r}$ with respect to O is

$$\mathbf{C}_P = \sum (\mathbf{p}_n \times \mathbf{F}_n)$$

$$= \sum (\mathbf{r}_n - \mathbf{r}) \times \mathbf{F}_n$$

$$= \sum [(\mathbf{r}_n \times \mathbf{F}_n) - (\mathbf{r} \times \mathbf{F}_n)] \quad \text{(distributive law for vector product)}$$

$$= \sum (\mathbf{r}_n \times \mathbf{F}_n) - \sum (\mathbf{r} \times \mathbf{F}_n)$$

$$= \mathbf{C}_O - \mathbf{r} \times \sum (\mathbf{F}_n)$$

$$\Rightarrow \quad \mathbf{C}_P = \mathbf{C}_O - \mathbf{r} \times \mathbf{F}$$

$$= \mathbf{C}_O \quad \text{(since } \mathbf{F} = 0)$$

When the resultant force is zero, you will get the same total torque about *any* point. You saw a special case of this earlier in the chapter, where two equal and opposite forces were considered. A system of forces with a zero resultant but a non-zero torque is said to reduce to a *couple*.

When the torque is also zero (i.e. a zero magnitude vector), the system of forces is in equilibrium. Thus a system of forces $\mathbf{F}_n$ is in equilibrium if and only if

$$\sum \mathbf{F}_n = 0 \quad \text{the resultant force is zero}$$

$$\text{and} \sum (\mathbf{r}_n \times \mathbf{F}_n) = 0 \quad \text{the resultant torque is zero about any point}$$

Equivalence of a system of forces

In the last example, the set of forces on the crate were shown to be *equivalent* to a couple. In general, two systems of forces are said to be equivalent if
(i) they have the same resultant;
(ii) they have the same torques about any given point.

Given (i) is true, you need to take moments about *only one point* to check (ii); generally the origin is easiest.

N O T E *All the foregoing is true only for* rigid *bodies. Take an example of two masses connected by an unstretched spring. If equal and opposite forces are applied to the masses they will move: the spring will extend. The system as a whole is not rigid if the masses can move relative to each other.*

EXAMPLE A force system acting on a rigid body consists of two forces $\mathbf{F}_1$, $\mathbf{F}_2$ where

$$\mathbf{F}_1 = 2\mathbf{i} - \mathbf{j} + 3\mathbf{k}$$

$$\mathbf{F}_2 = -3\mathbf{i} + 2\mathbf{k}$$

The forces act through points whose position vectors with respect to O are respectively $3\mathbf{j} + \mathbf{k}$ and $4\mathbf{i} + \mathbf{j} - \mathbf{k}$. This system is equivalent to a single force F acting through the point with position vector $\mathbf{i} + 2\mathbf{j} + 4\mathbf{k}$ together with a couple G. Find the magnitudes of F and G.

[Cambridge 1987]

Solution

Single force $\mathbf{F} = \mathbf{F}_1 + \mathbf{F}_2 = -\mathbf{i} - \mathbf{j} + 5\mathbf{k}$

$$\Rightarrow \quad |\mathbf{F}| = \sqrt{(1 + 1 + 25)} = \sqrt{27} = 3\sqrt{3}$$

Total torque of the system $\mathbf{F}_1$, $\mathbf{F}_2$ is

$$\mathbf{C} = \mathbf{r}_1 \times \mathbf{F}_1 + \mathbf{r}_2 \times \mathbf{F}_2 = \begin{vmatrix} \mathbf{i} & 0 & 2 \\ \mathbf{j} & 3 & -1 \\ \mathbf{k} & 1 & 3 \end{vmatrix} + \begin{vmatrix} \mathbf{i} & 4 & -3 \\ \mathbf{j} & 1 & 0 \\ \mathbf{k} & -1 & 2 \end{vmatrix}$$

$$= (10\mathbf{i} + 2\mathbf{j} - 6\mathbf{k}) + (2\mathbf{i} - 5\mathbf{j} + 3\mathbf{k})$$

$$= 12\mathbf{i} - 3\mathbf{j} - 3\mathbf{k}$$

The torque of **F** is

$$C_F = \begin{vmatrix} \mathbf{i} & 1 & -1 \\ \mathbf{j} & 2 & -1 \\ \mathbf{k} & 4 & 5 \end{vmatrix} = 14\mathbf{i} - 9\mathbf{j} + \mathbf{k}$$

The systems are equivalent, so the torque **C** is equal to C_F plus the couple **G**:

$$\mathbf{G} = \mathbf{C} - \mathbf{C}_F = (12\mathbf{i} - 3\mathbf{j} - 3\mathbf{k}) - (14\mathbf{i} - 9\mathbf{j} + \mathbf{k}) = -2\mathbf{i} + 6\mathbf{j} - 4\mathbf{k}$$

$$\Rightarrow \quad |\mathbf{G}| = \sqrt{(4 + 36 + 16)} = \sqrt{56}$$

Reduction of a system of forces

The last example demonstrated the equivalence of a system of forces and a force plus a couple. In other words the system was *reduced* to a force plus a couple. In general, a system of *any* number of forces can be reduced to a single force plus a couple. If $\sum \mathbf{F}_n = \mathbf{0}$, they will reduce to a couple alone as in the first example in this section or be in equilibrium.

Sometimes they can be reduced to a force alone. Suppose a system of forces $\mathbf{F}_1, \mathbf{F}_2, \ldots$ is equivalent to a single force **F** (figure 9.19). Then $\mathbf{F} = \mathbf{F}_1 + \mathbf{F}_2 + \ldots$.

The total torque of the $\mathbf{F}_n$ about O is

$$\mathbf{C} = \mathbf{r}_1 \times \mathbf{F}_1 + \mathbf{r}_2 \times \mathbf{F}_2 + \ldots$$

and since they are equivalent to **F**, this must be equal to the torque of **F** about O:

$$\mathbf{r} \times \mathbf{F} = \mathbf{C}.$$

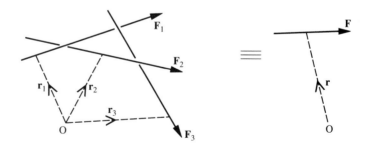

Figure 9.19

Hence, from the properties of the vector product, **F** is perpendicular to **C**. When a system of forces reduces to a single force, *the resultant force must be perpendicular to the sum of the torques of the original forces about any point*. This fact is used to help solve the example below.

Some special cases

There is an obvious case when the system can be reduced to a single force: when all the forces pass through the same point P. The system is equivalent to a single force $\mathbf{F} = \sum \mathbf{F}_n$ acting through the point P. The torques about the point P itself are all zero.

Also a set of *coplanar* forces, as in the two-dimensional problems you have met before, can always be reduced to a single force *or* a couple unless, of course, they are in equilibrium.

EXAMPLE

The forces $(3\mathbf{i}+\mathbf{j}+2\mathbf{k})$, $(6\mathbf{i}+p\mathbf{j}+\mathbf{k})$, $(2\mathbf{i}-2\mathbf{j}+2\mathbf{k})$, where p is a constant, act through the points with position vectors $(-\mathbf{i}+\mathbf{k})$, $(8\mathbf{i}+6\mathbf{j}+4\mathbf{k})$ and $(3\mathbf{j}+\mathbf{k})$ respectively, relative to an origin O.

(i) Find the value of p so that the above force system reduces to a single force **F**.

(ii) Find the torque of **F** about O.

(iii) A point on the line of action of **F** has co-ordinates $(x, 1, z)$. Find x and z and hence the vector equation of the line of action of **F**.

[AEB 1989]

Solution

(i) The resultant force is

$$\mathbf{F} = \sum \mathbf{F}_n = 11\mathbf{i} + (p-1)\mathbf{j} + 5\mathbf{k}$$

The total torque of the system about the origin is

$$\mathbf{C}_O = \sum (\mathbf{r}_n \times \mathbf{F}_n)$$

$$= \begin{vmatrix} \mathbf{i} & -1 & 3 \\ \mathbf{j} & 0 & 1 \\ \mathbf{k} & 1 & 2 \end{vmatrix} + \begin{vmatrix} \mathbf{i} & 8 & 6 \\ \mathbf{j} & 6 & p \\ \mathbf{k} & 4 & 1 \end{vmatrix} + \begin{vmatrix} \mathbf{i} & 0 & 2 \\ \mathbf{j} & 3 & -2 \\ \mathbf{k} & 1 & 2 \end{vmatrix}$$

$$= (-\mathbf{i} + 5\mathbf{j} - \mathbf{k}) + [(6 - 4p)\mathbf{i} + 16\mathbf{j} + (8p - 36)\mathbf{k}] + (8\mathbf{i} + 2\mathbf{j} - 6\mathbf{k})$$

$$= (13 - 4p)\mathbf{i} + 23\mathbf{j} + (8p - 43)\mathbf{k}$$

Since the system reduces to a single force, **F** is perpendicular to $\mathbf{C}_O$ so $\mathbf{F} . \mathbf{C}_O = 0$. Applying this scalar product to the components of **F** and $\mathbf{C}_O$ gives

$$11(13 - 4p) + 23(p - 1) + 5(+8p - 43) = 0$$

$$143 - 44p + 23p - 23 + 40p - 215 = 0$$

$$19p = 95$$

$$\Rightarrow \quad p = 5$$

Hence $\mathbf{F} = 11\mathbf{i} + 4\mathbf{j} + 5\mathbf{k}$.

(ii) Substituting for p in the value of $\mathbf{C}_O$ worked out above gives

$$\mathbf{C}_O = (13 - 4p)\mathbf{i} + 23\mathbf{j} + (-43 + 8p)\mathbf{k}$$

$$= -7\mathbf{i} + 23\mathbf{j} - 3\mathbf{k}$$

(iii) The torque of **F** about the origin must (since **F** is an equivalent system) be equal to $\mathbf{C}_O$, the torque of the original system. The torque of **F** about O is

$\mathbf{r} \times \mathbf{F}$, where $\mathbf{r} = x\mathbf{i} + \mathbf{j} + z\mathbf{k}$. So

$$-7\mathbf{i} + 23\mathbf{j} - 3\mathbf{k} = \begin{vmatrix} \mathbf{i} & x & 11 \\ \mathbf{j} & 1 & 4 \\ \mathbf{k} & z & 5 \end{vmatrix}$$

$$= (5 - 4z)\mathbf{i} + (11z - 5x)\mathbf{j} + (4x - 11)\mathbf{k}$$

This gives three equations:

$$5 - 4z = -7 \quad \Rightarrow \quad z = 3$$
$$4x - 11 = -3 \quad \Rightarrow \quad x = 2$$
$$11z - 5x = 23 \quad \text{(consistent with the other two)}$$

The point $2\mathbf{i} + \mathbf{j} + 3\mathbf{k}$ thus lies on the line of action of $\mathbf{F}$. The equation of the line of action of $\mathbf{F}$ is

$$\mathbf{r} = 2\mathbf{i} + \mathbf{j} + 3\mathbf{k} + \lambda(11\mathbf{i} + 4\mathbf{j} + 5\mathbf{k})$$

Exercise 9C

1. A force system on a rigid body consists of three forces:

   $\mathbf{F}_1 = -2\mathbf{i} + 4\mathbf{j} - 5\mathbf{k}$ through the point $(1, 1, 0)$
   $\mathbf{F}_2 = -\mathbf{i} - 3\mathbf{j} + 4\mathbf{k}$ through the point $(0, -2, 4)$
   $\mathbf{F}_3 = 3\mathbf{i} - \mathbf{j} + \mathbf{k}$ through the point $(4, 0, 1)$

   Prove the body is in equilibrium by showing that
   (i) the resultant force is zero;
   (ii) the resultant torque about the origin is zero.

2. A force system on a rigid body consists of three forces:

   $\mathbf{F}_1 = 2\mathbf{i} - \mathbf{j} - 3\mathbf{k}$ $\mathbf{F}_2 = \mathbf{i} + 4\mathbf{k}$ $\mathbf{F}_3 = -3\mathbf{i} + \mathbf{j} - \mathbf{k}$

   acting at the points with position vectors, $\mathbf{j} + \mathbf{k}$, $-2\mathbf{i} + 3\mathbf{j} + 15\mathbf{k}$, $2\mathbf{i} + \mathbf{j} + 9\mathbf{k}$, respectively. Prove that the body is in equilibrium.

3. The diagram shows a unit cube OABCPQRS. Forces of unit magnitude act along sides OA, AQ and QR and a further force of magnitude F acts along RO.

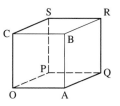

(i) Given that the system reduces to a couple, find the value of F.
(ii) Find the total torque about the origin.
(iii) Confirm that the same answer is given for the torque about the point R.

4. A force system acting on a rigid body consists of three forces $\mathbf{F}_1$, $\mathbf{F}_2$ and $\mathbf{F}_3$ where

   $$\mathbf{F}_1 = 2\mathbf{i} + \mathbf{k} \quad \mathbf{F}_2 = \mathbf{i} - \mathbf{j} + 3\mathbf{k} \quad \mathbf{F}_3 = X\mathbf{i} + Y\mathbf{j} + Z\mathbf{k}$$

   The force $\mathbf{F}_3$ acts through the origin O and $\mathbf{F}_1$ and $\mathbf{F}_2$ act through the points whose position vectors relative to O are $\mathbf{i} - \mathbf{j} + 2\mathbf{k}$ and $2\mathbf{i} - \mathbf{k}$ respectively. Given that the system reduces to a couple G, find the values of X, Y and Z and show that the magnitude of G is $\sqrt{20}$.

   [Cambridge 1987]

5. A force $\mathbf{F} = (\mathbf{i} + \mathbf{j} + 2\mathbf{k})$ acts through the origin. Show that this is equivalent to an equal force acting through the point with position vector $(\mathbf{i} + \mathbf{j} + \mathbf{k})$ together with a couple G. Find the value of G.

6. A force system acting on a rigid body consists of two forces $\mathbf{F}_1$ and $\mathbf{F}_2$ where

   $$\mathbf{F}_1 = \mathbf{j} + \mathbf{k} \quad \mathbf{F}_2 = 3\mathbf{i} - 2\mathbf{k}$$

   The forces act through points whose position vectors relative to the origin are $3\mathbf{j} + \mathbf{k}$ and $4\mathbf{i} + \mathbf{j} - \mathbf{k}$ respectively. This system is equivalent to a single force F acting

   M6

Exercise 9C continued

through the origin together with a couple G. Show the magnitudes of F and G are respectively √11 and √34.

7. A force system acting on a rigid body consists of three forces F_1, F_2 and F_3 where

$F_1 = 12i - 4j + 6k$ acting at the point with position vector $2i - 3j$

$F_2 = -3j + 2k$ acting at the point with position vector $i + j + k$

The force F_3 acts at the point with position vector $2i - k$. Given that the system reduces to a couple, find F_3 and the magnitude of the couple.

[Cambridge 1990]

8. A force system acting on a rigid body consists of two forces F_1 and F_2 where

$$F_1 = 2i - j + 3k \quad F_2 = -3i + 2k$$

The forces act through points whose position vectors relative to the origin are $3j + k$ and $4i + j - k$ respectively. This system reduces to a single force F acting through the point with position vector $i + 2j + 4k$ together with a couple G. Find the magnitudes of F and G.

[Cambridge 1987]

9. The diagram shows a light hollow cube ABCDEFGH of side 4 m. The faces ABCD, EGFH are horizontal and i, j, k denote unit vectors parallel to $\overrightarrow{AB}$, $\overrightarrow{AD}$, $\overrightarrow{AF}$ respectively, with k being vertically

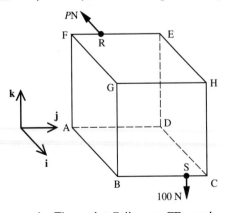

upwards. The point R lies on FE, such that FR = 1 m, and the point S lies on BC

such that BS = 3 m. A force of magnitude P N, parallel to $\overrightarrow{BA}$, acts at R and another force of magnitude 100 N, acts vertically downwards at S. Further forces $(X_A i + Z_A k)$ and $(X_D i + Z_D k)$ act at A and D respectively.

(i) Express the position vectors, relative to A, of D, R and S in the form $ai + bj + ck$, where a, b and c are constants to be determined for each point.

(ii) Given that the vector sum of all moments about A is zero, show that $P = 100$ N and find X_D and Z_D.

(iii) Given also that the vector sum of all forces is zero, find X_A and Z_A.

[AEB 1991]

10. A force F is applied first at a point A of a rigid body, and then at a point B of the body. Explain why, in general, the effect of the force will not be the same in both cases, and give the condition that would have to be satisfied for the effect to be the same.

In the diagram, the line OA is of length 0.4 m and is along the x axis. B is the mid-point of OA. The line BC is of length 0.15 m, perpendicular to OA and making an angle θ with the plane $y = 0$.

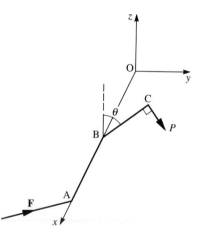

Unit vectors i, j, k are defined parallel to the x, y and z axes respectively.

(i) There is a force of magnitude P N acting at C perpendicular to BC and parallel to the plane $x = 0$. The sense of the force is such that it would produce a clockwise rotation when viewed from A to O. Express this force in the form $a\mathbf{j} + b\mathbf{k}$ and show that its moment about O is

$$-0.15\,P\mathbf{i} + (0.2\,P\sin\theta)\,\mathbf{j} + (0.2\,P\cos\theta)\,\mathbf{k}\ \text{N m}$$

(ii) There is also a force $\mathbf{F} = F_1\mathbf{i} + F_2\mathbf{j} + F_3\mathbf{k}$ N acting at A. Reduce the system of two forces to a force at O and a couple.

(iii) The structure OBCA is a simplified model of a screwdriver with a handle which is being used to tighten a screw into the plane $x = 0$ at O. The

component of the force $\mathbf{F}$ at A in the direction AO is 30 N and the other components of $\mathbf{F}$ should be adjusted so that the screw does not twist about Oy and Oz. This requires that there is no moment at O about these axes.

(a) Find, in terms of P and θ, the other components of the force at A such that the above condition is satisfied.

(b) Find also the least magnitude of P, given that the minimum magnitude of the couple required to turn the screw is 60 N m.

[Oxford 1993]

Use of vector products in rotation

Angular velocity vector

You know from watching a spinning top spin that its axis, unless perfectly vertical, describes a conical motion (known as *precession*). At any instant, the top is rotating with some angular speed about its axis—but the direction of that axis is continuously changing. The full analysis of the motion of a spinning top is beyond the scope of this book, but the example serves to show that rotation can be complicated. The *angular velocity vector*, introduced in this section, embodies both angular speed and axial direction and is an essential concept in modelling problems involving general rotation.

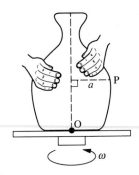

Figure 9.20

Figure 9.20 represents a pot being thrown on a potter's wheel. Each particle of clay is rotating in a circle about the axis of the wheel with an angular speed ω. A particle at a perpendicular distance a from the axis has a linear speed $a\omega$. O is any point on the axis.

The angular velocity vector $\boldsymbol{\omega}$ is defined as the vector whose magnitude is ω and which is in the direction of the axis. Imagine the clay being rotated in the same sense as when screwing up a right-handed screw, then the positive direction of $\boldsymbol{\omega}$ is taken as the direction the screw moves. In figure 9.21 $\boldsymbol{\omega}$ is upwards.

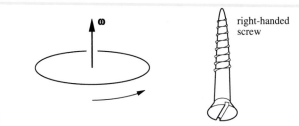

Figure 9.21

The definition of angular velocity as a *vector* thus incorporates not only the angular speed (as its magnitude), but also the direction of the axis of rotation (as its direction). However, as well as angular velocity, you need to specify a point O on the axis in order to describe the motion fully—just as a force **F** is not fully specified without also giving a point on its line of action. If the pot is moved slightly so that it is no longer located centrally on the wheel, it continues to rotate with the same angular velocity vector (assuming it somehow remains on the wheel) although the axis of rotation has changed relative to the pot.

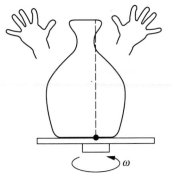

Figure 9.22

Linear velocity of a point

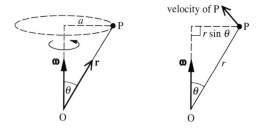

Figure 9.23

What is the linear velocity of a particle of clay P, with position vector $\mathbf{r}$ with respect to O? Its magnitude is $a\omega = \omega r \sin\theta$ and its direction is perpendicular to $\boldsymbol{\omega}$ and $\mathbf{r}$, so

$$\mathbf{v}_P = \boldsymbol{\omega} \times \mathbf{r}$$

Note how neatly the vector product works in giving a result that is independent of the position of O on the axis.

The ordering is important; $\boldsymbol{\omega}$ must come first to give the correct direction for the velocity.

EXAMPLE

A wheel is rotating with angular speed 4 rad s^{-1} in a right-handed sense about an axis through the origin which is in the direction of the vector $2\mathbf{i} + 6\mathbf{j} - 3\mathbf{k}$.
(i) Find its angular velocity.
(ii) Find the velocity of a point P on the wheel when P has co-ordinates $(0, 1, 0)$. (Linear units are in metres.)

Solution

(i) The magnitude of the vector $2\mathbf{i} + 6\mathbf{j} - 3\mathbf{k}$ is $\sqrt{(2^2 + 6^2 + 3^2)} = 7$, so a unit vector in the same direction is $\frac{1}{7}(2\mathbf{i} + 6\mathbf{j} - 3\mathbf{k})$. The angular speed is 4 rad s^{-1}, so the angular velocity vector is $\frac{4}{7}(2\mathbf{i} + 6\mathbf{j} - 3\mathbf{k})$ rad s^{-1}.

(ii) The position vector of P with respect to O is $\mathbf{p} = \mathbf{j}$. O is on the axis, so the velocity is

$$\boldsymbol{\omega} \times \mathbf{p} = \begin{vmatrix} \mathbf{i} & \frac{8}{7} & 0 \\ \mathbf{j} & \frac{24}{7} & 1 \\ \mathbf{k} & -\frac{12}{7} & 0 \end{vmatrix} = \frac{12}{7}\mathbf{i} + 0\mathbf{j} + \frac{8}{7}\mathbf{k}$$

The velocity is $\frac{12}{7}\mathbf{i} + \frac{8}{7}\mathbf{k}$ m s^{-1}.

Angular momentum

Chapter 6 introduced the concept of *angular momentum* of a rigid body rotating about an axis whose direction is fixed. Just as with angular velocity, angular momentum can be defined in vector form using a vector product (figure 9.24).

Figure 9.24

The diagram shows a particle of constant mass m moving with a velocity $\mathbf{v}$ with position $\mathbf{r}$ relative to some fixed origin O. It is under the influence of a force $\mathbf{F}$. By Newton's Second Law.

$$\mathbf{F} = \frac{d}{dt}(m\mathbf{v})$$

The torque of the force about O is:

$$\mathbf{r} \times \mathbf{F} = \mathbf{r} \times \frac{d}{dt}(m\mathbf{v})$$

This is the same as $\dfrac{d}{dt}(\mathbf{r} \times m\mathbf{v})$ since by the product rule

$$\frac{d}{dt}(\mathbf{r} \times m\mathbf{v}) = \frac{dr}{dt} \times m\mathbf{v} + \mathbf{r} \times \frac{d}{dt}(m\mathbf{v})$$

$$= (\mathbf{v} \times m\mathbf{v}) + \mathbf{r} \times \frac{d}{dt}(m\mathbf{v})$$

$$= \mathbf{r} \times \frac{d}{dt}(m\mathbf{v}) \quad (\mathbf{v} \times \mathbf{v} = \mathbf{0})$$

Hence
$$\mathbf{r} \times \mathbf{F} = \frac{d}{dt}(\mathbf{r} \times m\mathbf{v})$$

Now $m\mathbf{v}$ is the linear momentum of the particle, so the expression $\mathbf{r} \times m\mathbf{v}$ could be called the 'moment of the momentum'. It is a vector form of the angular momentum you met in Chapter 6 in connection with rotation about a fixed axis. The term *angular momentum* is used to mean either the vector or scalar form, depending on context.

N O T E

Vector angular momentum is most commonly denoted by the symbols $\mathbf{L}$, $\mathbf{H}$ or $\mathbf{h}$. This book will use $\mathbf{L}$.

Activity

Show that when a particle is moving in a circle about a point O, the *magnitude* of its vector angular momentum vector about O is the same as its scalar angular momentum about the axis of rotation through O (as defined in Chapter 6). Also show that the *direction* of the angular momentum vector is the same as the axis of rotation.

Rotational equivalent of Newton's Second Law

The equation $\mathbf{r} \times \mathbf{F} = \dfrac{d}{dt}(\mathbf{r} \times m\mathbf{v})$, the rotational equivalent of Newton's Second Law, states

> When a force is acting on a particle, the torque of the force about any point is equal to the rate of change of angular momentum of the particle about that point.

It may be summarised as $\mathbf{C} = \dfrac{d\mathbf{L}}{dt}$.

A particle moving in a straight line

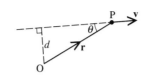

Figure 9.25

Figure 9.25 represents a particle P travelling at a constant velocity **v** (assumed in the plane of the page) in the absence of any force. O is any point also in the plane of the page and **r** is the position vector of P with respect to O. Then the angular momentum about O, $\mathbf{r} \times m\mathbf{v}$, is perpendicular to the plane (*into* the page, in the case of figure 9.25). Its magnitude is $mvr \sin\theta = mvd$, which is constant.

This is as expected; since there is no force, there is zero torque, and the angular momentum about *any* point is constant.

Angular momentum about an axis of rotation

In the case of rotation about a fixed axis, the component of the angular momentum vector along the axis can be shown to have the familiar form $I\omega$.

If O is a point on the axis, which is in the direction of the unit vector $\hat{\mathbf{n}}$, and $\boldsymbol{\omega}$ is the angular velocity about O then $\boldsymbol{\omega} = \omega\hat{\mathbf{n}}$. The momentum of a particle P with mass m and position vector $\mathbf{r}_P$ relative to O is

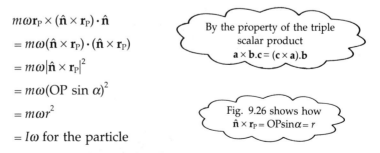

$$m\mathbf{v}_P = m\boldsymbol{\omega} \times \mathbf{r}_P$$
$$= m\omega(\hat{\mathbf{n}} \times \mathbf{r}_P)$$

Its angular momentum about O is:

$$\mathbf{r}_P \times m\mathbf{v}_P = m\omega\mathbf{r}_P \times (\hat{\mathbf{n}} \times \mathbf{r}_P)$$

Figure 9.26

The component of this angular momentum along the axis is the scalar product

$$m\omega\mathbf{r}_P \times (\hat{\mathbf{n}} \times \mathbf{r}_P) \cdot \hat{\mathbf{n}}$$
$$= m\omega(\hat{\mathbf{n}} \times \mathbf{r}_P) \cdot (\hat{\mathbf{n}} \times \mathbf{r}_P)$$
$$= m\omega|\hat{\mathbf{n}} \times \mathbf{r}_P|^2$$
$$= m\omega(OP \sin\alpha)^2$$
$$= m\omega r^2$$
$$= I\omega \text{ for the particle}$$

> By the property of the triple scalar product
> $$\mathbf{a} \times \mathbf{b}.\mathbf{c} = (\mathbf{c} \times \mathbf{a}).\mathbf{b}$$

> Fig. 9.26 shows how $\hat{\mathbf{n}} \times \mathbf{r}_P = OP\sin\alpha = r$

This is independent of the position of O on the axis. Hence the angular momentum about an axis as defined in Chapter 6 is now seen to be the component along that axis of the vector angular momentum about any point on the axis.

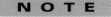

9

Mechanics 6

If you consider that the direction of the angular momentum vector $m\omega r_P \times (\hat{\mathbf{n}} \times \mathbf{r}_P)$ is perpendicular to both $\mathbf{r}_P$ and the normal to the plane OPC (so that it lies in this plane), you will realise that, for each individual particle, there is an additional component of angular momentum ($mr^2\omega \cot \alpha$) directed towards the centre of the circle in which it moves. When a rigid body is suitably symmetrical (e.g. about the axis), the sum of these components over all particles is zero. Otherwise an extra couple is required to ensure that the axis remains in a constant direction and does not wobble.

Central force

You met central force problems, exemplified by the motion of a planet about the sun, in *Mechanics 5*. The central force on a planet moving around the sun is directed towards the sun and thus has no torque about the sun. Hence the angular momentum of the planet about the sun is constant.

Figure 9.27

In figure 9.27, it is assumed that the sun S, the planet P and $\mathbf{v}$, the planet's velocity at some instant, are all in the plane of the page. The position vector of P with respect to S is $\mathbf{r}$. Then the angular momentum vector $\mathbf{r} \times m\mathbf{v}$ is a constant vector and, since m is constant, $\mathbf{r} \times \mathbf{v}$ is a constant vector.

(i)   This means that $\mathbf{r} \times \mathbf{v}$ has a constant direction (perpendicular to the page), so $\mathbf{r}$ and $\mathbf{v}$ must remain parallel to the page, i.e. the planet remains in a plane, as was shown in *Mechanics 5*.

(ii)

$$|\mathbf{r} \times \mathbf{v}| = rv \sin \phi$$

$$= rv_T \quad (v_T \text{ is transverse velocity})$$

$$= r^2\dot\theta \quad (\text{remember transverse velocity in polar co-ordinates})$$

Since $|\mathbf{r} \times \mathbf{v}|$ is constant, $r^2\dot\theta = $ constant h, say.

This result was proved in *Mechanics 5* and shown to be equivalent to Kepler's Second Law that a planet sweeps out equal areas in equal times. It is now seen to be a consequence of the constancy of angular momentum.

The symbol h is always used for this constant, even though this is not quite consistent with the widespread use (mentioned above) of $\mathbf{h}$ for angular momentum. The angular momentum of the planet (about the sun) has magnitude mh and has a constant direction perpendicular to the plane of the orbit.

Conservation of angular momentum

In the example of the planet there is no torque, and therefore no change in the angular momentum, about the sun. This is a generalisation (in vector form) of the *conservation of angular momentum* which you met in Chapter 6.

The total angular momentum of a rotating system under no external forces remains constant. It is a powerful principle in the analysis of rotating systems from atoms to galaxies.

EXAMPLE

A particle of mass m is attached to a string which passes through a hole in a smooth table and is held so that a length of string a remains on the table. The mass is set rotating in a circle about O with angular velocity ω (see diagram). The string is then gradually pulled until a length b remains on the table.

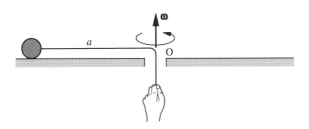

(i) What is the angular momentum about O of the particle when rotating in the circle radius a?
(ii) What is the angular velocity of the particle when the radius of the circle is b?
(iii) What is the increase in kinetic energy?
(iv) Work out the tension in the string when a length r remains on the table ($b \leqslant r \leqslant a$) and hence show that the increase in kinetic energy is equal to the work done in pulling against the tension in the string.

Solution

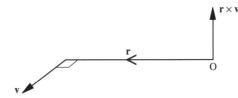

(i) The angular momentum about O is in the same direction as ω and has magnitude $ma^2\omega$. The angular momentum vector is thus $\mathbf{L} = ma^2\omega$.
(ii) Denote by ω_b the angular velocity of the particle when the radius of the circle is b. The only forces on the particle are the vertical ones which cancel out (no vertical motion) and the tension which passes through O. Thus angular momentum is conserved.

$$ma^2\omega = mb^2\omega_b$$

$$\Rightarrow \quad \omega_b = \frac{a^2}{b^2}\omega$$

(iii) Increase in kinetic energy

$$= \tfrac{1}{2}m(b^2\omega_b^2 - a^2\omega^2) \quad (\omega \text{ denotes magnitude of } \omega)$$

$$= \tfrac{1}{2}m\left(\frac{b^2a^4\omega^2}{b^4} - a^2\omega^2\right) \quad (\text{substituting for } \omega_b)$$

$$= \frac{ma^2\omega^2}{2b^2}(a^2 - b^2)$$

(iv) When the length of the string on the table is r, the angular speed is given by $\omega_r = \dfrac{a^2 \omega}{r^2}$.

Acceleration towards the centre is $r\omega_r^2 = \dfrac{a^4 \omega^2}{r^3}$.

Tension = mass × acceleration = $\dfrac{ma^4 \omega^2}{r^3}$.

Work done for a small *decrease* δr in r is $\left(\dfrac{ma^4 \omega^2}{r^3}\right)\delta r$.

So the total work is

$$\int_a^b -\frac{ma^4 \omega^2}{r^3}\, \mathrm{d}r = ma^4 \omega^2 \left[\frac{1}{2r^2}\right]_a^b$$

$$= \tfrac{1}{2}ma^4 \omega^2 \left(\frac{1}{b^2} - \frac{1}{a^2}\right)$$

$$= \frac{ma^2 \omega^2}{2b^2}(a^2 - b^2)$$

This is equal to the kinetic energy gained.

Total angular momentum of a body

You can find, by vector addition, the total angular momentum of all the particles in a body about some point. This gives the angular momentum of *the body* about that point. Suppose a typical particle in the body has mass m_p, position vector $\mathbf{r}_p$ with respect to a fixed origin O, and velocity $\mathbf{v}_p$. Then for each particle

torque of the force on it = rate of change of momentum

$$\mathbf{C}_p = \frac{\mathrm{d}}{\mathrm{d}t}(m_p \mathbf{r}_p \times \mathbf{v}_p) = \frac{\mathrm{d}}{\mathrm{d}t}\mathbf{L}_p \quad \text{(say)}$$

Summing over every particle in the body gives

$$\sum_{\text{all } p} \mathbf{C}_p = \sum_p \frac{\mathrm{d}}{\mathrm{d}t}\mathbf{L}_p = \frac{\mathrm{d}}{\mathrm{d}t}\sum_p \mathbf{L}_p$$

$$= \frac{\mathrm{d}}{\mathrm{d}t}\mathbf{L}$$

where $\mathbf{L}$ is the total angular momentum.

In other words, the total torque of forces on the body is the rate of change of total angular momentum.

Activity

Show that when a body is rotating with angular speed $\dot\theta$ about a fixed axis through a point O, the component *along the axis* of its total angular momentum vector about O is equal to $I\dot\theta$, i.e. the scalar angular momentum.

NOTE

In a rigid body all the particles which make it up exert internal forces on each other: a particle m_i moves because a neighbouring particle m_j acts on it. But by Newton's Third Law m_i exerts an equal and opposite force on m_j. When you add up the torques of these forces about any point, they cancel out. Hence only external forces need be considered in working out the total torque.

Angular momentum about a moving point

The formula

torque of the force on P = rate of change of angular momentum of P

can be written $\qquad \mathbf{r} \times \mathbf{F} = \dfrac{\mathrm{d}}{\mathrm{d}t}(\mathbf{r} \times m\dot{\mathbf{r}})$

It is valid only if the vector $\mathbf{r}$, the position vector of P, is relative to a *fixed* origin. However, both the torque $\mathbf{r} \times \mathbf{F}$ and the angular momentum $\dfrac{\mathrm{d}}{\mathrm{d}t}(\mathbf{r} \times m\dot{\mathbf{r}})$ can be defined relative to a *moving* point A. This is very useful in the analysis of the general motion of a rigid body.

Suppose P is a particle with, at time t, position vector $\mathbf{r}_p$ relative to a fixed origin O. P is acted on by a force $\mathbf{F}_p$. A is a moving point with position vector $\mathbf{r}_A$. Denote the position vector of P *relative to* A by $\boldsymbol{\rho}_p$, i.e. $\boldsymbol{\rho}_p = \mathbf{r}_p - \mathbf{r}_A$ (figure 9.28).

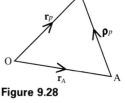

Figure 9.28

Newton's Second Law gives

$$\mathbf{F}_p = m\ddot{\mathbf{r}}_p = m(\ddot{\mathbf{r}}_A + \ddot{\boldsymbol{\rho}}_p)$$

The torque about the point A is

$$\boldsymbol{\rho}_p \times \mathbf{F}_p = \boldsymbol{\rho}_p \times m(\ddot{\mathbf{r}}_A + \ddot{\boldsymbol{\rho}}_p)$$

$$= \boldsymbol{\rho}_p \times m\ddot{\mathbf{r}}_A + \boldsymbol{\rho}_p \times m\ddot{\boldsymbol{\rho}}_p \qquad \qquad ①$$

Now the rate of change of angular momentum about A is

$$\dfrac{\mathrm{d}}{\mathrm{d}t}(\boldsymbol{\rho}_p \times m\dot{\boldsymbol{\rho}}_p) = \dot{\boldsymbol{\rho}}_p \times m\dot{\boldsymbol{\rho}}_p + \boldsymbol{\rho}_p \times m\ddot{\boldsymbol{\rho}}_p \quad \text{(by the product rule} - m \text{ constant)}$$

$$= \boldsymbol{\rho}_p \times m\ddot{\boldsymbol{\rho}}_p \quad \text{(since a vector crossed with itself is zero)}$$

Hence from ①

$$\boldsymbol{\rho}_p \times \mathbf{F}_p = m\boldsymbol{\rho}_p \times \ddot{\mathbf{r}}_A + \dfrac{\mathrm{d}}{\mathrm{d}t}(\boldsymbol{\rho}_p \times m\dot{\boldsymbol{\rho}}_p)$$

This can be interpreted as

torque about A $= m\boldsymbol{\rho}_p \times \ddot{\mathbf{r}}_A +$ rate of change of angular momentum about A

Thus relative to an *accelerating* point A, the relationship between torque and rate of change of momentum is complicated by an extra term. This term disappears when A is fixed or moving with constant velocity. This equation can also be summed for all particles of a body, and then the extra terms cancel out in the special case when A is G, the centre of mass of the body.

Angular momentum about the centre of mass of a moving body

Consider now a *moving* body made up of a number of particles. Let m_p be the mass of a typical particle with position vector $\boldsymbol{\rho}_p$ with respect to the centre of mass G, which itself has position vector $\mathbf{r}_G$ with respect to a fixed origin O. By the result in the last section, for each particle

torque about G $= m_p\boldsymbol{\rho}_p \times \ddot{\mathbf{r}}_G +$ rate of change of angular momentum about G

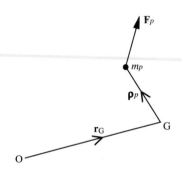

Figure 9.29

Summing over every particle gives

$$\text{total torque about G} = \sum_{\text{all } p} m_p\boldsymbol{\rho}_p \times \ddot{\mathbf{r}}_G + \frac{d\mathbf{L}}{dt}$$

where $\mathbf{L}$ is the *total* angular momentum about G.

Now

$$\sum_{\text{all } p} m_p\boldsymbol{\rho}_p \times \ddot{\mathbf{r}}_G = \left(\sum_{\text{all } p} m_p\boldsymbol{\rho}_p\right) \times \ddot{\mathbf{r}}_G$$

since $\ddot{\mathbf{r}}_G$ is the same for each element of the sum. But $(\sum_{\text{all } p} m_p\boldsymbol{\rho}_p) = \mathbf{0}$ by the definition of the centre of mass, since $\boldsymbol{\rho}_p$ is the position of particle m_p relative to the centre of mass. So as stated in Chapter 6, the relationship

total torque about centre of mass =

rate of change of angular momentum about centre of mass

holds even when the centre of mass of the body is moving.

1. The diagram shows a wheel rotating anticlockwise about the origin in the xy plane at 5 rad s^{-1}. Write down the vector angular velocity in terms of the usual vectors $\mathbf{i}, \mathbf{j}, \mathbf{k}$ ($\mathbf{k}$ coming out of the page). Using $\mathbf{v} = \boldsymbol{\omega} \times \mathbf{r}$, find the velocity of the point on the wheel with position vector $2\mathbf{j}$ and confirm this is what you would expect.

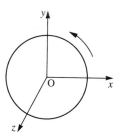

2. A wheel is rotating about an axis through the origin. Its angular velocity is $\boldsymbol{\omega} = \mathbf{i} + \mathbf{j} + \mathbf{k}$. At one instant a point P on the wheel has position vector $\mathbf{r} = \mathbf{i} + \mathbf{j} + 2\mathbf{k}$ with respect to the origin. Use $\mathbf{v} = \boldsymbol{\omega} \times \mathbf{r}$ to find the velocity and hence the speed of P at this moment.

3. A gyroscope is rotating at 21 rad s^{-1} about a fixed axis in a direction $2\mathbf{i} - \mathbf{j} - 2\mathbf{k}$.
(i) Find the constant angular velocity vector of the gyroscope.
(ii) Find the velocity of a point P on the gyroscope which momentarily has position vector $\mathbf{i} - \mathbf{j} - \mathbf{k}$ with respect to a point O on the axis.

4. One component of a double-star system has mass m and at time t has position vector

$$\mathbf{r}_1 = a \cos \omega t\, \mathbf{i} + a \sin \omega t\, \mathbf{j} + ut\mathbf{k}$$

where u is constant with respect to a fixed origin. The other has mass $2m$ and position vector

$$\mathbf{r}_2 = \tfrac{1}{2} a \cos (\omega t + \pi)\, \mathbf{i} + \tfrac{1}{2} a \sin (\omega t + \pi)\, \mathbf{j} + ut\mathbf{k}$$

(i) Find the velocity of each star.
(ii) Work out the angular momentum vector of each star about the origin.
(iii) Show that the total angular momentum is constant and in the direction of $\mathbf{k}$.
(iv) Give a general description of the motion of the system.

5. The position vector $\mathbf{r}$, relative to an origin O, of a particle of mass m at time t is given by

$$\mathbf{r} = (t^2 - 1)\mathbf{i} + 2t\mathbf{j} + t\mathbf{k}$$

(i) Describe, using one or more sketches, the path of the particle.
(ii) Find the angular momentum $\mathbf{h}$ of the particle about O at time t and its time derivative $\mathrm{d}\mathbf{h}/\mathrm{d}t$.
(iii) Find the force $\mathbf{F}$ acting on the particle and its torque about O.

6. A vase is turning on a potter's wheel with a constant angular velocity vector $\boldsymbol{\omega}$. A particle of clay P of mass m has position vector $\mathbf{r}$ with respect to O, a point on the axis of rotation.

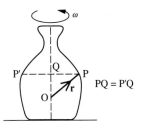

(i) Write down the vector expression for the velocity of P.
(ii) Show that the angular momentum of P about O is $m|\mathbf{r}|^2 \boldsymbol{\omega} - m(\mathbf{r}.\boldsymbol{\omega})\mathbf{r}$. Deduce that the angular momentum is parallel to $\boldsymbol{\omega}$ only if O is the nearest point on the axis of rotation to P.
Note that for any vectors $\mathbf{a}, \mathbf{b}, \mathbf{c}$: $\mathbf{a} \times (\mathbf{b} \times \mathbf{c}) = (\mathbf{a}.\mathbf{c})\mathbf{b} - (\mathbf{a}.\mathbf{b})\mathbf{c}$.
(iii) P' is a particle of the same mass as P, diametrically opposite the axis of rotation from P. Show that the sum of the angular momentum of P and P' about O is parallel to the axis. Deduce that the total angular momentum of the vase about O is parallel to the axis when the vase has rotational symmetry about the axis.

7. (i) Define the angular momentum L about the origin O of a particle of mass m,

position vector **r** moving with velocity **v**.
A system of n particles with total mass
M is defined by masses m_i, $i = 1, 2, \ldots n$,
with position vectors $\mathbf{r}_i$ and velocities $\mathbf{v}_i$.
The centre of mass of the system is at a
point G with position vector $\bar{\mathbf{r}}$ and
moves with velocity $\bar{\mathbf{v}}$.

(ii) Show that the angular momentum $\mathbf{L_O}$
of the system of particles about the
origin O is given by

$$\mathbf{L_O} = \sum_{i=1}^{n} m_i \mathbf{r}_i \times \mathbf{v}_i$$

(iii) Write down an expression for the
angular momentum $\mathbf{L_{OG}}$ about the
origin O of the total mass M of the
particles acting as if they were all
located at the mass centre G.

(iv) Find an expression for the angular
momentum $\mathbf{L_G}$ of the system of
particles in their motion relative to
G about G.

(v) By writing $\mathbf{r}_i$ as $\bar{\mathbf{r}} + (\mathbf{r}_i - \bar{\mathbf{r}})$ and $\mathbf{v}_i$ as
$\bar{\mathbf{v}} + (\mathbf{v}_i - \bar{\mathbf{v}})$ in the expression for $\mathbf{L_O}$ show
that $\mathbf{L_O} = \mathbf{L_{OG}} + \mathbf{L_G}$.

[MEI 1994]

8. The diagrams show a gyroscope consisting
of a disc centre P and radius a, spinning at
ω rad s^{-1} about an axis OP where OP $= d$.

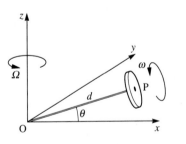

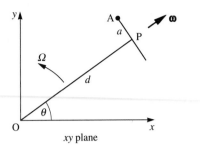

xy plane

The axis itself remains in the *xy* plane but is
rotating with angular speed Ω about Oz.
Hence the angular velocity vector of OP
about O is $\mathbf{\Omega} = \Omega\mathbf{k}$.
At one instant the axis OP makes an angle
θ with the x axis. Thus $\boldsymbol{\omega} = \omega(\cos\theta\mathbf{i} + \sin\theta\mathbf{j})$ is
a vector with magnitude ω in the direction
of OP. At this instant a point A on the rim
of the disc is in the *xy* plane. Due to the
spinning of the disc it has a component of
velocity $a\omega$ in the **k** direction.

(i) Show that **r**, the position vector of A
relative to O is

$$\mathbf{r} = (d\cos\theta - a\sin\theta)\mathbf{i} + (d\sin\theta + a\cos\theta)\mathbf{j}$$

(ii) Differentiate to find the component of
velocity of A in the *xy* plane. Hence
show that the velocity of A is

$$(-d\sin\theta - a\cos\theta)\Omega\mathbf{i}$$
$$+ (d\cos\theta - a\sin\theta)\Omega\mathbf{j} + a\omega\mathbf{k}$$

(iii) Confirm that this is equal to $(\boldsymbol{\omega} + \boldsymbol{\Omega}) \times \mathbf{r}$,
where **r** is the position vector of A with
respect to O.

(iv) What is the angular momentum vector
of a particle of mass m at A?

KEY POINTS

- Given two vectors **a** and **b**, inclined at an angle θ, the vector product $\mathbf{a} \times \mathbf{b}$ is defined as a *vector* which
  - * is perpendicular to both **a** and **b**;
  - * has a magnitude $|\mathbf{a}||\mathbf{b}|\sin\theta$
  - * has a direction such that **a**, **b** and $\mathbf{a} \times \mathbf{b}$ form a right-handed set of vectors.

- For any vectors **a**, **b**, **c**

  $$\mathbf{a} \times \mathbf{b} = -(\mathbf{b} \times \mathbf{a})$$

  $$(\mathbf{a} \times \mathbf{b}).\mathbf{c} = (\mathbf{b} \times \mathbf{c}).\mathbf{a} = (\mathbf{c} \times \mathbf{a}).\mathbf{b}$$

  $$\mathbf{a} \times (\mathbf{b} + \mathbf{c}) = (\mathbf{a} \times \mathbf{b}) + (\mathbf{a} \times \mathbf{c})$$

- Given two vectors

  $$\mathbf{a} = a_1\mathbf{i} + a_2\mathbf{j} + a_3\mathbf{k} \quad \text{and} \quad \mathbf{b} = b_1\mathbf{i} + b_2\mathbf{j} + b_3\mathbf{k}$$

  then

  $$\mathbf{a} \times \mathbf{b} = \begin{vmatrix} \mathbf{i} & a_1 & b_1 \\ \mathbf{j} & a_2 & b_2 \\ \mathbf{k} & a_3 & b_3 \end{vmatrix} = (a_2b_3 - a_3b_2)\mathbf{i} + (a_3b_1 - a_1b_3)\mathbf{j} + (a_1b_2 - a_2b_1)\mathbf{k}$$

- Where two vectors, **F** and **G**, depend on variable t, their vector and scalar products can be differentiated using the product rule:

  $$\frac{d}{dt}(\mathbf{F}.\mathbf{G}) = \frac{d\mathbf{F}}{dt}.\mathbf{G} + \mathbf{F}.\frac{d\mathbf{G}}{dt}$$

  $$\frac{d}{dt}(\mathbf{F} \times \mathbf{G}) = \frac{d\mathbf{F}}{dt} \times \mathbf{G} + \mathbf{F} \times \frac{d\mathbf{G}}{dt}$$

- The *vector moment* or *torque* of a force **F** about a point O is defined as

  $$\mathbf{C} = \mathbf{r} \times \mathbf{F}$$

  where **r** is the position vector of any point R on the line of action of the force. The moment of **F** about an *axis* through O is then

  $$\hat{\mathbf{n}}.\mathbf{C} = \hat{\mathbf{n}}.(\mathbf{r} \times \mathbf{F})$$

  where $\hat{\mathbf{n}}$ is the unit vector in the direction of the axis.

- A *couple* can be represented by equal and opposite forces **F** acting through A and $-\mathbf{F}$ acting through B. The torque of a *couple* about any point is the same, $\overrightarrow{BA} \times \mathbf{F}$.

- A system of forces $\mathbf{F}_i$ is in equilibrium if and only if

  $$\sum \mathbf{F}_i = 0 \qquad \text{the resultant force is zero}$$

  and $\quad \sum (\mathbf{r}_i \times \mathbf{F}_i) = 0 \qquad$ the resultant torque is zero

- Two systems of forces are said to be *equivalent* if
  * they have the same resultant; and
  * they have equivalent torques about a point.
  In general a system of *any* number of forces can be reduced to a single force plus a couple, either of which may be zero.

- When a body is rotating about an axis, the *angular velocity vector* $\boldsymbol{\omega}$ associated with a point O on that axis is defined as the vector whose magnitude is ω and which is in the direction of the axis, in a right-handed screw sense. The velocity of a particle with position vector $\mathbf{r}$ with respect to O is

$$\mathbf{v}_p = \boldsymbol{\omega} \times \mathbf{r}$$

- When a particle of mass m is moving with velocity $\mathbf{v}$ and has position vector $\mathbf{r}$ with respect to a point O, the *moment of the momentum* or *angular momentum* of the particle about O is

$$\mathbf{L} = \mathbf{r} \times m\mathbf{v} = \mathbf{r} \times m\dot{\mathbf{r}}$$

- The torque about a fixed point O of the total force on the particle is equal to the rate of change of total angular momentum about that point

$$\mathbf{C} = \frac{d\mathbf{L}}{dt}$$

- If there is no torque, as with a central force, angular momentum is conserved.

- When a particle is rotating with angular speed ω at a distance r from a fixed axis of rotation, the *angular momentum about the axis* is $mr^2\omega$. This is the same as $I\omega$, where I is the moment of inertia of the particle about the axis.

- The angular momentum of a body made up of a number of particles is the vector sum of the individual angular momentum. The total external torque on a body about *any fixed point* is equal to the rate of change of angular momentum about that point.

- The preceding result also applies when the point is the centre of mass, even if it is moving.

Answers
Mechanics 5

Exercise 1A

1. $27\sqrt{5}$ or 60.37
2. $3t^2\mathbf{i} + 2t\mathbf{j}$, $t^3\mathbf{i} + (t^2 + c)\mathbf{j}$ where c is constant
3. (i) $3\mathbf{i} - 4\mathbf{j}$ (ii) $\mathbf{r} = (\mathbf{i} + 2\mathbf{j}) + \lambda(3\mathbf{i} - 4\mathbf{j})$
   (iii) 10 units clockwise
4. Velocity $8\mathbf{i} + 6\mathbf{j}$ m s^{-1}, kinetic energy 50 J
5. (i) $\mathbf{r} = (\mathbf{i} + \mathbf{j} + \mathbf{k}) + t(\mathbf{i} + 2\mathbf{j} - \mathbf{k})$ (iii) 2 s
6. (i) $m(2\mathbf{i} - \sin t\mathbf{j})$ (ii) $m(4t - \sin t\cos t)$
   (iii) $m(2t^2 - \frac{1}{2}\sin^2 t)$
8. (i) $\frac{1}{5}mg\,(4 - t)$
9. (i) Acceleration $2\cos t\mathbf{i} + 2\sin t\mathbf{j}$, velocity $2\sin t\mathbf{i} - 2\cos t\mathbf{j}$
   (iii) $2(1 - \cos t)\mathbf{i} - 2\sin t\mathbf{j}$
   (iv) Centre $(2, 0)$, radius 2
10. (i) $3\cos 3t\mathbf{i} + 3\sin 3t\mathbf{j}$ (ii) 3 m s^{-1}
11. (ii) Minimum speed $2a\omega$ (iii) $t = \pi/2\omega$

Exercise 2A

1. (i) 15 km h^{-1} due W
   (ii) 105 km h^{-1} due W
   (iii) 75 km h^{-1} at 307°
   (iv) 42.5 km h^{-1} at 318.5°
   (v) 98.8 km h^{-1} at 253°
2. (i) 30 km h^{-1} due N (ii) 8.4 m
   (iii) 1.008 s (iv) 0.144 s
3. 62°
4. 0.728 m s^{-1}
5. On a bearing of 124.4°
6. (i) Upstream at $\arccos(u/v)$ to the bank
   (ii) au/v
7. (i) (a) 15 s (b) 3 s
   (ii) 11.72 s; turbo goes on accelerating; no
8. (ii) 1.21 s
9. (ii) $\frac{1}{2}(a - \mu g)T^2$.
   (iii) It continues slipping but slows down to the speed of the truck
   (v) 5.16 m s^{-2}
10. (i) 0.72 (ii) 6.43 (iii) 250/3
11. (ii) 14.14 km h^{-1} (iii) 17 km h^{-1} at 152°
   (iv) From 171.4°
14. (i) 1.25 km h^{-1} (ii) 2.505 (iii) 42 min

Exercise 2B

1. (ii) $t = 2$ (iii) $11\mathbf{i} + 3\mathbf{j}$
2. (i) 3.32 knots (ii) 1.01 nautical miles
3. (iii) 44.5 km h^{-1}, 47 min

4. (i) $-12(\mathbf{i} + \mathbf{j})$
   (ii) $(4800 - 12t)\mathbf{i} + (12\,000 - 24t)\mathbf{j}$
   (iii) $(4800 - 12t)^2 + (12\,000 - 24t)^2$
   (iv) 1.07 km
5. (i) 12.30, 7.79 km (ii) (a) $2.5\mathbf{i} - 2\mathbf{j}$
   (b) 13 00 (c) 7.2 km
6. (i) $\mathbf{r}_A = -9\mathbf{i} - 5\mathbf{j} + t_1(9\mathbf{i} + 7\mathbf{j})$,
   $\mathbf{r}_B = 6\mathbf{i} + 5\mathbf{j} + t_2(7\mathbf{i} + 6\mathbf{j})$, $27\mathbf{i} + 23\mathbf{j}$
   (ii) $15\mathbf{i} + 10\mathbf{j} - t(2\mathbf{i} + \mathbf{j})$, 2.24 km
7. 7.62 km; $6\mathbf{i} + 8\mathbf{j}$
8. (i) $3nu\mathbf{i} + (5nu - gt)\mathbf{j}$,
   $(3n + 4)u\mathbf{i} + (5n - 3)u\mathbf{j}$ (ii) $n = 4$
   (iii) $-12u^2/g$, $9u^2/2g$
9. (i) $a\sin\omega t\mathbf{i} - a\cos\omega t\mathbf{j} + 3h\mathbf{k}$
   (iii) $V\cos\theta\mathbf{j} - V\sin\theta\mathbf{k}$,
   $a\mathbf{i} + VT\cos\theta\mathbf{j} + (2h - VT\sin\theta)\mathbf{k}$
10. (ii) $400p^2 + (9 - 25p)^2$, $p_{\mathrm{m}} = \frac{9}{41}$ (iii) $41\mathbf{j}$
12. (i) $\mathbf{r}_{\mathrm{m}} = \frac{3}{2}(\cos\pi t\mathbf{i} + \sin\pi t\mathbf{j})$,
   $\mathbf{r}_{\mathrm{e}} = \cos 2\pi t\mathbf{i} + \sin 2\pi t\mathbf{j}$
   (ii)
   $(\frac{3}{2}\cos\pi t - \cos 2\pi t)\mathbf{i} + (\frac{3}{2}\sin\pi t - \sin 2\pi t)\mathbf{j}$
   (iv)

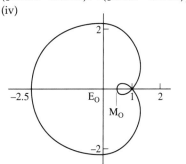

13. (i) Yes
   (ii) (d)

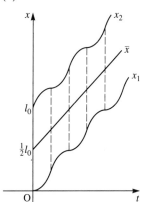

Exercise 3A

1. In the following answers, k is an arbitrary constant
   (i) $s = \frac{1}{2}v + k$ (ii) $v = ke^{2t}$
   (iii) $v = -\frac{2}{3}\cos 3t + k$
   (iv) $v = 1/(t+k)$ (v) $v = ke^{-s}$
   (vi) $v = \sqrt{[4s(1-s)+k]}$
   (vii) $s = -\frac{1}{3}\ln k(3v+2)$

2. (i) $\frac{16}{3}$ m s^{-1} (ii) $\frac{1}{2}$ m s^{-1}
   (iii) $\sqrt{2\ln 2}$ or 1.177 m s^{-1}
   (iv) $7\frac{1}{2}$ s   (v) $\pi/4$ or 0.785 s
   (vi) $\frac{1}{3}\ln\frac{5}{2}$ or 0.305 s
   (vii) $\frac{1}{2}\ln 3$ or 0.549 s
   (viii) $\frac{1}{2}\ln\frac{4}{3}$ or 0.144 m

3. (i) $v\dfrac{dv}{ds} = \dfrac{1}{v+2}$ (iii) $\dfrac{dv}{dt} \approx \dfrac{1}{v+2}$
   (v) $s = -2t + \frac{1}{3}(4+2t)^{3/2} - \frac{8}{3}$

4. (ii) $v = \pm\omega\sqrt{a^2 - s^2}$

5. (i) $\dfrac{dv}{dt} = \dfrac{10}{v}$
   (ii) $v = \sqrt{(20t+25)}$ (iv) $s = \frac{1}{30}(v^3 - 125)$
   (v) Start from $v\dfrac{dv}{ds} = \dfrac{10}{v}$

Exercise 3B

1. (i) $\dfrac{g}{100} - kv^2 = \dfrac{1}{100}\dfrac{dv}{dt}$, $v_T = \dfrac{1}{10}\sqrt{\left(\dfrac{g}{k}\right)}$

   (ii) $k = \dfrac{g}{22\,500} = 4.36 \times 10^{-4}$
   (iii) $15/\sqrt{2}$ or 10.61 m s^{-1}

2. (i) $m = 3.4 \times 10^{-15}$ kg
   (ii) $m\dfrac{dv}{dt} = mg - 3.1 \times 10^{-10}\,v$,
   $v_T = 1.1 \times 10^{-4}$ m s^{-1}
   (iii) 1.5 micron

3. (i) $40\dfrac{dv}{dt} = -kv$ (ii) $Ue^{-k/40}$ (iv) 18.6 m

5. (i) Weight $\frac{4}{3}\pi R^3 \rho_w g$,
   $2R^2(\rho_g - \rho_w)g - 9\eta v = 2R^2\rho_g\dfrac{dv}{dt}$
   (iii) Terminal velocity of 1 cm marble $\frac{1}{9}$ m s^{-1}, of 2 cm marble $\frac{4}{9}$ m s^{-1}
   (iv) $\frac{1}{60}\ln 100 = 0.077$ s

6. (i) $k = mg/40$ (ii) $F = mg\sin 2° + mgv/40$
   (iii) $m\dfrac{dv}{dt} = -(mg\sin 2° + mgv/40)$
   (iv) 66 m

7. (i) $\dfrac{dv}{dx} = g\left(\dfrac{1}{v} - \dfrac{1}{V}\right)$

8. (iii) $\dfrac{UV}{\sqrt{U^2 + V^2}}$
   (iv) $v = V\tan\left(\alpha - \dfrac{gt}{V}\right)$ where
   $\alpha = \arctan\left(\dfrac{U}{V}\right)$

Exercise 3C

1. (i) $\mathrm{M}^{-1}\mathrm{L}^3\mathrm{T}^{-2}$ (ii) $a_h = GM/(R+h)^2$, $a_0 = g$
   (v) 1.6%

2. (ii) $R^2 g/r^2$

3. (i) $\sqrt{(k/R)}$ (ii) $k = 4.9 \times 10^{12}$, 1673 m s^{-1}

4. (i) $(r/R)^3 M$ (ii) GMr/R^3
   (iii) Period $2\pi\sqrt{(R^3/GM)}$

5. (i) $\dfrac{GM_m}{(d-x)^2} - \dfrac{GM_e}{x^2}$
   (iii)
   $$v = \sqrt{\left[2\left(\dfrac{GM_m}{d-x} + \dfrac{GM_e}{x} - \dfrac{GM_m}{d-R} - \dfrac{GM_e}{R}\right) + u^2\right]}$$
   (iv) $v^2 > 0$ when $x = d\left[1 + \sqrt{\left(\dfrac{M_m}{M_e}\right)}\right]^{-1}$

Exercise 3D

1. $\frac{1}{6}$ J

2. (i) 75 J   (ii) $\sqrt{15}$ or 3.87 m s^{-1}
   (iii) $10\sqrt{15}$ or 38.7 N s
   (iv) $10\sqrt{15}$ or 38.7 N s

3. $\sqrt{(40/3)}$ or 3.65 m s^{-1}
   $10\sqrt{(40/3)}$ or 36.5 N s

4. (i) $\sqrt{(2gl)}$

5. (i) $\frac{1}{2}mu^2 - mgR$ (ii) 11.3 km s^{-1}

6. (ii) $\frac{1}{3}\sqrt{\dfrac{2}{G}}(h^{3/2} - R^{3/2})$

7. (iii) $t = 16\ln\left(\dfrac{1600}{1600 - v^2}\right)$, 23.2 s

8. (i) $7g/80$ or 0.858 m s^{-2} (ii) 313 m

9. (i) $\frac{1}{2}mga$ (ii) $\sqrt{(ga)}$

11. (i) $p + \dfrac{q}{v}$ (ii) $p + \dfrac{q}{v} = m\dfrac{dv}{dt}$

12. (ii) 2002 s   (iv) Note: fastest when $v_1 = v_2$

Exercise 3E

2. (i) 40 s

(iii) $\mathbf{v} = -a\omega\sin\omega t\,\mathbf{i} - a\omega\sin 2\omega t\,\mathbf{j}$
$\mathbf{a} = -\omega^2 a\cos\omega t\,\mathbf{i} - 2a\omega^2\cos 2\omega t\,\mathbf{j}$

(iv) $t = 5.8$ or 14.2,
max speed $\pi a/16 = 0.196a$

3. (ii) $\dfrac{1}{k}\ln\left(1 + \dfrac{kV}{g}\right)$ (iii) $\dfrac{UV}{g + kV}$

4. $x = \dfrac{u}{k}(1 - e^{-kt})$, $y = \tfrac{1}{2}gt^2$

5. (ii) $R_{max} = \dfrac{V^2}{g}\left(1 + \dfrac{n^2V^2}{g^2}\right)^{-\frac{1}{2}}$

$\arctan\left(1 + \dfrac{n^2V^2}{g^2}\right)^{-1/2}$

6. (i) mg/V

Exercise 3F

2. (i) $-\tfrac{2}{3}$ (ii) $\tfrac{2}{3}$ (iii) $\tfrac{14}{15}$
Work done is zero between any two
points on circumference

3. (i) $\tfrac{1}{2}\ln 10$ (ii) $\tfrac{1}{2}\ln 10$

4. Work done 280π. Starting point makes
no difference: going round in the
opposite direction would change the
sign of the answer.

5. (iii) $-k\ln x(+\text{const})$

Exercise 4A

1. (i) $r = a$ (ii) $\theta = \alpha$ (iii) $r = c\,\text{cosec}\,\theta$
(iv) $r = d\sin\beta\,\text{cosec}\,(\beta - \theta)$

2. (i)

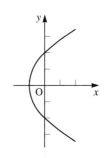

(iii) Point P: tangent to curve is
perpendicular to radius vector OP

3. (i) $-\tfrac{\pi}{4} \leqslant \theta \leqslant \tfrac{\pi}{4}, \tfrac{3\pi}{4} \leqslant \theta \leqslant \pi, -\pi < \theta \leqslant -\tfrac{3\pi}{4}$

(iv)

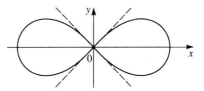

4. (i) $\hat{\mathbf{r}} = \cos\theta\,\mathbf{i} + \sin\theta\,\mathbf{j}$, $\hat{\boldsymbol{\theta}} = -\sin\theta\,\mathbf{i} + \cos\theta\,\mathbf{j}$

5. (i) $3\mathbf{j}$ (ii) $5\hat{\mathbf{r}}$ (iii) $\tfrac{3}{5}(4\hat{\mathbf{r}} + 3\hat{\boldsymbol{\theta}})$
(iv) $\hat{\mathbf{r}} = \tfrac{3}{5}\mathbf{i} + \tfrac{4}{5}\mathbf{j}$, $\hat{\boldsymbol{\theta}} = -\tfrac{4}{5}\mathbf{i} + \tfrac{3}{5}\mathbf{j}$

6. (i)

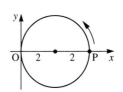

(ii) $\pi/4$, $\pi t/8$ (iii) $4\cos(\pi t/8)$

7. (i) $-2\hat{\boldsymbol{\theta}}$ (ii) $\sqrt{2}(\hat{\mathbf{r}} - \hat{\boldsymbol{\theta}})$ (iii) $2\hat{\mathbf{r}}$

Exercise 4B

2. (i) ωe^θ, ωe^θ, $\sqrt{2}\omega e^\theta$

(ii) $-3\omega\sin\theta$, $3\omega\cos\theta$, 3ω

(iii) $-\omega\sin\theta$, $\omega(1 + \cos\theta)$, $\omega\sqrt{2(1 + \cos\theta)}$

(iv) $2\omega\sin\theta(1 + \cos\theta)^{-2}$, $2\omega(1 + \cos\theta)^{-1}$,
$2\sqrt{2}\omega(1 + \cos\theta)^{-3/2}$

3. (i) $-2\theta\omega^2$, $4\omega^2$ (ii) 0, $2\omega^2 e^\theta$

(iii) $-6\omega^2\sin\theta$, $6\omega^2\cos\theta$

(iv) $\omega^2(2\sin\theta - 1)$, $-2\omega^2\cos\theta$

4. $0.6\,g \approx 6\,\text{m s}^{-2}$

5. (ii) $r = Re^{-\theta\cot\alpha}$

6. (i) $\ddot{r} - r\omega^2 = 0$ (iii) $r = e^{\omega t} + e^{-\omega t}$

(iv)

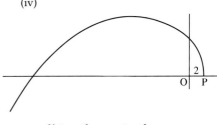

7. (ii) $ae^{2\theta}(\ddot{\theta} + 4\dot{\theta}^2)$ (iii) $\ddot{\theta} - \dot{\theta}^2 = 0$

8. $32a^4\lambda^2/r^5$

Exercise 4C

4. (ii) $\sqrt{(k/R)}$ (iii) $18R/7$

6. (iv) $h = \sqrt{(\frac{1}{2}k)}$

7. (i) $\dfrac{h}{r}$ (ii) $h\sqrt{\left(\dfrac{e^2}{l^2}\sin^2\theta + \dfrac{1}{r^2}\right)}$

9. (i) $\dot{r} = \dfrac{er^2\sin\theta}{l}\dot{\theta}$

(ii) $\dot{r} = \dfrac{he\sin\theta}{l}$

(iii) $-\dfrac{h^2}{lr^2}$

10. (iv) $\dfrac{d^2u}{d\theta^2} + u = -\dfrac{k}{h^2}$

11. (i) $V = \sqrt{(k/R)}$ (ii) $\frac{1}{2}\sqrt{kR}$ (iii) $-mk/2R$
(iv) $(2+\sqrt{3})V, (2-\sqrt{3})V$

12. (i) $\alpha V(H+R)$ (iii) $\alpha^2 > 2R/(H+2R)$

Exercise 5A

1. (i) $2ma^2, \frac{2}{3}Ma^2; 5ma^2, \frac{5}{3}Ma^2$
(ii) $2ma^2, \frac{1}{2}Ma^2; ma^2, \frac{1}{4}Ma^2; 4ma^2, Ma^2$
(iii) $\frac{1}{2}ma^2, \frac{1}{6}Ma^2; \frac{3}{4}ma^2, \frac{1}{4}Ma^2; 2ma^2, \frac{2}{3}Ma^2;$
$ma^2, \frac{1}{3}Ma^2$
(iv) $2ma^2, \frac{2}{5}Ma^2; \frac{11}{4}ma^2, \frac{11}{20}Ma^2$

2. (i) 0.3125 kg m^2 (ii) 1172 J

3. 11.85 J

4. (i) 25 J (ii) 70.7 rad s^{-1}

5. (i) $\rho\delta x$ (ii) $x^2\rho\delta x$ (iii) $\displaystyle\int_0^{2a} x^2\rho\,dx = \frac{8}{3}\rho a^3$

(v) $\frac{4}{3}Ma^2$

6. (i) 4.29 rad s^{-1} (ii) 6.86 m s^{-1}

7. (i) 7.47 kg m^2 (ii) 23.3 J

8. (i) $\pi\rho y^2\delta x$ (iii) $\frac{8}{15}\rho\pi r^5$ (v) $\displaystyle\int\frac{1}{2}\rho\pi y^4\,dx$

9. (i) 9.83×10^{37} kg m^2
(ii) 2.6×10^{29} J, overestimate
(iii) 2.9×10^{23} J (iv) 3.75×10^{28} J
(v) approx $7{:}1$

10. (i) $M/\pi r^2h$ (ii) $4.41h$ (iii) $0.824Mr^2$
(iv) $1.65{:}1$

Exercise 5B

1. (i) 0.0241 kg m^2 (ii) 105 rad s^{-2}
(iii) 2.53 N m (iv) 1.19 kJ

2. (i) 5.4 kg m^2 (ii) 10.5 J (iii) 1.97 rad s^{-1}

3. (i) 4.608×10^{-3} N m
(ii) A 1.91×10^{-3} J, B 0.869 J, C $234\,000$ J
(iii) $1{:}454{:}1.22\times10^8$

4. (i) 250 rad s^{-2} (ii) 0.8 rad (iii) 150 N
(iv) 167 N; 2.4 rad

5. (i) 0.0573 J (iii) 0.209 rad s^{-2},
0.098 kg m^2
(iv) 0.18 kg m^2
Much of the mass is near the centre of
the wheel so reducing the moment of
inertia.

6. 0.054 kg m^2 (iii) 0.9 s

7. (i) $2000v$ rad s^{-1} (ii) 0.603 m s^{-1}

8. (ii) $\frac{1}{8}Mg$

Exercise 5C

1. (i) (a) $6ma^2$ (b) $3ma^2$ (c) $3ma^2$
(iii) (a) $13ma^2$ (b) $\frac{33}{4}ma^2$

2. (ii) $\frac{1}{3}n(4n+1)Ma^2$

3. See table on page 305–7

4. $\frac{1}{2}Mr^2$
(i)

(a)	$\frac{1}{4}Mr^2$	$\frac{1}{6}Mr^2$	$\frac{1}{8}Mr^2$	$\frac{1}{10}Mr^2$	$\frac{1}{12}Mr^2$
(b)	$\frac{1}{2}mr^2$	$\frac{1}{2}mr^2$	$\frac{1}{2}mr^2$	$\frac{1}{2}mr^2$	$\frac{1}{2}mr^2$

(ii) $\frac{1}{4}Mr^2$. The first is $\frac{1}{8}Mr^2 = \frac{1}{4}mr^2$ and the
third is $\frac{1}{16}Mr^2 = \frac{1}{4}mr^2$. In the other three
diagrams some of the sectors are not
distributed about the axis in the same
way as the shaded sector.

5. (ii) $\sum m_p(y_p^2 + z_p^2), \sum m_p(x_p^2 + z_p^2)$
(iii) $I = I_x = I_y = I_z$ by symmetry

6. (i) $\rho\delta x$ (ii) $\rho\delta x(x\sin\alpha)^2$ (iii) $\frac{2}{3}\rho\sin^2\alpha a^3$
(iv) $\frac{1}{3}Ma^2\sin^2\alpha$ (v) $\frac{1}{3}Ma^2\sin^2\alpha$
(vi) The moment of inertia will be
multiplied by $\sin^2\alpha$

7. (i) $2y\sigma\delta x$ (ii) $\frac{2}{3}\sigma y^3\delta x$ (iii) $\dfrac{2\sigma a^3}{3h^3}\displaystyle\int_0^h x^3\,dx$

(iv) $\frac{1}{6}Ma^2$
(vi) $\frac{1}{2}Mh^2$ (vii) $\frac{1}{6}M(a^2+3h^2)$
(viii) $\frac{1}{18}M(3a^2+h^2)$

9. (i) $\pi y^2\rho\delta x$ (iii) $\displaystyle\int_0^h \frac{1}{2}\pi\rho y^4\,dx$ (iv) $\frac{1}{10}\pi\rho r^4 h$

(v) $\frac{3}{10}Mr^2$

10. (i) $2\pi ab^2\rho$ (ii) $\frac{4}{3}Mab$

11. (i) $\frac{2}{5}Mb^2$ (ii) $\frac{2}{5}Ma^2$ These are equal to
the moments of inertia of spheres of
radius b and a about their diameters

12. $\frac{3}{20}M(r^2+4h^2)$

13. (i) $\sum my^2$
(iii) (a) $\frac{1}{3}M(a^2+b^2+3ac+3c^2)$
(b) $\frac{1}{3}M(b^2\cos^2\theta+a^2+3ac+3c^2)$

14. (iii) 8.816×10^{37} kg m^2

Answers to Question 3, Exercise 5C

Uniform body of mass M	Axis	Moment of inertia
Hoop of radius r	Through centre $\perp$ to plane	Mr^2
Hollow cylinder radius r	Through centre $\perp$ to circular cross section	Mr^2
Thin ring radius r	Diameter	$\frac{1}{2}Mr^2$
Thin ring radius r	Through edge $\perp$ to plane	$2Mr^2$
Thin ring radius r	Tangent	$\frac{3}{2}Mr^2$
Disc of radius r	Through centre $\perp$ to plane	$\frac{1}{2}Mr^2$
Solid cylinder radius r	Through centre $\perp$ to circular cross section	$\frac{1}{2}Mr^2$

Uniform body of mass M	Axis	Moment of inertia
Thin disc radius r	Diameter	$\frac{1}{4}Mr^2$
Thin disc radius r	Through edge $\perp$ to disc	$\frac{3}{2}Mr^2$
Thin disc radius r	Tangent	$\frac{5}{4}Mr^2$
Thin rod	$//$ to rod at distance d	Md^2
Thin rod of length $2l$	Through centre $\perp$ to rod	$\frac{1}{3}Ml^2$
Thin rod of length $2l$	Through end $\perp$ to rod	$\frac{4}{3}Ml^2$
Rectangular lamina	In plane, through centre $\perp$ to sides length $2l$	$\frac{1}{3}Ml^2$
Rectangular lamina	Edge $\perp$ to sides length $2l$	$\frac{4}{3}Ml^2$

Uniform body of mass M	Axis	Moment of inertia
Rectangular lamina sides $2a$ and $2b$	Through centre $\perp$ to plane	$\frac{1}{3}M(a^2+b^2)$

Rectangular block sides $2a$, $2b$ and $2c$	Through centre $//$ to sides of length $2c$	$\frac{1}{3}M(a^2+b^2)$

Solid sphere radius r	Diameter	$\frac{2}{5}Mr^2$

Solid sphere radius r	Tangent	$\frac{7}{5}Mr^2$

Hollow sphere radius r	Diameter	$\frac{2}{3}Mr^2$

Hollow sphere radius r	Tangent	$\frac{5}{3}Mr^2$

Exercise 5D

1. Rod (i) a^2 (ii) $\frac{2}{3}a$ (iii) $\frac{1}{2}a^4 = \frac{1}{2}Ma^2$
   Disc (i) $\frac{2}{3}\pi ka^3$ (ii) At O
   (iii) $\frac{2}{5}\pi ka^5 = \frac{3}{5}Ma^2$

   Rod (i) $\dfrac{a(2p + aq)}{2}$ (ii) $\dfrac{a(3p + 2aq)}{3(2p + aq)}$

   (iii) $\dfrac{a^3(4p + 3aq)}{12} = \dfrac{Ma^2}{6}\dfrac{(4p + 3aq)}{(2p + aq)}$

   Pipe (i) $3\pi rkh^2$ (ii) $\frac{5}{9}h$ (iii) $3\pi r^3kh^2 = Mr^2$
2. (i) $\ddot{\theta} = -(3g/4a)\sin\theta$ (ii) $2\pi\sqrt{(4a/3g)}$
   (iii) $4a/3$
3. (i) $(61/3)ml^2$ (ii) $2\pi\sqrt{(61l/21g)}$
   (iii) 0.342 m

4. (i) $3\rho - 2\rho y/h$ (ii) $2\pi a^2\rho h$ (iii) $\frac{5}{12}h$
   (iv) $\pi a^4\rho h = \frac{1}{2}Ma^2$
5. (iv) C, $\sqrt{(35ga)}$
6. (i) $1.04g/l$
   (ii) $\sqrt{(4.16gl)}$
7. (i) $143ma^2$ (ii) $\dfrac{\sqrt{(124g)}}{\sqrt{(143a)}}$
8. (i) $\frac{1}{2}ma^2$ (ii) 19.9 cm
9. (ii) $\frac{5}{3}Ml^2$ (iii) $2\pi\sqrt{\left(\dfrac{5kl}{2\sqrt{3}g}\right)}$
10. (ii) $4\pi a\sqrt{(37/k)}$
11. (iv) $\frac{15}{8}mg\cos\theta$ (v) $16°$
12. (i) $\frac{3}{13}(a + 4x)$ (ii) $6m(3a^2 + 2x^2)$

Mechanics 6

Exercise 6A

1. (i) $0.4M$ (ii) (a) $\frac{5}{2}g$ (b) $\frac{5}{4}g$
   (iii) (a) $\frac{9}{4}Mg$ (b) $\frac{13}{8}Mg$
2. (i) $\frac{\omega}{2}$ (ii) $\frac{3}{4}Ma\omega$
   (iii) $\frac{1}{4}a\omega^2 - g(1 - \cos\theta)$ (iv) $-\frac{1}{2}g\sin\theta$
   (vi) $\frac{1}{8}Ma\omega^2 - \frac{1}{2}Mg(1 - 3\cos\theta)$, $\frac{3}{4}Mg\sin\theta$
   (viii) $2\sqrt{(g/a)}$
3. (ii) $9J/7ml$ (iii) $2m\sqrt{(7gl)}/3\sqrt{3}$
4. (i) $ma^2\omega/3x$ (ii) $ma\omega(2a - 3x)/6x$
5. (i) No centripetal force   (ii) $\frac{3}{2}Ma^2\omega$
   (iii) No moment about axis
6. (i) $pqa\omega/(p + q)$
   (ii) $p\omega/(p + q)$, $ap\omega/b(p + q)$
   (iii) Reactions at pivot produce
   impulsive couple $J(a + b)$
   (iv) $pqa^2\omega^2/2(p + q)$
7. $\sqrt{7}a$
8. (iii) 228 (3sf)
10. (ii) $(3g\cos\theta)/4b$
    (iii) $X = -\frac{1}{4}Mg\cos\theta$, $Y = \frac{5}{2}Mg\sin\theta$
    (iv) 0, $\frac{5}{2}Mg$; $Y = \frac{5}{2}\times$ value at rest
11. (ii) $5g\cos\theta$ (iii) $1.5mg\sin\theta$, $0.75mg\cos\theta$
12. (ii) $\frac{1}{3}mg$ $(7\cos\theta - 2\sqrt{2})$, $\frac{1}{3}mg\sin\theta$
14. (ii) $\frac{9}{16}g\sin\theta$
    (iii) $\frac{1}{8}mg(27 - 43\cos\theta)$, $\frac{5}{16}mg\sin\theta$

Exercise 6B

1. (i) (a) $\frac{7}{10}Mv^2$ (b) $\frac{3}{4}mr^2\omega^2$
   (ii) $\sqrt{(\frac{10}{7}gd\sin\alpha)}$ $\sqrt{(\frac{4}{3}gd\sin\alpha)}$
2. (i) $Mr\omega$ (ii) $\frac{2}{3}r$
3. (i) $\ddot{x} = a\ddot{\theta}$
   (ii) $Fa = \frac{2}{5}ma^2\ddot{\theta}$, $mg\sin\alpha - F = m\ddot{x}$
   (iii) $\frac{5}{7}g\sin\alpha$, $\sqrt{\left(\dfrac{14s}{5g\sin\alpha}\right)}$
4. (i) $\sqrt{(10gy/7)}$ (ii) $\sqrt{(2gy)}$
5. (iii) $r\sqrt{\left(\dfrac{(2Mgdr^2\sin\alpha)}{I + Mr^2}\right)}$
6. (i) Greater acceleration for smaller I
   (ii) $(m + M)R^2$
   (iv) $g/3$, the solid
7. (i) $ma(\omega_1 + \omega_2)$
8. (i) $\frac{5}{12}Ma^2$ (ii) $(1 - k)V$, $6kV/5a$
   (iii) $5a\pi(1 - k)/6k$
9. (i) $3V/l$, V (ii) $\dot{x} = -l(\sin\theta)\dot{\theta}$
   (iv) $\sqrt{(2gl/3)}$
10. (i) u (ii) $3u/10r$ (iii) $7u/10$
    (iv) $\frac{1}{2}mg$, $\frac{1}{2}g$, $u - \frac{1}{2}gt$
    (v) $5g/4r$, $(3u/10r) + (5g/4r)t$
    (vii) $9u^2/25g$
11. (ii) $u + gt(\sin\alpha - \mu\cos\alpha)$
    (iii) $2u/g(7\mu\cos\alpha - 2\sin\alpha)$
12. (ii) $a^2\omega^2/2\mu g$ (iii) $\frac{9}{4}ma^2\omega^2$ (iv) $u < a\omega$
13. $a\ddot{\theta} = \frac{3}{4}g\sin\theta$, $\arccos\frac{2}{3}$
14. (ii) $\sqrt{(10gl/7a^2)}$ (iv) $\frac{1}{7}(5g - 2f)$
15. (ii) $3d/2$ (iii) $\mu > f/3g$

Mechanics 5

Exercise 7A

1. (i) $-mgx \sin \theta$ (ii) $mg(x-a)^2/a$
   (iii) $a(1 + \frac{1}{2}\sin\theta)$
   (iv) Period $\pi\sqrt{(2a/g)}$
2. (i) $-mgx$, $\lambda x^2/a$ (ii) $2\pi\sqrt{(ma/2\lambda)}$
3. (i) $\lambda a(1-\cos\theta)^2 - mga\sin\theta$
4. (i) $mga\cos 2\theta$, $\frac{1}{2}\lambda a(4\sin\theta - 1)^2$
   (ii) $\theta = \pi/2$ (unstable when $\lambda > mg/3$),
   $\theta = \arcsin\left(\dfrac{\lambda}{4\lambda - mg}\right)$ provided $\lambda > mg/3$
   (stable)
5. Angle DBA $= 2\arcsin\left[\dfrac{\lambda}{2(\lambda - 2mg)}\right]$
6. $\theta = 0$ (stable) and $\theta = \pi/3$ (unstable)
   where θ is the inclination of OP to the
   vertical
7. (i) $\frac{1}{2}mgd[k + (k-1)\sin 2\theta]$ (ii) $k = 1$
   (iii) $k > 1$
8. (i) $a\dot\theta^2(ka - mg) + 2mga$ (PE zero level
   through O)
   (ii) $\ddot\theta = -\dfrac{(ka - mg)}{2ma}\theta$
9. (i) $x = \pm(2b/a)^{1/6}$ (ii) $\frac{\pi}{3}m^{1/2}(2b)^{2/3}a^{-7/6}$
   (iii)

Force is repulsive when
atoms closer than
$(2b/a)^{1/6}$ otherwise
attractive.

10. (i) $\lambda < mgd/r$ (ii) $\lambda = mgd/r$
    (iii) $\lambda > mgd/r$ (iv) None
11. (ii) $\theta = \arccos(11\sqrt{3}/20) = 0.31$
    (iii) Unstable
13. $-2(M+m)ga\cos\frac{1}{2}\theta + mga\cos\theta$ (relative
    to centre of larger cylinder);
    equilibrium when $\theta = 0$ or
    $\cos\frac{1}{2}\theta = \frac{1}{2}[1 + (M/m)]$; initial position
    stable when $M > m$

Exercise 8A

1. (ii) $v = \dfrac{F}{Mk}(e^{kt} - 1)$
2. $\dfrac{F}{k}\ln\left(\dfrac{M}{M - m_0}\right)$
3. (ii) $v^2 = \dfrac{2000P}{r}\ln\left(\dfrac{M_0}{M_0 - rt}\right)$
4. (ii) $F = (M - ks)v\dfrac{dv}{ds}$
5. (ii) $\dfrac{P}{v} = M\left(1 - \dfrac{t}{2T}\right)^3\dfrac{dv}{dt}$
6. (i) $k = \dfrac{1}{T}\ln 2$ (ii) $Mge^{kt} = \dfrac{d}{dt}(Me^{kt}v)$
   (iv) $\dfrac{gT^2}{\ln 2}\left(1 - \dfrac{1}{2\ln 2}\right)$
8. (i) Dimensions of k are MT^{-1}
   (iii) $v = \dfrac{g}{3km^2}(m^3 - M^3)$
11. (ii) $\dfrac{c}{k}[(1 + k^2t^2)^{\frac{1}{2}} - 1]$

Exercise 8B

1. (ii) In order, expulsion speeds in m s^{-1}
   are: 617, 602, 598, 803, 803, 798, 806;
   probably two different fuels
2. (ii) 1290 m s^{-1}
4. (i) $m\dfrac{dv}{dt} = ku$
6. (i) $M_0 - kt$, $\alpha M_0/k$
   (iv) $R_1 = M_0/(M_0 - \alpha M_1)$, $R_2 = 1/(1 - \alpha)$
   (v) $c\ln R_1$ (vi) $c\ln R_1 R_2$
7. Speed $c\ln\left(\dfrac{1}{1 - \alpha}\right) - \dfrac{g\alpha M}{k}$; altitude
   $\dfrac{Mc}{k}[(1 - \alpha)\ln(1 - \alpha) + \alpha] - \frac{1}{2}g\dfrac{\alpha^2 M^2}{k^2}$

Exercise 9A

1. (i) $-5\mathbf{i} - 2\mathbf{j} - 3\mathbf{k}$ (ii) $5\mathbf{i} + 2\mathbf{j} + 3\mathbf{k}$
   (iii) $10\mathbf{i} + 4\mathbf{j} + 6\mathbf{k}$
3. $-21\mathbf{i} + 10\mathbf{j} - 24\mathbf{k}$ (or any multiple)
4. $\sqrt{3}\mathbf{i} + \sqrt{3}\mathbf{j} + \sqrt{3}\mathbf{k}$
5. Sine Rule
6. (i) $-2\mathbf{j} + 2\mathbf{k}$, $-2\mathbf{i} + 2\mathbf{j} - 4\mathbf{k}$, $2\mathbf{i} + 2\mathbf{k}$
   (ii) $2\sqrt{3}$
9. (i) $-\cos 2t$, $2\sin 2t$
   (ii) $-t\cos t\mathbf{i} - t\sin t\mathbf{j} + \sin 2t\mathbf{k}$,
   $(t\sin t - \cos t)\mathbf{i} - (t\cos t + \sin t)\mathbf{j} + 2\cos 2t\mathbf{k}$

Exercise 9B

1. $-\mathbf{i} - \mathbf{j}$
2. $2\mathbf{i} + \mathbf{j}$, $-4\mathbf{i} + 2\mathbf{j}$
3. (i) $-2\mathbf{i} + \mathbf{j} - \mathbf{k}$ (ii) $2\mathbf{i} - \mathbf{j}$
4. (i) $\mathbf{i} - \mathbf{k}$ (ii) $-\mathbf{j} - \mathbf{k}$
5. (i) $\dfrac{9}{\sqrt{14}}(-\mathbf{i} + 2\mathbf{j} + \mathbf{k})$
   (ii) Zero (force passes through point)
6. $(r - 1)\mathbf{i} + (1 - r)\mathbf{j} + (q - p - 1)\mathbf{k}$,
   $(2 - r)\mathbf{i} + (r - 2)\mathbf{j} + (p - q - 1)\mathbf{k}$,
   sum $\mathbf{i} - \mathbf{j} - 2\mathbf{k}$
7. (i) $-4\mathbf{i} + 10\mathbf{j} - 3\mathbf{k}$ (ii) $22/\sqrt{29}$
8. (i) $1/\sqrt{3}\,(-\mathbf{i} - \mathbf{j} + \mathbf{k})$, $10\sqrt{3}\,(-\mathbf{i} - \mathbf{j} + \mathbf{k})$
   (ii) $10\sqrt{3}\,(\mathbf{i} - \mathbf{j})$
   (iii) $10\sqrt{3}$ N m (iv) 60 N m
9. (i) Force $\dfrac{T}{\sqrt{(3 - 2\sin\theta)}}$
   $[-\cos\theta\mathbf{i} + \mathbf{j} + (1 - \sin\theta)\mathbf{k}]$
   (ii) $\dfrac{4T(-\sin\theta\mathbf{i} - \cos\theta\mathbf{j} + \cos\theta\mathbf{k})}{\sqrt{(3 - 2\sin\theta)}}$

Exercise 9C

3. (i) $\sqrt{3}$ (ii) $\mathbf{i} - \mathbf{j} + \mathbf{k}$
4. $X = -3$, $Y = 1$, $Z = -4$
5. $-\mathbf{i} + \mathbf{j}$
7. $-12\mathbf{i} + 7\mathbf{j} - 8\mathbf{k}$; couple $-6\mathbf{i} + 14\mathbf{j} + 39\mathbf{k}$,
   magnitude $\sqrt{(1753)}$
8. $|\mathbf{F}| = 3\sqrt{3}$, $|\mathbf{G}| = \sqrt{56}$
9. (i) D: $4\mathbf{j}$, R: $\mathbf{j} + 4\mathbf{k}$, S: $4\mathbf{i} + 3\mathbf{j}$
   (ii) $X_D = -25$, $Z_D = 75$
   (iii) $X_A = 125$, $Z_A = 25$
10. (i) $\mathbf{P} = (P\cos\theta)\mathbf{j} - (P\sin\theta)\mathbf{k}$
    (ii) Equivalent to force
    $F_1\mathbf{i} + (F_2 + P\cos\theta)\mathbf{j}$
    $+ (F_3 - P\sin\theta)\mathbf{k}$ at O with a couple
    $-0.15P\mathbf{i} + (0.2P\sin\theta - 0.4F_3)\mathbf{j}$
    $+ (0.2P\cos\theta + 0.4F_2)\mathbf{k}$
    (iii) (a) $F_2 = -\tfrac{1}{2}P\cos\theta$, $F_3 = \tfrac{1}{2}P\sin\theta$
    (b) $P = 400$ N

Exercise 9D

1. $\boldsymbol{\omega} = 5\mathbf{k}$; $\mathbf{v} = -10\mathbf{i}$
2. $\mathbf{v} = \mathbf{i} - \mathbf{j}$, speed $\sqrt{2}$
3. (i) $\boldsymbol{\omega} = 7(2\mathbf{i} - \mathbf{j} - 2\mathbf{k})$ (ii) $-7\mathbf{i} - 7\mathbf{k}$
4. (iii) $-(\omega a\sin\omega t)\mathbf{i} + (\omega a\cos\omega t)\mathbf{j} + u\mathbf{k}$,
   $(\tfrac{1}{2}\omega a\sin\omega t)\mathbf{i} - (\tfrac{1}{2}\omega a\cos\omega t)\mathbf{j} + u\mathbf{k}$
   (ii) Star 1: $mua(\sin\omega t - \omega t\cos\omega t)\mathbf{i}$
   $- mua(\omega t\sin\omega t + \cos\omega t)\mathbf{j} + ma^2\omega\mathbf{k}$
   Star 2: $-mua\,(\sin\omega t + \omega t\cos\omega t)\mathbf{i}$
   $+ mua(\omega t\sin\omega t + \cos\omega t)\mathbf{j} + \tfrac{1}{2}ma^2\omega\mathbf{k}$
   (iii) Total $\tfrac{3}{2}a^2\omega m\mathbf{k}$
   (iv) Stars circling about a common centre, which itself is moving with constant velocity perpendicular to the plane of the circles
5. (i) A parabola

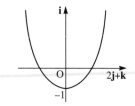

   (ii) $\mathbf{h} = m(t^2 + 1)\mathbf{j} - 2m(t^2 + 1)\mathbf{k}$,
   $\dfrac{d\mathbf{h}}{dt} = 2mt(\mathbf{j} - 2\mathbf{k})$
   (iii) $\mathbf{F} = 2m\mathbf{i}$, moment $2mt(\mathbf{j} - 2\mathbf{k})$
6. (i) $\mathbf{v} = \boldsymbol{\omega} \times \mathbf{r}$
7. (i) $\mathbf{L} = m\mathbf{r} \times \mathbf{v}$ (iii) $\mathbf{L}_{OG} = M\bar{\mathbf{r}} \times \bar{\mathbf{v}}$
   (iv) $\mathbf{L}_G = \sum [m_i(\mathbf{r}_i - \bar{\mathbf{r}}) \times (\mathbf{v}_i - \bar{\mathbf{v}})]$
8. (iv) $m[a\omega(d\sin\theta + a\cos\theta)\mathbf{i}$
   $+ a\omega(a\sin\theta - d\cos\theta)\mathbf{j} + \Omega(a^2 + d^2)\mathbf{k}]$

Mechanics 5

Index